Biopolymers and Biodegradable Polymers: Synthesis, Properties, Application and Degradation Behavior

Biopolymers and Biodegradable Polymers: Synthesis, Properties, Application and Degradation Behavior

Guest Editors

Vishal Gavande
Vasi Shaikh
Won-Ki Lee

MDPI

Basel • Beijing • Wuhan • Barcelona • Belgrade • Novi Sad • Cluj • Manchester

Guest Editors

Vishal Gavande
Department of Material Science and Engineering
Saarland University
Saarbrucken
Germany

Vasi Shaikh
Department of Chemistry
Dr. Vishawanath Karad MIT World Peace University
Pune
India

Won-Ki Lee
Department of Polymer Engineering
Pukyong National University
Busan
Korea, South

Editorial Office

MDPI AG
Grosspeteranlage 5
4052 Basel, Switzerland

This is a reprint of the Special Issue, published open access by the journal *Polymers* (ISSN 2073-4360), freely accessible at: www.mdpi.com/journal/polymers/special_issues/YBWH9Z54TG.

For citation purposes, cite each article independently as indicated on the article page online and using the guide below:

Lastname, A.A.; Lastname, B.B. Article Title. *Journal Name* **Year**, *Volume Number*, Page Range.

ISBN 978-3-7258-3860-8 (Hbk)
ISBN 978-3-7258-3859-2 (PDF)
https://doi.org/10.3390/books978-3-7258-3859-2

Contents

About the Editors

Vishal Gavande

Dr. Vishal Madhukar Gavande is a postdoctoral research scientist specializing in polymer engineering and materials science, recognized for his contributions to the development and characterization of advanced polymer-based materials. His academic and research career reflects a strong focus on innovative polymer technologies with applications ranging from coatings to biodegradable composites. Dr. Gavande's research interests are extensive, including polymers, nanocomposites, electrospinning, nanofibers, biopolymers, composites, coatings, flexible electronics, wearable sensors, ultra-high-molecular-weight polypropylene, polymer degradation, hydrogels, and 3D printing, among others. His dedication to advancing polymer science is evident through his extensive research, editorial endeavors, and commitment to fostering innovation in the field.

Vasi Shaikh

Professor Vasi Ahmed Ebrahim Shaikh is a distinguished academic and researcher at MIT World Peace University (MIT-WPU) in Pune, India, with over 30 years of experience in teaching and research across science and engineering disciplines. Based in the School of Polymer Engineering, he has built a robust reputation as a leader in polymer chemistry and materials science, contributing significantly to both academia and industry-oriented education. At MIT-WPU, Prof. Shaikh serves as a Professor and has held pivotal roles such as Associate Dean and Head of the M.Sc. Industrial Polymer Chemistry program, which he helped conceptualize and launch in 2020. Under his leadership, the program has grown to enroll about 60 students annually, maintaining high academic standards while boosting admissions and placements. Beyond research, Prof. Shaikh is recognized for his teaching excellence, earning accolades such as the Best Teacher Award from Savitribai Phule Pune University and the Ideal Teacher Award from MIT, Pune. His innovative practices have enhanced student employability, blending rigorous academics with practical skills.

Won-Ki Lee

Professor Won-Ki Lee is a faculty member at Pukyong National University (PKNU) in Busan, South Korea, known for his contributions to the field of polymer engineering. While specific details about his personal background are limited, his academic and research profile aligns with PKNU's strong reputation in science and technology, particularly in areas like materials science and engineering.

At PKNU, Prof. Lee is affiliated with the Department of Polymer Engineering, where he has likely played a significant role in advancing research and education related to polymer materials. His work intersects with the university's broader focus on innovative technologies, which is evident in PKNU's collaborations with industry and its emphasis on applied sciences. Given his field, his research probably explores topics such as polymer composites, sustainable materials, and advanced manufacturing techniques—areas that are critical to both academic inquiry and industrial applications.

Prof. Lee's academic journey likely includes advanced degrees in polymer science or a related discipline, possibly from a prestigious institution, though exact details are not widely publicized. His presence at PKNU suggests he contributes to a legacy of practical, industry-oriented research. PKNU's Yongdang and Daeyeon campuses, where such work is conducted, provide a robust environment for his endeavors, equipped with facilities like the Engineering Library and research centers focused on cutting-edge materials.

Preface

The burgeoning global awareness of environmental sustainability has propelled the field of biopolymers and biodegradable polymers to the forefront of materials science and engineering. This reprint, "Biopolymers and Biodegradable Polymers: Synthesis, Properties, Application, and Degradation Behavior," aims to provide a comprehensive and up-to-date exploration of this critical area. Its scope encompasses the fundamental principles of biopolymer synthesis, the intricate relationships between their molecular structure and material properties, the diverse applications that leverage their unique characteristics, and the crucial aspect of their degradation behavior.

My motivation for writing this work stems from the urgent need for accessible and consolidated knowledge regarding these eco-friendly materials. The rapid advancements in biopolymer research have created a vast body of literature, often fragmented and dispersed. This book seeks to bridge that gap, offering a single, authoritative resource for researchers, students, and industry professionals, offering both fundamental knowledge and the latest developments in the field. The chapters within this volume explore a spectrum of topics, from the stabilization of earth-based pavements with bio-based binders to the intricate molecular design of sustainable polyesters and polyurethanes. We delve into the fabrication of shape- and temperature-stable nonwoven structures, the characterization of chitosan-based thin films, and the genetic underpinnings of polyhydroxybutyrate synthesis. Furthermore, this reprint explores advanced analytical techniques, such as deep learning for cell migration studies, and investigates the crystallization behavior of polylactic acid.

The practical applications of biopolymers are equally diverse, encompassing ternary blends for tailored material properties, starch-based flame retardants, and the revolutionary use of polyetheretherketone in restorative dentistry. Finally, we examine the critical role of biomaterials in guided tissue and bone regeneration for periodontal treatments, highlighting the clinical relevance of biopolymer research.

This reprint is intended for a broad audience, including researchers, engineers, material scientists, and clinicians seeking to understand the latest developments in biopolymers and their applications. It is the product of contributions from leading experts in their respective fields, and I am deeply grateful for their insightful work. I would also like to express my sincere appreciation to the editors, reviewers, colleagues, and research assistants. Their invaluable insights and support have been instrumental in shaping the content and ensuring the accuracy of this reprint.

It is my sincere hope that this reprint will catalyze further innovation and applications of biopolymers and biodegradable polymers, contributing to a more sustainable and environmentally responsible future.

Vishal Gavande, Vasi Shaikh, and Won-Ki Lee
Guest Editors

polymers

MDPI

Article

Investigation of Mechanical Properties and Microstructural Characteristics of Earth-Based Pavements Stabilised with Various Bio-Based Binders

Nuriye Kabakuş [1,*] and Yeşim Tarhan [2]

[1] Department of Emergency and Disaster Management, Faculty of Applied Sciences, Atatürk University, Erzurum 25240, Türkiye
[2] Technical Sciences Vocational School, Ardahan University, Ardahan 75000, Türkiye; yesimtarhan@ardahan.edu.tr
* Correspondence: nsirin@atauni.edu.tr

Abstract: For centuries, earthen materials have regained popularity because of the high carbon emissions caused by the construction sector. Although earth-based materials possess superior properties, such as recyclability, easy accessibility, affordability, and high thermal conductivity, they are not without drawbacks. They are, for instance, relatively weak and sensitive to water, and their physical and chemical properties can vary considerably depending on the source from which they are obtained. Stabilisation is often used to overcome these drawbacks. In this study, natural earth-based materials were stabilised with biopolymers of organic origin, such as alginate, Arabic gum, xanthan gum, and locust bean gum, to preserve their natural properties. To produce the samples, the earth material used in the road sub-base layer was mixed with kaolin clay and silica sand, and the mixtures were prepared by substituting biopolymer materials with clay at a ratio of 0.1%. After determining the fresh unit volume weights, spreading diameters (flow table test), penetration depths (fall cone test), and air content of the mixtures, the flexural and compressive strengths of the cured specimens were measured. In addition, scanning electron microscopy (SEM) and X-ray diffraction (XRD) analyses were performed to determine the microstructural characteristics. According to the 28-day compressive strength results, the mix with xanthan gum was found to be almost twice as strong as the other mixes. It has been concluded that biopolymer-stabilised earth mixtures can be used as a fill material in buildings where high strength is not required, or as a paving material on low-traffic roads.

Keywords: earth roads; biobased materials; xanthan gum; new-generation pavement

Academic Editors: Vishal Gavande, Vasi Shaikh and Won-Ki Lee

Received: 14 February 2025
Revised: 13 March 2025
Accepted: 22 March 2025
Published: 24 March 2025

Citation: Kabakuş, N.; Tarhan, Y. Investigation of Mechanical Properties and Microstructural Characteristics of Earth-Based Pavements Stabilised with Various Bio-Based Binders. *Polymers* **2025**, *17*, 864. https://doi.org/10.3390/polym17070864

1. Introduction

Earth-based natural materials have been used widely in the construction industry since ancient times [1]. The use of energy-intensive materials, such as cement and bitumen, in this sector has increased environmental problems. With growing interest in sustainable building materials, earthen mixtures have once again come to the fore [2]. Currently, scientists are exploring various methods to enhance the strength and durability of earth-based mixtures, including stabilization techniques [3–5] and fiber reinforcement [6]. Additionally, advancements in construction technologies, such as 3D printing, are gaining traction due to their potential to optimise material usage, reduce waste, and incorporate sustainable materials like recycled plastics and bio-based composites, thereby contributing to circular economy principles and sustainable building practices [7–10]. The increasing demand for sustainable building materials has accelerated efforts to standardise earth-based mixtures' design, application, and testing. Earth-based materials are often used in the construction of unpaved roads [11–13], road sub-base layers [14], cycle paths, pedestrian roads [12], berms, and some non-high-strength structures [15].

As demand increases, studies and strategies are being developed to replace depleted petroleum resources with renewable resources that offer environmental and economic benefits. In this context, the new goal in pavement engineering is to reduce the use of bituminous materials and emphasise the use of renewable resources, such as biopolymers. Biopolymers are preferred binders for earth-based materials because they are natural materials derived from plants, animals, and microorganisms, have a low environmental impact, and contribute to strength. These materials have been successfully used in the construction sector to stabilise various building and road materials.

The stabilisation of earth-based materials with biopolymers is an emerging field, driven by the need for sustainable construction practices and the reduction in environmental impacts associated with traditional stabilisers, such as cement and lime [16,17]. Biopolymers, which are environmentally friendly and derived from natural sources, offer a promising alternative with the potential to improve the mechanical properties of earth structures while reducing carbon emissions [18]. However, the use of biopolymers is challenging and still needs to be improved. Poor water resistance and susceptibility to biodegradation can limit their effectiveness and durability [19]. Research is ongoing to improve these properties through novel treatments, such as acetylation and the addition of inorganic minerals. Studies have also shown that biopolymers, such as casein and sodium caseinate, can significantly improve the compressive strength of treated sand, with factors, such as curing time and temperature, playing a role in the stabilisation process [16,20].

The stabilisation of the earth is one of the primary methods for correcting the expanding earth in response to moisture changes and improving its engineering quality [20]. Although there are many studies on soil stabilisation materials and techniques in the literature, there are only a limited number of studies on stabilisers (especially biopolymers) and techniques recommended for earth-based pavements [21,22].

This study developed an earth-based unpaved road mix using biopolymers as binders as an alternative to flexible and rigid pavements that use bitumen and cement as binders for low-volume road pavements. The engineering properties of mixtures obtained by adding biopolymers, which are also natural binders, to natural and locally available earth and clay mixtures were investigated. The biopolymers used in this study, which investigated the most effective type of biopolymer that could increase the strength and durability of this earth-based mix, were selected from water-soluble gums [23]. Arabic gum (AG) from the resin group, alginate (A) from seaweed extracts, locust bean gum (LBG) from seed extracts, and xanthan gum (KG) from microbial gums were used. In this way, unlike studies in literature, the effect of different categories of gums in earth-based mixtures was investigated.

The water solubility of biopolymers is a key factor influencing their performance as earth-stabilising agents. When dissolved in water, these biopolymers form hydrogels or viscous solutions that enhance soil cohesion, modify rheological properties, and improve mechanical stability [24]. Their capacity to interact with soil particles and pore water enables them to function as binding agents, thereby reducing erosion, regulating moisture fluctuations, and increasing soil strength [25]. Among water-soluble biopolymers, xanthan gum, guar gum, alginate, and Arabic gum have demonstrated significant potential for soil stabilisation due to their capacity to form cross-linked polymer networks. In particular, xanthan gum and guar gum create three-dimensional hydrogel structures that improve the mechanical properties and durability of soil, making them highly suitable for road stabilisation [26]. Xanthan gum, in particular, has been shown to enhance water retention, reduce evaporation and percolation, and mitigate soil erosion—critical factors in maintaining road stability [23].

This study investigates the potential of biopolymer stabilisation in soil mixes by incorporating four types of biopolymers—alginate, gum arabic, xanthan gum, and locust bean gum—into a road sub-base soil mix with sand and kaolin clay. The primary objective is to evaluate the feasibility of these stabilised mixtures as paving materials for low-traffic roads. Fresh state properties were assessed and hardened specimens were tested for mechanical strength and durability. In addition, the microstructural analysis provided insight into the stabilisation mechanisms. The results contribute to a broader understanding of biopolymer-stabilised soil materials and their role in sustainable road construction.

2. Materials and Methods

2.1. Materials and Specimen Preparation

In this study, a mixture of soil, sand, and clay stabilised with four types of biopolymers to determine the most effective biopolymer for enhancing the mechanical performance of an earth-based mixture. The primary soil used in the mixture was classified as silty sand (SM) according to the Unified Soil Classification System (USCS), with a unit volume weight of 2.35 g/cm^3. This material, sourced from a road sub-base layer, was crushed and sieved to a 0–2 mm particle size to ensure uniformity. The silica sand component, with a bulk density of 2.47 g/cm^3, was included in the 0–1 mm size range to enhance gradation. Kaolin clay, supplied by ATA KİMYA (the company in Ankara, Türkiye) commercial limited company as a dry, ground material packaged in 20 kg bags, was incorporated as binder. Figure 1 presents images of all raw materials used in this study.

Figure 1. Road subgrade earth, sand, and kaolin clay, from left to right.

The grain diameter distribution and chemical composition of kaolin clay were provided by ATA KİMYA (the company in Ankara, Türkiye), while the geotechnical properties were determined through experimental tests conducted at the Soil Mechanics Laboratory of Atatürk University's Faculty of Engineering. Hydrometer analysis confirmed that 100% of kaolin particles passed through the 0.075 mm sieve. The chemical composition, analysed via X-ray fluorescence spectrometry (XRF), and the geotechnical properties, obtained from both laboratory experiments and the manufacturer's published data, are presented in Table 1.

Table 1. Chemical and geotechnical properties of kaolin clay [27,28].

Chemical Properties	Value (%)	Geotechnical Properties	Value (%)
SiO_2	69.10	Grain size < 0.002 mm, %	28
Al_2O_3	15.20	Grain size < 0.075 mm, %	100
Fe_2O_3	0.20	Specific gravity	2.63
CaO	0.10	Liquid limit, %	49
SO_3	4.29	Plastic limit, %	26
Na_2O	0.03	Plasticity index, %	23
K_2O	11.07	Optimum moisture content, %	25
CR_2O_3	0.01	Maximum dry unit weight, kN/m^3	13.2
		Unconfined compressive strength, kPa	307

Alginate (A), Arabic gum (AG), xanthan gum (KG), and locust bean gum (LBG) were added to the control mixture (C) to stabilise the earth-based mixture. The properties of the biopolymers are listed in Table 2.

Table 2. Properties of biopolymers used.

Product Properties	LBG	A	AG	KG
CAS number	9000-40-2	14984-39-5	9000-01-5	11138-66-2
Appearance/ texture/smell	Pale white, no smell	White, powder, no smell	White, powder, slightly smelly	White, solid, no smell
pH	5.4–7	6.0–8.0	4.1–4.8	6–8
Density	$1.2 \pm 0.1\ g/cm^3$	$1.6\ g/cm^3$	$1.15\ g/cm^3$	$1.5\ g/cm^3$
Molecular weight	226.66 g/mol	216.121 g/mol	180.41 g/mol	1016.8 g/mol
Molecular formula	$C_{10}H_{11}CIN_{202}$	$C_6H_7O_6Na$	$C_{12}H_{36}$	$(C_{35}H_{49}O_{29})_n$
E-code	E410	E401	E414	E415

Dry mixes were prepared using 40% by weight road sub-base, 30% sand, and 30% clay. Tap water was added to 20% of the total mix to ensure a suitable consistency for workability in the control mix. To stabilise the mix, biopolymers were incorporated at 0.1% by weight of clay, proportioned to the total clay content. The biopolymers were first dissolved in water to form a slightly gel-like consistency before being incorporated into the dry mix.

In preliminary trials, the biopolymer ratio was initially set at 1% of the clay content, but this resulted in excessive gelation, making the mixture too dense and highly viscous, preventing proper mixing with the dry components. A subsequent 0.5% substitution was tested, but the mix still had a solidified consistency that prevented proper moulding. Based on these observations, a 0.1% biopolymer level was selected as it provided a workable consistency that allowed for proper mixing and moulding. The final compositions of the mixes used in this study are shown in Table 3.

Table 3. Earth-based mix designs stabilised with biopolymers.

Sample Code	Material Amounts (%)							
	Earth	Sand	Clay	Water	A	AG	KG	LBG
C	40	30	30	20	-	-	-	-
A	40	30	29.9	20	0.1	-	-	-
AG	40	30	29.9	20	-	0.1	-	-
KG	40	30	29.9	20	-	-	0.1	-
LBG	40	30	29.9	20	-	-	-	0.1

A 5-litre laboratory mortar mixer was used to prepare the mixtures. First, the dry components were mixed for 1 min, then the biopolymer solution—prepared by dissolving the biopolymer in water—was added. The mixture was then mixed at low speed for 1 min, followed by a 1 min rest period, and finally mixed at high speed for 2 min. The biopolymers were first dissolved in water for 2 min using a magnetic mixer before being incorporated into the soil mixture.

For mechanical testing, $4 \times 4 \times 16$ cm^3 beam specimens were prepared for flexural strength evaluation, while $5 \times 5 \times 5$ cm^3 cube specimens were cast for compressive strength testing. Due to the gel-forming nature of biopolymers, specimens were manually placed in moulds rather than compacted, as immediate compaction could result in uneven density distribution and affect mechanical performance. The samples were left undisturbed for 1 day at 21 ± 2 °C and 50% relative humidity under laboratory conditions. After demoulding, specimens were oven-dried at 60 °C for 1 day to ensure controlled moisture removal and to maintain consistency across all specimens. This temperature was chosen to prevent excessive shrinkage while promoting uniform drying. Whilst thermal drying can increase the compressive strength of clay containing materials, the effect was consistent across all mixtures tested, ensuring reliable comparisons. The samples were then cured for 28 days at 23 ± 2 °C and 65% relative humidity to allow further strength development under controlled conditions.

2.2. Experimental Program

The unit volume weights, air content [29], spreading diameter (measured using a flow table test to evaluate the viscosity and workability of the mixtures) [30], and penetration depth (measured using a fall cone test to determine the thixotropic behaviour of the mixtures) [31] were evaluated. The flow table test, typically used for cementitious materials, was used in this study to evaluate the workability of biopolymer-stabilised mixtures. Due to the gel-forming nature of biopolymers, conventional soil workability tests do not adequately capture the changes in mix consistency. The flow table test provided a controlled assessment of the flowability and deformation characteristics of the mixes, ensuring a standardised comparison between different formulations.

The cured specimens were subjected to three-point flexure at 28 days and uniaxial compression tests at 28 and 56 days to evaluate their mechanical performance over time. The loading rate was set at 0.004 MPa/s for the flexure test and 0.04 MPa/s for the compression test.

Capillary water absorption tests were performed to assess the durability of the samples according to the methodology outlined in [32]. This test evaluates the water absorption capacity of the material over time, expressed as a percentage increase in mass relative to the initial dry weight. To ensure consistency, the samples were dried in an oven at 40 ± 2 °C for seven days until a stable mass was reached, maintaining a uniform moisture content prior to testing. After drying, the sides of the samples were sealed with tape to prevent lateral water absorption. The capillary water absorption test was carried out by placing the samples on steel rods in a tray, ensuring that only the lower surface (up to a height of 5 mm) was in contact with tap water. Water uptake was recorded as a percentage increase in mass relative to the initial dry weight to ensure consistency with standardised reporting methods. Results are presented as percentage mass gain (%) to allow comparability with similar studies. Pictures of the test set-up are shown in Figure 2.

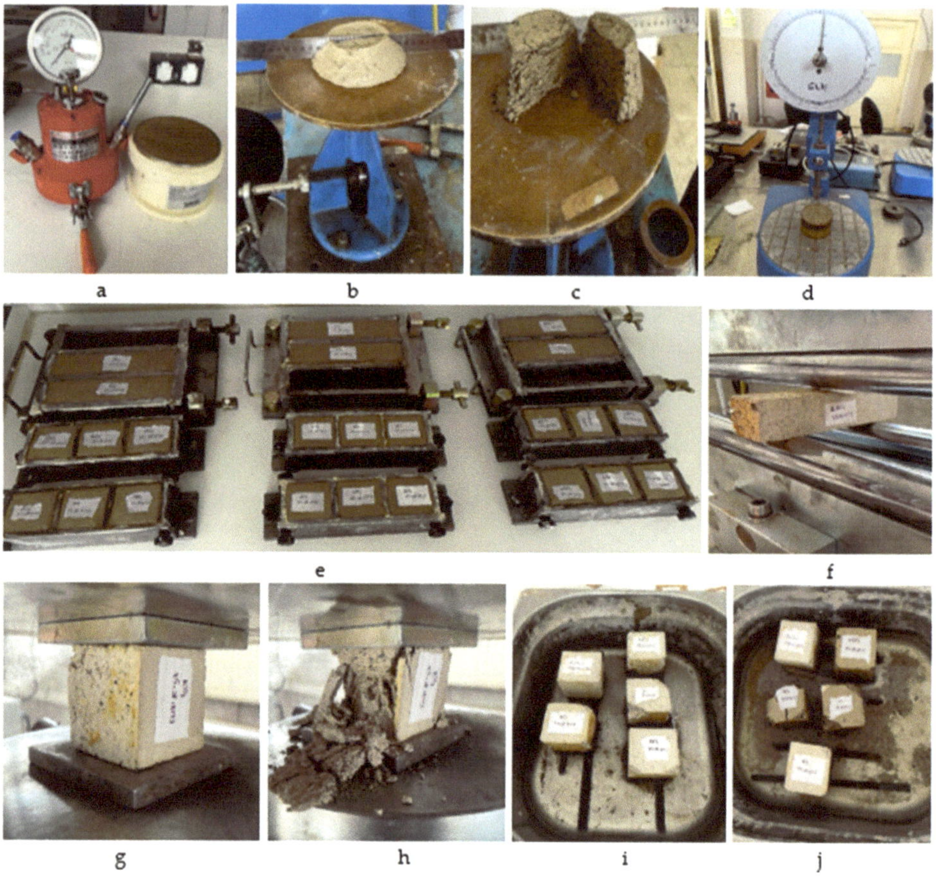

Figure 2. Tests applied: (**a**) Air content determination; (**b**,**c**) Flow table; (**d**) Fall cone; (**e**) Sample preparation; (**f**) Flexural strength; (**g**,**h**) Compressive strength; and (**i**,**j**) Capillary water absorption.

After 28 days of curing, all mechanical tests, including evaluation of compressive and flexural strength, were performed. However, SEM and XRD analyses were carried out on specimens taken from the specimens after 56 days. These analyses were carried out at Atatürk University DAYTAM to investigate the microstructural properties. SEM images were obtained using a Zeiss Sigma 300 scanning electron microscope. XRD analyses were performed using a Malvern PANalytical EMPYREAN X-ray diffractometer (manufactured in Almelo, The Netherlands) with Cu-α radiation (λ = 1.54 Å) on an X-ray diffractometer. The scan range (2θ) was 10–90° with a scan speed of 4°/min, an operating voltage of 5 kV, and an operating current of 40 mA.

3. Results and Discussion

3.1. Fresh State Results

This study investigated the potential of stabilising earth-based construction materials with biopolymer materials to improve their mechanical properties. The results of the fresh unit volume weight, air content, spreading diameter, and penetration depth tests of the mixtures are plotted in Figure 3. In the graph in Figure 3, all fresh-state test results are plotted together for comparison.

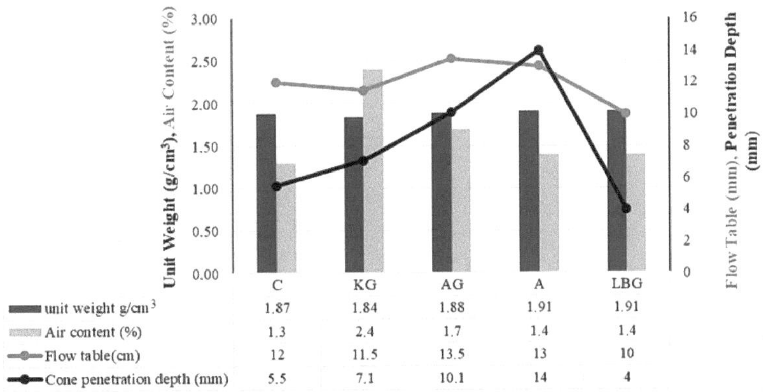

Figure 3. Fresh-state test results.

Figure 3 shows that the unit volume weights of all mixes are below 2 g/cm^3, and the values show a very close distribution. In this study, the mix with the lowest unit volume weight and the highest air content was observed in the xanthan gum-stabilised group. In terms of penetration values, the highest penetration depth was obtained in the group with alginate, while the lowest value was recorded in the group with locust bean gum. Based on the flow table results, the lowest flow value was recorded in the group in which locust bean gum was added, indicating that the mixture to which locust bean gum was added had a more solid-like consistency and, therefore, a higher unit volume weight compared to the other mixtures. Thus, the falling cone and flow table experiments supported each other. As can be seen from the graph, the results from the flow table and the penetration depth are generally in agreement, indicating that these two tests are effective tools for determining the rheological properties of the mixtures, that is, their consistency. Mixing biopolymers with water resulted in a gel-like consistency, increasing the viscosity of the mixtures, and consequently resulting in a more solid structure.

3.2. Hardened State Results

The 28-day flexural strengths of the control group and all the groups to which the biopolymer stabiliser was added are shown in Figure 4.

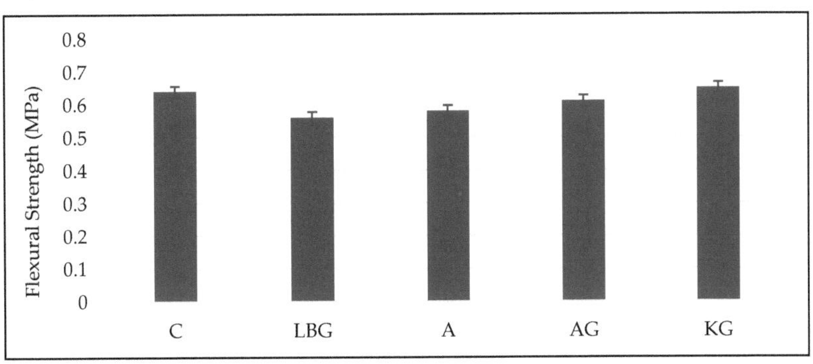

Figure 4. Flexural strength test results.

Analysis of the flexural strength results (Figure 4) shows that the highest flexural strength was achieved in the KG group, reaching 0.65 MPa. The stabilisation process with KG increased the flexural strength of the earth-based specimens by approximately 2%.

Although this increase may seem small, even small improvements are significant given the inherently brittle nature of earth-based materials.

In contrast, biopolymers other than KG had a negative effect on flexural strength. Among the formulations tested, the KG group exhibited the highest consistency, as evidenced by its low unit volume weight, high air content, and cohesive behaviour. As can be seen in Figure 2c, the KG mix exhibited superior consistency compared to the other formulations, despite its lower unit volume weight. This suggests that KG improves the bonding between the soil materials, contributing to higher flexural strength.

The flexural strength test results support this observation, showing that the KG group had a greater flexural strength capacity than the other groups. This finding reinforces the conclusion that KG acts as an effective binder, improving both cohesion, and mechanical performance.

The 28 and 56-day compressive strength values of the earth-based specimens are shown in Figure 5. In addition, the increasing trends representing the relationship between the specimens in the figure are shown separately for the 28-day and 56-day compressive strength results.

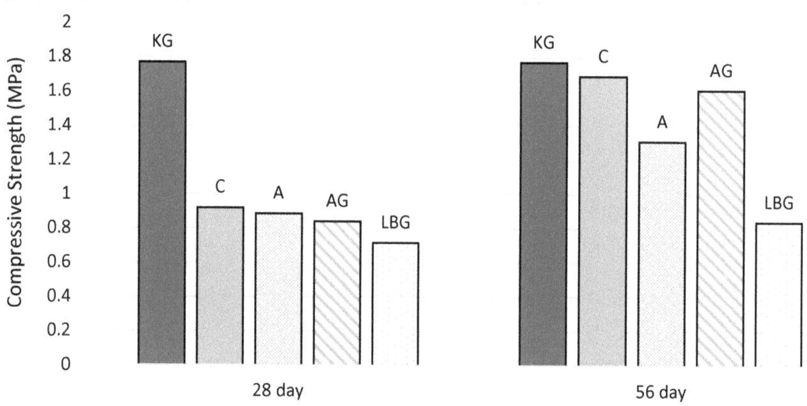

Figure 5. Compressive strength test results for 28 and 56 days.

According to the results presented in Figure 5, the compressive strength at 56 days increased across nearly all groups compared to 28 days, indicating that the soil-based specimens continued to gain strength over time. This increase was particularly pronounced in the control group without biopolymers, which exhibited an 84% strength increase compared to 28 days. The increase in strength is likely due to densification, im-proved interparticle bonding, and void reductions over time, which have been identified in soil mechanics studies as key contributors to the strength development of soil-based materials [33–36]. Studies show that biopolymers, such as chitosan, xanthan gum, and guar gum, improve interparticle bonding and significantly increase soil strength [37,38]. For example, the chitosan biopolymer increased the compressive strength of sand by up to 320 kPa and improved the cohesion by 34.2 kPa [38], while the combination of xanthan gum and guar gum resulted in a greater shear strength than individual treatments [39].

The KG group exhibited the highest compressive strength at both 28 and 56 days, with almost identical values at both time points. This suggests that xanthan gum promoted early strength development by improving cohesion, water retention, and interparticle bonding through its hydrogel-forming ability. Unlike hydraulic binders, such as cement, xanthan gum does not undergo hydration reactions but instead forms a three-dimensional polymeric network that fills soil pores, improves load distribution, and reduces voids, leading to improved mechanical performance. Its hydrogel formation plays a critical role in soil stabilisation by cementing soil particles together, significantly improving inter-particle cohesion [40,41]. Studies have shown that even a 1% xanthan content can increase cohesion by 3.8 to 14 times compared to untreated soil [42]. In addition, its water-retaining properties help to retain moisture in the soil pores, further contributing to its stabilising effect [43].

In contrast, the LBG group exhibited the lowest compressive strength, even lower than the control group, indicating that LBG was ineffective in improving the strength of earth-based mixtures. Cheng and Geng [44] compared the effectiveness of different biopolymers in improving the unconfined compressive strength (UCS) of clay and found that sodium alginate provided the highest UCS improvement, while LBG was not specifically identified as an effective stabiliser.

The AG (Arabic gum) group showed a gradual increase in strength, reaching values close to those of the control group at 28 days and approximately 90% of its maximum strength at 56 days. However, as its compressive strength remained lower than that of the control mix, it cannot be classified as a strength-enhancing binder. While Brzyski [45] reported that Arabic gum at concentrations of 3% and 5% significantly increased the flexural strength (by ~300%) and compressive strength (by 25% and 60% respectively) in lime–metakaolin pastes, the results of this study suggest that the effectiveness of AG as a binder is limited for the soil-based mixtures tested. Similarly, the A group (alginate) showed a 30% increase in strength at 56 days compared to 28 days, indicating that while it improved in strength over time, its performance was less effective than that of the AG group.

The earth-based samples were subjected to capillary water absorption tests for durability testing. The results of the capillary water absorption tests are presented in Table 4.

Table 4. Capillary water absorption test results.

Sample	Beginning	2 min Later Weight (g)	6 min Later Weight (g)	30 min. Later Weight (g)	Mass Change (%)
C	204.19	206.63	206.39	195.39	(−) 4.31
A	188.11	188.76	190.43	170.61	(−) 9.3
LBG	190.62	203.04	206.06	219.12 **	(+) 14.95
AG	198.48	185.44	178.36	*	(−) 10.14
KG	195.60	201.98	203.79	215.33	(+) 10.09

* Disintegrated before leaving water. ** Water level reached top of sample.

During the capillary water absorption test, the KG and LBG samples continued to gain weight due to water absorption after immersion, while the C, A, and AG samples began to dissolve and lose weight within minutes.

For the KG and LBG samples, capillary water reached the top surface within 30 min, and both remained intact during this period. However, after 30 min, they began to disperse, and the test was terminated. Despite the negative effect on mechanical properties, locust bean gum (LBG) exhibited temporary resistance to water penetration before eventually disintegrating. The poorest water resistance was observed in the AG group, where samples disintegrated and lost material immediately upon immersion.

KG improved both mechanical properties and durability, improving also resistance to water absorption.

3.3. Micro Structural Analysis Results

The graphs obtained from the XRD analysis of the samples taken after the 56-day compressive strength test are shown in Figure 6.

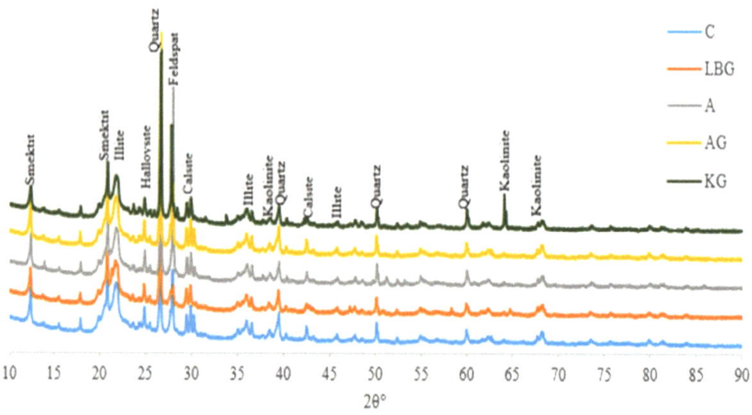

Figure 6. XRD graphs of earth-based mixtures stabilised with biopolymers.

The XRD analysis of the mixtures (Figure 6) revealed the presence of ceramic-based minerals, including quartz, kaolinite, feldspar, smectite, and illite, with no significant differences observed between the control and biopolymer-modified samples. The absence of biopolymer-related peaks in the XRD spectra is consistent with the findings in the literature, which highlight that XRD primarily detects crystalline structures and has a limited sensitivity to non-crystalline or amorphous organic compounds [46]. Research has demonstrated that biopolymers introduced at low substitution rates ($\leq$1%) frequently fall below the detection threshold of XRD due to their weak diffraction signals and amorphous nature [47,48].

Furthermore, the presence of biopolymers may be diminished by high-energy irradiation, thereby further reducing their detectability in XRD analysis [37]. Despite their absence in XRD spectra, biopolymers have been documented to improve soil cohesion and mechanical performance through physical interactions, such as hydrogel formation, interparticle bonding, and moisture retention, rather than through crystalline phase alterations [39,40,49].

These enhancements occur without compromising the crystalline structure of soil minerals. Instead, biopolymers form fibrous and reticulated networks that fill voids, bond soil particles, and enhance interparticle cohesion, leading to increased strength, water stability, and erosion resistance [40].

The SEM images used to visualise the microvoids and bond structures formed in the control, LBG, A, AG, and KG group samples are shown in Figure 7.

The SEM analysis (Figure 7) confirmed the presence of clay minerals in all mixtures. However, due to the low substitution rate (0.1%) of biopolymers, no pronounced biopolymer bonds with earth-based materials were observed. This finding supports the XRD and compressive strength results, suggesting that the biopolymers affected soil properties through physical rather than chemical interactions.

The LBG and KG groups exhibited a more crystalline mineral structure, while other groups exhibited a more reticulated morphology. A closer examination of the KG sample at 100× magnification (lower right-hand corner of Figure 7) revealed large capillaries and macrovoids, a feature observed in all groups. The high percentage of voids was identified as a key factor contributing to the poor mechanical and durability performance of the earth-based mixes.

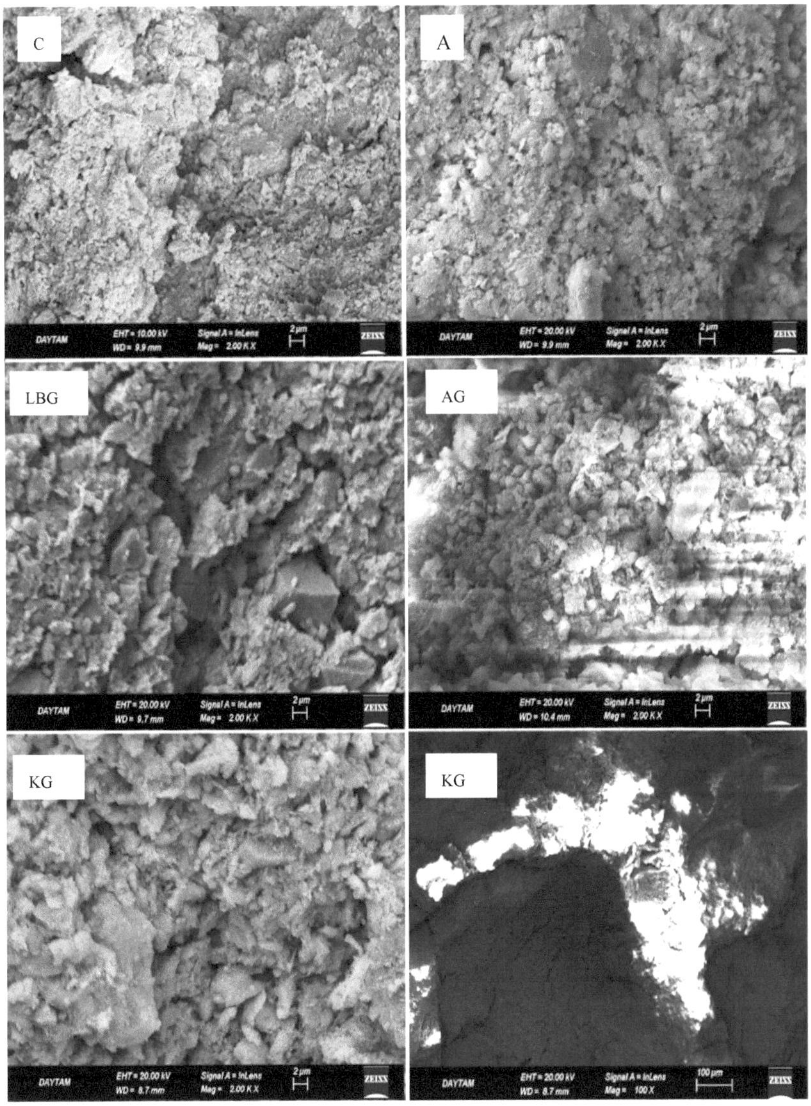

Figure 7. SEM images of the mixtures.

4. Conclusions

This study investigated the potential for improving the mechanical properties of earth-based building materials by stabilising them with four types of biopolymer materials (xanthan gum, Arabic gum, alginate and locust bean gum).

The addition of the biopolymer reduced the unit volume of the mixture. The mixtures stabilised with xanthan gum reached a more solid consistency despite having a low unit volume weight and high air content. When biopolymers were mixed with water, they changed to gel consistency and increased the viscosity of the mixtures.

The incorporation of biopolymers improved both flexural and compressive strength, demonstrating their potential as binders in soil-based materials. Among the formulations tested, xanthan gum exhibited the highest strength values at both 28 and 56 days and showed the greatest resistance to capillary water absorption.

Microstructural analyses (XRD and SEM) confirmed the presence of crystalline minerals but did not detect biopolymer components, probably due to their non-crystalline nature and low substitution rate (0.1%). While the SEM analysis revealed differences in bonding structures and the presence of macrovoids, further investigation is needed to better understand biopolymer–soil interactions at the molecular level.

Therefore, Fourier Transform Infrared (FTIR) spectroscopy is recommended as it can identify functional groups in biopolymers and detect potential chemical interactions between biopolymer molecules and soil particles that are not captured by SEM or XRD analyses.

In conclusion, this study showed that the mechanical and durability properties of earth-based building materials can be improved when stabilised with biopolymers. Of the biopolymer types investigated in this study, xanthan gum improved the properties of the earth-based mix more than the other biopolymers. Biopolymers can be effective components in the development of sustainable building materials and can potentially reduce environmental impact and improve material performance. Based on the results of this study investigating the effective biopolymer type, it is recommended that future research should investigate the optimum biopolymer substitution ratio and the results of advanced day strengths (90, 180, and 360).

While biopolymer-based stabilisation is an emerging and promising approach to sustainable construction, a universally accepted standard for evaluating its performance has yet to be established. However, this study provides a quantitative assessment of mechanical and durability properties, consistent with previous research on biopolymer-stabilised soils. The results show that xanthan gum significantly improves cohesion, flexural strength, and water resistance, highlighting its potential for soil-based stabilisation. Future research should prioritise the development of standardised test methods to enable wider adoption and comparative analyses within the construction industry.

Author Contributions: Conceptualisation, Y.T. and N.K.; methodology, Y.T. and N.K.; validation, Y.T. and N.K.; investigation, Y.T. and N.K.; resources, Y.T.; data curation, N.K.; writing—original draft preparation, Y.T. and N.K.; writing—review and editing, Y.T. and N.K.; visualisation, Y.T. and N.K.; supervision, Y.T. and N.K. All authors have read and agreed to the published version of the manuscript.

Funding: This research was funded by the Coordinatorship of Scientific Research Projects (BAP) of Atatürk University, grant number FKP-2023-13046.

Institutional Review Board Statement: Not applicable.

Informed Consent Statement: Not applicable.

Data Availability Statement: All data supporting the reported results of this study are included within the article.

Conflicts of Interest: The authors declare no conflicts of interest.

References

1. Rael, R. Earth Architecture. Available online: https://books.google.com.tr/books?hl=tr&lr=&id=BsLAeifqe4EC&oi=fnd&pg=PA1&dq=Rael,+R.+(2009).+Earth+Architecture,+Princeton+Architectural+Press,+New+York.+USA.&ots=bMaUeGeTBG&sig=T7hqfxToPaKl4d7FV-mCeMILXTY&redir_esc=y#v=onepage&q&f=false (accessed on 13 March 2025).
2. Galan-Marin, C.; Rivera-Gomez, C.; Garcia-Martinez, A. Use of Natural-Fiber Bio-Composites in Construction versus Traditional Solutions: Operational and Embodied Energy Assessment. *Materials* **2016**, *9*, 465. [CrossRef] [PubMed]
3. Afrin, H. A Review on Different Types Soil Stabilization Techniques. *Int. J. Transp. Eng. Technol.* **2017**, *3*, 19. [CrossRef]
4. Van Damme, H.; Houben, H. Earth Concrete. Stabilization Revisited. *Cem. Concr. Res.* **2018**, *114*, 90–102. [CrossRef]

5. Zuber, S.Z.S.; Kamarudin, H.; Abdullah, M.M.A.B.; Binhussain, M. Review on Soil Stabilization Techniques. *Aust. J. Basic. Appl. Sci.* **2013**, *7*, 258–265.
6. Gomes, M.I.; Faria, P.; Gonçalves, T.D. Earth-Based Mortars for Repair and Protection of Rammed Earth Walls. Stabilization with Mineral Binders and Fibers. *J. Clean. Prod.* **2018**, *172*, 2401–2414. [CrossRef]
7. Chang, R.; Antwi-Afari, M.F. Critical Success Factors for Implementing 3D Printing Technology in Construction Projects: Academics and Construction Practitioners' Perspectives. *Constr. Innov.* 2023, *ahead-of-print*. [CrossRef]
8. Singh, N.; Colangelo, F.; Farina, I. Sustainable Non-Conventional Concrete 3D Printing—A Review. *Sustainability* **2023**, *15*, 10121. [CrossRef]
9. Zaneldin, E.; Ahmed, W.; Alharbi, N.; Alkaabi, R.; Alnuaimi, M. Potential Construction Applications of Sustainable 3D Printed Elements Made from Plastic Waste. *Mater. Sci. Forum* **2023**, *1082*, 277–283. [CrossRef]
10. Sinka, M.; Spurina, E.; Korjakins, A.; Bajare, D. Hempcrete—CO_2 Neutral Wall Solutions for 3D Printing. *Environ. Clim. Technol.* **2022**, *26*, 742–753. [CrossRef]
11. Chamorro, A.; Tighe, S. Development and Application of a Sustainable Management System for Unpaved Rural Road Networks. *Transp. Res. Rec.* **2019**, *2673*, 891–901. [CrossRef]
12. Ramdas, V.M.; Mandree, P.; Mgangira, M.; Mukaratirwa, S.; Lalloo, R.; Ramchuran, S. Review of Current and Future Bio-Based Stabilisation Products (Enzymatic and Polymeric) for Road Construction Materials. *Transp. Geotech.* **2021**, *27*, 100458. [CrossRef]
13. Tarhan, Y.; Kabakuş, N. Enhancing Sustainable Road Construction: Evaluation of the Mechanical and Durability Properties of Stabilized Earth-Based Pavement Materials. *Sustainability* **2024**, *16*, 10784. [CrossRef]
14. Antunes, V.; Simão, N.; Freire, A.C. A Soil-Cement Formulation for Road Pavement Base and Sub Base Layers: A Case Study. *Transp. Infrastruct. Geotechnol.* **2017**, *4*, 126–141. [CrossRef]
15. Muguda, S.; Booth, S.J.; Hughes, P.N.; Augarde, C.E.; Perlot, C.; Bruno, A.W.; Gallipoli, D. Mechanical Properties of Biopolymer-Stabilised Soil-Based Construction Materials. *Geotech. Lett.* **2017**, *7*, 309–314. [CrossRef]
16. Smith, C.; McGregor, J.; Martsinovich, N.; Armistead, S.; Yu, X.; Siripanich, N. Enhancing the Water and Biodegradation Resistance of Biopolymer Stabilised Soils—Design Concepts. *RILEM Bookseries* **2023**, *45*, 293–302. [CrossRef]
17. Ilham, M.; Baba, K.; Simou, S. Naturally Strengthening Rammed Earth: The Promising Potential of Biopolymers. Google Books. Available online: https://books.google.com.tr/books?hl=tr&lr=&id=vKAAEQAAQBAJ&oi=fnd&pg=PA233&dq=16.%09Masrour,+I.;+Baba,+K.;+Simou,+S.+Naturally+Strengthening+Rammed+Earth:+The+Promising+Potential+of+Biopolymers.+Materials+Research+&ots=f4fj_iYSF5&sig=tKDmfs6eT4i8c9sBf6ewJ29dEkE&redir_esc=y#v=onepage&q&f=false (accessed on 13 March 2025).
18. Losini, A.E.; Grillet, A.C.; Bellotto, M.; Woloszyn, M.; Dotelli, G. Natural Additives and Biopolymers for Raw Earth Construction Stabilization—A Review. *Constr. Build. Mater.* **2021**, *304*, 124507. [CrossRef]
19. George, A.; Sanjay, M.R.; Srisuk, R.; Parameswaranpillai, J.; Siengchin, S. A Comprehensive Review on Chemical Properties and Applications of Biopolymers and Their Composites. *Int. J. Biol. Macromol.* **2020**, *154*, 329–338. [CrossRef]
20. Fatehi, H.; Abtahi, S.M.; Hashemolhosseini, H.; Hejazi, S.M. A Novel Study on Using Protein Based Biopolymers in Soil Strengthening. *Constr. Build. Mater.* **2018**, *167*, 813–821. [CrossRef]
21. Jones, D. Assessment of the Potential for Salt Damage to Asphalt Surfacings When Chloride-Treated Unpaved Roads Are Upgraded to a Sealed Standard. *Transp. Res. Rec.* **2019**, *2673*, 774–782. [CrossRef]
22. Cabalar, A.F.; Akbulut, N.; Demir, S.; Yildiz, O. Use of a Biopolymer for Road Pavement Subgrade. *Sustainability* **2023**, *15*, 8231. [CrossRef]
23. Berninger, T.; Dietz, N.; González López, Ó. Water-Soluble Polymers in Agriculture: Xanthan Gum as Eco-Friendly Alternative to Synthetics. *Microb. Biotechnol.* **2021**, *14*, 1881–1896. [CrossRef] [PubMed]
24. Fatehi, H.; Ong, D.E.L.; Yu, J.; Chang, I. Biopolymers as Green Binders for Soil Improvement in Geotechnical Applications: A Review. *Geosciences* **2021**, *11*, 291. [CrossRef]
25. Arabani, M.; Shalchian, M.M.; Baghbani, A. A State-of-the-Art Review on Interactive Mechanisms and Multi-Scale Application of Biopolymers (BPs) in Geo-Improvement and Vegetation Growth. *J. Environ. Manag.* **2024**, *358*, 120905. [CrossRef] [PubMed]
26. Muguda, S.; Hughes, P.N.; Augarde, C.E.; Perlot, C.; Walter Bruno, A.; Gallipoli, D. Cross-Linking of Biopolymers for Stabilizing Earthen Construction Materials. *Build. Res. Inf.* **2022**, *50*, 502–514. [CrossRef]
27. Ana Sayfa. Ata Kimya. Yapı Kimyasalları ve Boya Sanayisine Hammadde Tedariği. Available online: http://www.atakimya.com.tr/ (accessed on 13 March 2025).
28. Artuk, F. Lif ve uçucu kül katkili polimer içi boş kaziklarin yumuşak kil zeminlerdeki davranişinin deneysel olarak araştirilmasi. Master's Thesis, Atatürk University, Erzurum, Türkiye, 2016.
29. *TS EN 12350-7/AC TS EN 12350-7/AC*; Beton—Taze Beton Deneyleri—Bölüm 7: Hava Içeriğinin Tayini—Basınç Yöntemleri (Testing Fresh Concrete—Part 7: Air Content—Pressure Methods). TSE: Ankara, Turkey, 2022. Available online: https://intweb.tse.org.tr/Standard/Standard/Standard.aspx?081118051115108051104119110104055047105102120088111043113104073086065089087052072056121054068120 (accessed on 4 January 2024).

30. *TS EN 12350-5 TS EN 12350-5*; Beton—Taze Beton Deneyleri—Bölüm 5: Yayılma Tablası Deneyi (Testing Fresh Concrete—Part 5: Flow Table Test). TSE: Ankara, Turkey, 2010. Available online: https://intweb.tse.org.tr/standard/standard/Standard.aspx?081118051115108051104119110104055047105102120088111043113104073082112065070113087101115117109052 (accessed on 4 January 2024).
31. *CEN ISO 17892-6 ISO 17892-6:2017*; Geotechnical Investigation and Testing—Laboratory Testing of Soil—Part 6: Fall Cone Test. ISO: Geneva, Switzerland, 2017. Available online: https://www.iso.org/standard/55248.html (accessed on 4 January 2024).
32. *TS EN 13057 TS EN 13057*; Beton Yapılar—Koruma ve Tamir Için Mamul ve Sistemler—Deney Metotları—Kılcal Su Emmeye Direncin Tayini (Products and Systems for the Protection and Repair of Concrete Structures—Test Methods—Determination of Resistance of Capillary Absorption). TSE: Ankara, Turkey, 2002. Available online: https://intweb.tse.org.tr/standard/standard/Standard.aspx?081118051115108051104119110104055047105102120088111043113104073081068111056079084053120078107114 (accessed on 4 January 2024).
33. Giroud, J.P.; Bonaparte, R.; Beech, J.F.; Gross, B.A. Design of Soil Layer-Geosynthetic Systems Overlying Voids. *Geotext. Geomembr.* **1990**, *9*, 11–50. [CrossRef]
34. Arabani, M.; Shalchian, M.M. A Review of the Use of Bio-Based Substances in Soil Stabilization. *Environ. Dev. Sustain.* **2024**, *26*, 13685–13737. [CrossRef]
35. Olarte, J.; Dashti, S.; Liel, A.B. Can Ground Densification Improve Seismic Performance of the Soil-Foundation-Structure System on Liquefiable Soils? *Earthq. Eng. Struct. Dyn.* **2018**, *47*, 1193–1211. [CrossRef]
36. Shah, S.S.A.; Turrakheil, K.S.; Naveed, M. Impact of Wetting and Drying Cycles on the Hydromechanical Properties of Soil and Implications on Slope Stability. *Atmosphere* **2024**, *15*, 1368. [CrossRef]
37. Shariatmadari, N.; Reza, M.; Tasuji, A.; Ghadir, P.; Javadi, A.A. Experimental Study on the Effect of Chitosan Biopolymer on Sandy Soil Stabilization. *E3S Web Conf.* **2020**, *195*, 06007. [CrossRef]
38. Amiri Tasuji, M.R.; Ghadir, P.; Hosseini, A.; Javadi, A.A.; Habibnejad Korayem, A.; Ranjbar, N. Experimental Investigation of Sandy Soil Stabilization Using Chitosan Biopolymer. *Transp. Geotech.* **2024**, *46*, 101266. [CrossRef]
39. Wan, J.; Ouyang, F.; Xiao, H.; Wang, L.; Tao, G. Experimental Study on the Physical and Mechanical Properties of Modified Clay Using Xanthan Gum and Guar Gum Composite Materials. *Sustainability* **2024**, *16*, 5432. [CrossRef]
40. Rong, X.; Deng, S.; Liang, B.; Zhuang, J.; Yu, Y.; Wu, Z. Mechanical Behavior and Strengthening Mechanism of Loess Stabilized with Xanthan Gum and Guar Gum Biopolymers. *Mater. Res. Express* **2024**, *11*, 105305. [CrossRef]
41. Chen, C.; Wu, L.; Harbottle, M. Exploring the Effect of Biopolymers in Near-Surface Soils Using Xanthan Gum–Modified Sand under Shear. *Can. Geotech. J.* **2019**, *57*, 1109–1118. [CrossRef]
42. Lee, S.; Chang, I.; Chung, M.K.; Kim, Y.; Kee, J. Geotechnical Shear Behavior of Xanthan Gum Biopolymer Treated Sand from Direct Shear Testing. *Geomech. Eng.* **2017**, *12*, 831–847. [CrossRef]
43. Chang, I.; Kwon, Y.M.; Im, J.; Cho, G.C. Soil Consistency and Interparticle Characteristics of Xanthan Gum Biopolymer–Containing Soils with Pore-Fluid Variation. *Can. Geotech. J.* **2019**, *56*, 1206–1213. [CrossRef]
44. Cheng, Z.; Geng, X. Investigation of Unconfined Compressive Strength for Biopolymer Treated Clay. *Constr. Build. Mater.* **2023**, *385*, 131458. [CrossRef]
45. Brzyski, P. The Influence of Gum Arabic Admixture on the Mechanical Properties of Lime-Metakaolin Paste Used as Binder in Hemp Concrete. *Materials* **2021**, *14*, 6775. [CrossRef] [PubMed]
46. Bagherinia, M. Mechanical, Durability, and Microstructure of Soft Clay Stabilised with Anionic Biopolymer. *Constr. Build. Mater.* **2024**, *417*, 135343. [CrossRef]
47. Mitikie, B.B.; Geleta, D.G. Enhancement of Concrete Performance with Xanthan Gum: Potential of Natural Biopolymer. *Discov. Civil Eng.* **2025**, *2*, 41. [CrossRef]
48. de Matos, P.R.; Andrade Neto, J.S.; Jansen, D.; De la Torre, A.G.; Kirchheim, A.P.; Campos, C.E.M. In-Situ Laboratory X-Ray Diffraction Applied to Assess Cement Hydration. *Cem. Concr. Res.* **2022**, *162*, 106988. [CrossRef]
49. Torres-Huerta, A.M.; Palma-Ramírez, D.; Domínguez-Crespo, M.A.; Del Angel-López, D.; De La Fuente, D. Comparative Assessment of Miscibility and Degradability on PET/PLA and PET/Chitosan Blends. *Eur. Polym. J.* **2014**, *61*, 285–299. [CrossRef]

polymers

MDPI

Article

Functionalization of Phenolic Aldehydes for the Preparation of Sustainable Polyesters and Polyurethanes

Rachele N. Carafa [1], Brigida V. Fernandes [1], Clara Repiquet [2], Sidrah Rana [3], Daniel A. Foucher [1] and Guerino G. Sacripante [1,*]

[1] Department of Chemistry and Biology, Toronto Metropolitan University, 350 Victoria St., Toronto, ON M5B 2K3, Canada; rachele.carafa@torontomu.ca (R.N.C.); brigida.fernandes@torontomu.ca (B.V.F.); daniel.foucher@torontomu.ca (D.A.F.)
[2] Graduate School of Chemistry and Engineering, Ecole Nationale Supérieure de Chimie de Lille (ENSCL) Centrale Lille Institut, Cité Scientifique, CS 20048, 59651 Villaneuve d'Ascq, France; clara.repiquet@enscl.centralelille.fr
[3] Department of Chemistry, University of Toronto, 80 St. George St., Toronto, ON M5S 1A1, Canada; sidrah.rana@mail.utoronto.ca
* Correspondence: gsacripante@torontomu.ca

Abstract: Biobased organic diols derived from the phenolic aldehyde by-products in the depolymerization of lignin (4-hydroxybenzaldehyde, vanillin, and syringaldehyde) for the synthesis of polyesters and polyurethanes is described. Methods to prepare lignin-based diols involved a two-step synthetic route using either a hydroxy alkylation and aldehyde reduction or an aldehyde reduction and Williamson–Ether substitution. The preparation of five polyesters (PEs) and ten polyurethanes (PUs) from lignin-based diols was also performed and their physical and thermal properties were analyzed. DSC analysis confirmed the amorphous nature of all synthesized polymers, and GPC analysis revealed broad dispersities and high molecular weights. Two PE polyols were also derived from a vanillin-based diol at concentrations of 10 and 25 wt% for their usage in sustainable PU foams. PU foams were prepared from these polyols, where it was found that only the foam containing the 10 wt% formulation was suitable for mechanical testing. The PU foam samples were found to have good hardness and tensile strengths compared to both control foams, showing potential for the incorporation of biobased polyols for PU foam formation.

Keywords: vanillin; 4-hydroxybenzaldehyde; syringaldehyde; lignin; biobased; polyesters; polyurethanes; polyester polyols; polyurethane foams

Academic Editors: Vishal Gavande, Vasi Shaikh and Won-Ki Lee

Received: 31 January 2025
Revised: 22 February 2025
Accepted: 25 February 2025
Published: 27 February 2025

Citation: Carafa, R.N.; Fernandes, B.V.; Repiquet, C.; Rana, S.; Foucher, D.A.; Sacripante, G.G. Functionalization of Phenolic Aldehydes for the Preparation of Sustainable Polyesters and Polyurethanes. *Polymers* **2025**, *17*, 643. https://doi.org/10.3390/polym17050643

1. Introduction

Polymers are an important part of everyday life, with applications such as packaging, clothing, medical, and electronics, to name a few [1,2]. More recently, the polymer industry has embraced the move towards sustainability to lessen the dependency on fossil fuel-based resins [3]. In particular, the use of renewable feedstocks, such as biomass, as an alternative to fossil fuels is promising [3,4]. Lignocellulose is an abundant and renewable biomass source that has the potential to produce sustainable chemicals and materials [5]. Lignin, one of the three main components of lignocellulose, is an aromatic macromolecule comprising three phenylpropanoid structures, *p*-coumaryl alcohol, coniferyl alcohol, and sinapyl alcohol, which then give rise to soluble small molecules, *p*-hydroxyphenyl (H), guaiacyl (G), and syringyl (S) units [6,7]. Vanillin is a G-type phenolic aldehyde derived from various biomass sources, including oil, woody biomass such as lignin, or orchid pods, and is used as a vanilla flavouring agent in the cosmetic and food industry [8]. 4-hydroxybenzaldehyde is an H-type phenolic aldehyde that is also one of the main components of the vanilla flavour in vanilla pods [9]. Syringaldehyde is an S-type phenolic aldehyde that can be isolated from Klason lignin by treating spruce or maple wood with alkali solutions [10]. All three compounds are obtainable by depolymerization methods of lignin, thereby making them potential candidates for the synthesis of biobased polymers [11].

Polyurethanes (PUs) are a class of polymers that can be used in various markets for applications as foams, coatings, and elastomers, among others [12]. PUs are synthesized from the polyaddition reaction between the hydroxyl (OH) groups of a diol or polyol with the isocyanate (N=C=O) groups of diisocyanates [13]. Most polyols used for PU synthesis typically have low molecular weights (M_W's) and comprise either polyesters (PEs) or polyethers [13,14]. Both types of polyols are used for PU foam synthesis, with PE polyols offering the advantage of being more stable towards oxidation but more susceptible to hydrolysis and the reverse being true for polyether polyols [15]. Since the bulk of these components used to prepare PUs are derived from petrochemicals, more sustainable approaches to PUs have been investigated, including the use of biobased diisocyanates and the incorporation of polyols obtained from renewable feedstocks [16,17].

PEs are a class of polymers that are typically used for plastic packaging or in clothing fibers [4,18]. PEs are synthesized by the polycondensation reaction between the carboxylic acid or ester groups of diacids with the OH groups of diols or polyols, resulting in the formation of water or methanol (MeOH) as a by-product [19]. Like PUs, there has been recent interest in finding more sustainable sources of PEs [2]. Aliphatic PEs have gained considerable attention since they have biodegradable properties and are susceptible to hydrolysis [20,21]. Since most aliphatic PEs have limited uses due to low softening points and melt temperatures, the incorporation of a rigid benzene ring in their structure to form aliphatic–aromatic PEs improves both their mechanical properties and thermal stability compared to aliphatic PEs alone [21].

While lignin itself has been utilized in the direct synthesis of PUs [22,23], PEs [20], or polyols [24,25], the incorporation of lignin-derived phenolic aldehydes for these applications is of interest. One method to utilize these phenolic aldehydes is by first converting them to a diol. A study by Zhao et al. successfully made a vanillin-based diol to synthesize both PEs and PUs [26]. Here, the phenol group on vanillin was first hydroxy alkylated using ethylene carbonate in the presence of tetrabutylammonium iodide (TBAI) as a phase-transfer catalyst to yield 4-(2-hydroxyethoxy)-3-methoxybenzaldehyde (HMBD) at elevated temperatures (110 °C for 24 h). The aldehyde group was then reduced using sodium borohydride ($NaBH_4$) in MeOH at room temperature (RT) to produce the desired diol 2-(4-(hydroxymethyl)-2-methoxyphenoxy) ethan-1-ol (HMEO). PEs were synthesized using either oxalyl chloride, succinyl chloride (SC), terephthaloyl chloride or 2,5-furandicarbonyl dichloride as the diacyl chlorides in anhydrous pyridine and tetrahydrofuran (THF). PUs were synthesized using either hexamethylene diisocyanate (HDI), isophorone diisocyanate (IPDI), or methylene diphenyl diisocyanate (MDI) in anhydrous THF with 1,8-Diazabicyclo(5.4.0)undec-7-ene (DBU) as a catalyst [26]. Canceill et al. synthesized an alternative diol based on vanillin for their use as cryptophanes [27]. In this study, vanillin was reduced to vanillyl alcohol and then dimerized using various dihalides in ethanol (EtOH) under reflux for 3 h [27]. Similar derivatives of these diols have also been synthesized from 4-hydroxybenzaldehyde [28,29] and syringaldehyde [30].

The main goal of this study is to prepare PEs and PUs from these lignin-derived phenolic aldehydes. In the first step, diols from these compounds are synthesized in one of two ways (Scheme 1): a hydroxy alkylation and aldehyde reduction or an aldehyde reduction and Williamson–Ether substitution. PUs are synthesized using either MDI or 1,5-pentamethylene diisocyanate (PDI) (Scheme 2a) while PEs are synthesized using SC

(Scheme 2b). In lieu of an additional diacid chloride for PE synthesis, one of the vanillin-based diols is utilized in the synthesis of a low M_W PE polyol to be used in PU foam formation using succinic acid (SA) and 1,3-propanediol (PDO) (Scheme 3). The properties of these polymers are also investigated.

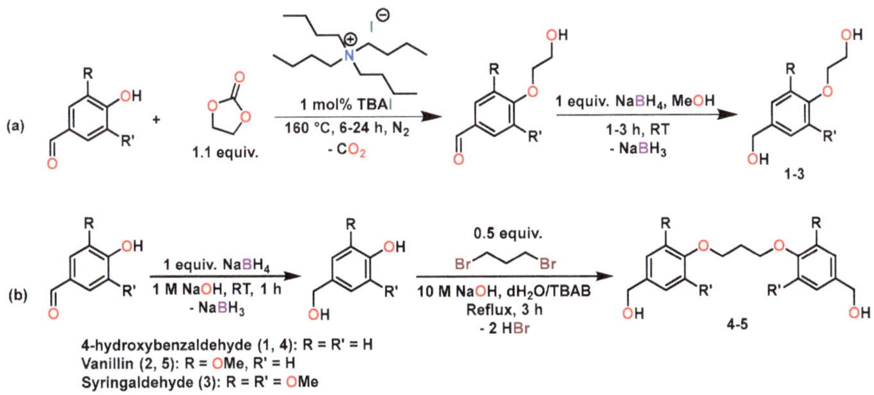

Scheme 1. Synthetic routes for the preparation of organic diols from phenolic aldehydes for polymer synthesis: (**a**) hydroxy alkylation with ethylene carbonate and aldehyde reduction with $NaBH_4$. (**b**) Aldehyde reduction with $NaBH_4$ and Williamson–Ether synthesis with 1,3-dibromopropane.

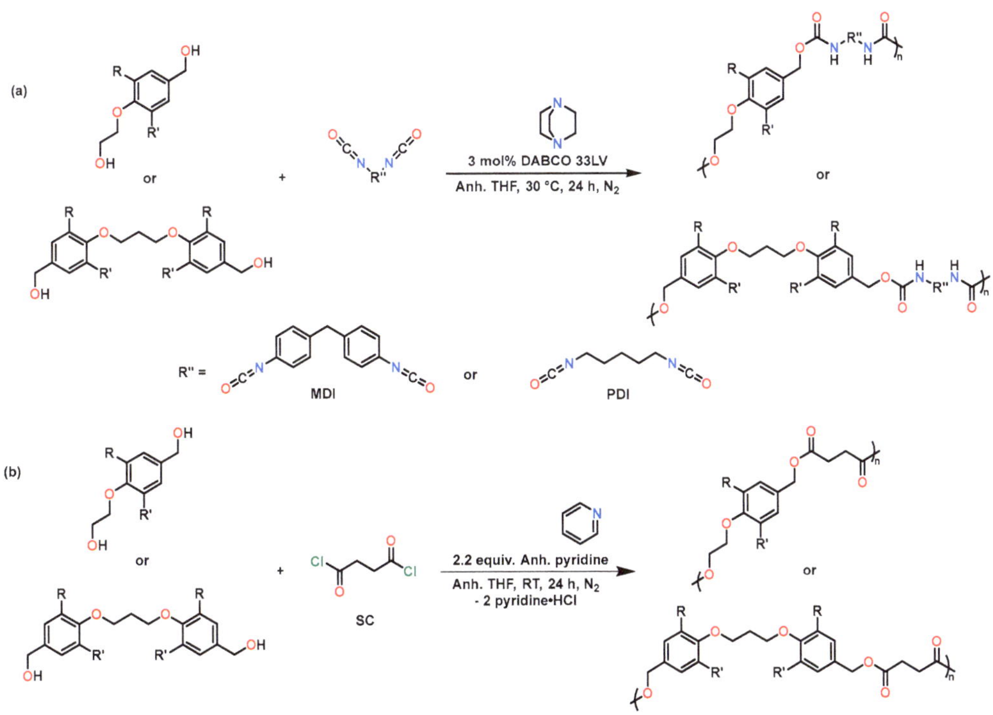

Scheme 2. Synthetic routes for (**a**) PUs from MDI or PDI and (**b**) PEs from SC.

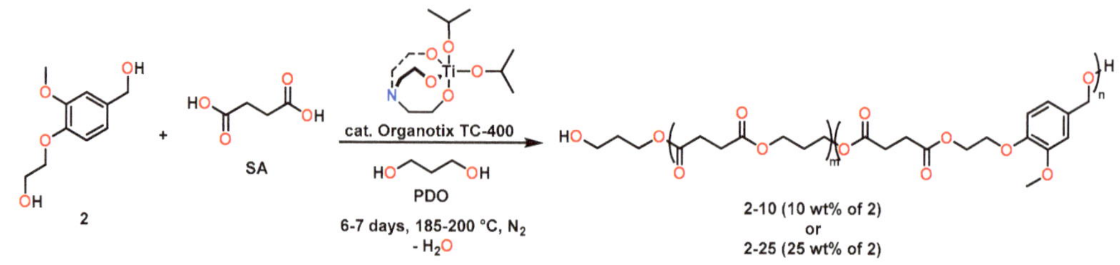

Scheme 3. Synthesis of a PE polyol from diol **2**.

2. Materials and Methods

2.1. Materials

Vanillin, 4-hydroxybenzaldehyde, syringaldehyde, $NaBH_4$, TBAI, tetrabutylammonium bromide (TBAB), SC, and anhydrous pyridine were purchased from MilliporeSigma Canada in Oakville, ON. Ethylene carbonate and 1,3-dibromopropane were purchased from Alfa Aesar in Ward Hill, MA. MDI, PDI, 1,4-diazabicyclo[2.2.2]octane (DABCO 33LV), SA, PDO, titanium (IV) triethanolaminato isopropoxide (8.2 Ti wt%) (Organotix TC-400), triethyl citrate (TEC), Tegostab B4113, and orange dye paste (Dye 2012C) were supplied by Evoco Ltd in Toronto, ON. Anhydrous THF was dried using an SPS solvent purification solvent. All other solvents were reagent grade and used as received. All reactions carried out in a nitrogen environment were performed using standard Schlenk line techniques.

2.2. General Synthesis of Phenoxy Diols (Diols **1**–**3**)

In a 100 mL round bottom flask, phenolic aldehyde (5.00–10.00 g), ethylene carbonate (1.1 equiv.), and TBAI (1 mol%) were added. The reaction was refluxed at 160 °C under N_2 and the reaction was monitored by thin layer chromatography (TLC) until completion (6–24 h). After cooling to RT, the material solidified in a quantitative yield and was used without further purification. Afterwards, the hydroxy alkylated aldehyde was dissolved in MeOH (50–100 mL) and cooled to 0 °C in an ice bath. $NaBH_4$ (1 equiv.) was slowly added to the solution and the reaction was monitored by TLC until completion (1–3 h). The solvent was removed via rotary evaporation to yield a waxy solid in a quantitative yield that was used without further purification.

2.3. General Synthesis of Bis(Phenolic Alcohols) (Diols **4**–**5**)

In a 125 mL Erlenmeyer flask, 5.00–10.00 g of vanillin or 4-hydroxybenzaldehyde were dissolved in NaOH (1 M, 50–100 mL) and cooled to 0 °C on an ice bath. $NaBH_4$ (1 equiv.) was slowly added to the solution with cooling and the reaction was stirred for 1 h at RT. HCl (3 M) was added dropwise until a slightly acidic pH was reached, after which a white precipitate formed. The solution was left to sit until crystallization was complete, then it was vacuum filtered and washed with ~15 mL of cold H_2O (Yield = 58–65%). The reduced alcohol (2.00 g) was then dissolved in a saturated solution of TBAB in water (7 mL). NaOH (10 M, 1.3 mL) was added along with 1,3-dibromopropane (0.5 equiv.) and the mixture was refluxed for 3 h. The solution was cooled to RT and the product was isolated by vacuum filtration and washed with ~15 mL of DCM to obtain a beige or white powder (Yield = 96%).

2.4. General Preparation of PUs

The diol (1.00 g) was added to a 100 mL 3-neck round bottom flask along with anhydrous THF (20 mL) and DABCO (3 mol% based on diol amount). MDI or PDI (1 equiv.) was added under N_2, and the reaction was stirred at 30 °C for 24 h, after which the mixture was precipitated or washed in cold MeOH (MDI) or hexanes (PDI) to remove any unreacted diol or diisocyanate.

2.5. General Preparation of PEs

The diol (1.00 g) was added to a 100 mL 3-neck round bottom flask along with anhydrous THF (20 mL). Anhydrous pyridine (2.2 equiv.) was added dropwise to the stirred solution at 0 °C under N_2 followed by a dropwise addition of SC (1 equiv.), after which the by-product, pyridine hydrochloride, immediately precipitated out of solution as a pink powder. The reaction was warmed to RT and stirred for 24 h. The by-product was filtered off via vacuum filtration and discarded. The brown filtrate was transferred to a round bottom flask and solvent removed via rotary evaporation. The product was stored in a desiccator to form either brown oil or powder.

2.6. General Preparation of PE Polyols

For exact amounts of materials and reaction conditions used, please refer to the supplementary section (Tables S3 and S4). To a 500 mL 3-neck round bottom flask equipped with a heating mantle, temperature probe, nitrogen outlet, and a water condenser, an appropriate amount of PDO was added. A mechanical stirrer was used to mix the solution at ~200 rpm and the temperature set to 100 °C. SA was added slowly, followed by diol **2**. Aluminum foil was wrapped around the flask and the nitrogen flow set to 0.25 L/min. The water condenser was turned on to remove any excess water from the reaction mixture. After an hour, the condenser was turned off and the temperature was adjusted to 185 °C. AV and viscosity measurements were recorded twice a day and when necessary, additional TC-400 or PDO were added. The reaction was stopped when the AV measurements were found to remain constant (~6–7 days), and the polyol transferred into a 500 mL plastic container.

2.7. Preparation of PU Foams

2.7.1. PU Foam Premix

To a 500 mL container, 120.00 g of the polyol, 36.00 g of TEC, 4.80 g of PDO, 1.01 g of dH_2O, 0.66 g of Tegostab B4113, 0.96 g of DABCO 33 LV, and 4.46 g of orange dye paste (Dye 2012C) was added. The solution was then mixed for 2 min using a homogenizer at 1800 rpm. The premix (~95 g) was poured into an 18 L cup for plaque formation while ~57 g was poured into a 12 L cup for free rise analysis.

2.7.2. PU Foam Free Rises

The premix was stirred for 30 s at 1800 rpm, then ~24 g of MDI was injected via a syringe. The mixture was homogenized for an additional 5 s, and ~30 g of this mixture was poured into a small paper cup and the cream time, rise time, and tack free time were analyzed.

2.7.3. PU Foam Plaque Moulds

The premix was stirred for 30 s at 1800 rpm, then ~39 g of MDI was injected via a syringe. The mixture was homogenized for an additional 5 s, and ~125 g of this mixture was poured into a plaque mould with a volume of 311 cm^3 and kept at a temperature of 50–55 °C. The mixture was allowed to cure in the mould for 20 min, yielding a PU elastomer foam having a plaque density of ~0.32 g/cm^3.

2.8. Characterization Methods

For the diols, PEs, and select PUs, characterization by ^1H and ^{13}C nuclear magnetic resonance (NMR) were carried out on a 400 MHz Bruker Avance II spectrometer using deuterated acetone (Acetone-d_6), deuterated chloroform ($CDCl_3$), deuterated methanol

(CD_3OD), or deuterated dimethyl sulfoxide (DMSO-d_6) as the solvent. ^{1}H NMR spectra (400 MHz) and ^{13}C NMR spectra (101 MHz) were referenced against the internal standard tetramethylsilane (TMS). For the remaining PUs, ^{1}H and ^{13}C NMR were carried out on a 600 MHz Bruker Avance II spectrometer using DMSO-d_6 or $CDCl_3$ as the solvent. Fourier Transform Infrared Spectroscopy (FTIR) analysis was performed on an Agilent Cary 630 FTIR Spectrometer. Differential Scanning Calorimetry (DSC) was completed on a DSC Q20 TA Instrument at a heating rate of 10 °C/min in an aluminum Tzero pan with approximately 10 mg of sample under a nitrogen atmosphere with a flow rate of 20 mL/min. The third cycle was used for sample analysis. Thermal gravimetric analysis (TGA) was conducted on a PerkinElmer TGA 4000 under a nitrogen atmosphere with a flow rate of 40 mL/min and a heating rate of 10 °C/min in a Platinum HT pan with approximately 10 mg of sample. Elemental analysis was performed at McMaster University in Hamilton, Ontario. GPC of the PUs was determined using a TOSOH GPC-WS instrument equipped with two TSKgel a-M 13 mm mixed bed columns, a refractive index, and viscometer detector in conjunction with a Wyatt DAWN HELOS multi-angle light-scattering detector. DMSO at 70 °C was used as the eluent (0.5 mL/min, 70 °C, 1.00 mg/mL). Prior to injection into the GPC, each sample was filtered through a 0.22 mm PTFE filter to remove any particulates that could clog or disrupt the GPC column. Viscosity was measured using a CAP 2000+ Viscometer at 70 °C at a speed of 50 rpm. Acid value (AV) measurements were performed by titration with a 0.1 N solution of potassium hydroxide (KOH) in EtOH and 3–5 drops of phenolphthalein as an indicator. Samples were first dissolved in THF, and the volume of titrant used to cause a color change was recorded to determine the AV. The OH number measurements were performed on a Metrohm 848 Titrino plus following ASTM E1899-08 [31].

2.9. Mechanical Testing of PU Foams

The ADMET eXpert 7601 Tensile Tester was used to evaluate split tear strength, and the Instron 34SC-1 was used to measure tensile strength, tensile elongation, and Die C tear strength. Tensile test samples of the PU foams were prepared in dog-bone shapes using a die cutter. The samples had dimensions in accordance with one or more of ASTM D412, ASTM D3574-17, and SATRA TM-2 standards [32,33]. Each sample was placed between clamps of the tensile tester, and the appropriate force was applied to the sample at a particular rate to measure the characteristics and properties of the PU foams. PU foam hardness was measured on an Asker C scale.

3. Results

3.1. Synthesis and Characterization of Organic Diols

3.1.1. Phenoxy Diols

4-hydroxybenzaldeyde, vanillin, and syringaldehyde were subjected to a hydroxy alkylation reaction following a modified literature procedure (Scheme 1a) [34]. A slight excess of ethylene carbonate was utilized with catalytic amounts of TBAI under solvent-less conditions to synthesize the target intermediates [4-(2-hydroxyethoxy)benzaldehyde (HEB) from 4-hydroxybenzaldehyde, HMBD from vanillin, and 4-(2-hydroxyethoxy)-3,5-dimethoxybenzaldehyde (HEDB)] from syringaldehyde. The reaction conditions proposed by Sacripante et al. on rosin acids indicated complete conversion after 6 h [34]. Monitoring the reaction by TLC showed that vanillin was fully converted after this allotted time, whereas 4-hydroxybenzaldehyde and syringaldehyde were fully converted after 24 h. The ^{1}H NMR spectra for all three intermediates revealed two triplets in the 3–4 ppm range (Figures S1, S5 and S9), while the ^{13}C NMR spectra displayed two new signals at ~60 and 70 ppm (Figures S2, S6 and S10), indicative of the newly formed methylene groups. Afterwards, the aldehyde was reduced to a primary alcohol following the procedure by

Zhao et al. [26]. The intermediates were dissolved in MeOH, and the solution was cooled to 0 °C. One equivalent of $NaBH_4$ was then added before slowly warming the solution to RT. A TLC indicated that HEB and HMBD fully reacted after only 1 h to form diols **1** and **2**, respectively, while HEDB required 3 h to form diol **3**. ^{1}H NMR analysis showed that the aldehyde proton resonances at ~9.9 ppm disappeared completely, and a new resonance at ~4.4 ppm appeared, representing the two protons on the primary alcohol (Figures S3, S7 and S11). ^{13}C NMR analysis also confirmed the disappearance of the aldehyde signal at ~190 ppm and a new signal at ~63 ppm (Figures S4, S8 and S12). ^{1}H and ^{13}C NMR spectroscopy confirmed the structures of the intermediates and diols and agreed with previously reported literature values [26,28,30].

3.1.2. Bis(Phenolic Alcohols)

4-hydroxybenzaldehyde and vanillin were first reduced using $NaBH_4$ in NaOH, followed by an acid workup with HCl to form 4-hydroxybenzyl alcohol (HBA) and vanillyl alcohol (VA), respectively (Scheme 1b) [35]. Similar to diols **1** and **2**, the ^{1}H NMR spectra for HBA and VA revealed the absence of the aldehyde protons and a new resonance at ~4.5 ppm (Figures S13 and S17), while the ^{13}C NMR spectra revealed a new resonance at ~63 ppm (Figures S14 and S18), indicative of the newly formed alcohol, which agrees with previously reported data [35,36]. The reduced alcohols were then dimerized following a modified procedure by Canceill et al. [27]. The dihalide used was 1,3-dibromopropane with a concentrated NaOH solution, along with a saturated solution of TBAB and H_2O. Diols **4** and **5** were successfully synthesized after refluxing the mixtures for 3 h. ^{1}H NMR analysis showed two new signals: a triplet at ~4.1 ppm for the two methylene protons near the oxygen atoms and a quintet at ~2.1 ppm for the protons in between the methylene protons (Figures S15 and S19). ^{13}C NMR analysis also confirmed the structures of **4** and **5** with the addition of carbon signals at ~65 and 29 ppm (Figures S16 and S20). The results are in agreement with previously reported literature values for diols **4** and **5** [27,29]. FTIR analysis of all five diols also showed the presence of the O-H stretch (Figure S21) to further confirm their successful synthesis.

3.2. Synthesis and Characterization of PUs

For PU synthesis, diols **1–5** underwent a polyaddition reaction following a modified literature procedure with a diisocyanate using DABCO 33LV as the catalyst and anhydrous THF as the solvent (Scheme 2a) [26]. These conditions were utilized due to the ability to synthesize thermoplastic PUs at lower temperatures, allow for proper mixing of all components on a small scale, and explore the properties of these PUs for their potential application in PU foams. MDI and PDI were selected as the diisocyanates since MDI is a commonly used petroleum-derived aromatic diisocyanate while PDI is a biobased aliphatic diisocyanate [37]. The resulting polymers were either white or beige powders after precipitation. The yields ranged from 26 to 98%, where it was observed that the PUs synthesized from MDI typically had higher yields compared to PDI. The overall properties of the PUs are summarized in Table 1.

^{1}H NMR analysis of PU-1M, PU-2M, PU-3M, PU-4M, and PU-5M showed an increase in the integration of the aromatic protons in between 6.8 ppm and 7.3 ppm, confirming the presence of the MDI backbone (Figures S22, S26, S30, S34 and S38). There was also a broad singlet at ~3.6 ppm representative of the methylene bridge. Due to the poor solubility of select samples, ^{13}C NMR analysis did not always reveal all the respective signals for the polymers. However, they all showed the presence of aromatic carbons (Figures S23, S27, S31, S35 and S39). ^{1}H NMR analysis of PU-1P, PU-2P, PU-3P, PU-4P, and PU-5P showed new signals at ~1.3 ppm and ~2.9 ppm, which correspond to the aliphatic protons from

PDI (Figures S24, S28, S32, S36 and S40). For ^{13}C NMR analysis, new resonances between 20 ppm and 30 ppm were typically observed, which were indicative of the PDI backbone (Figures S25, S29, S33, S37 and S41). FTIR analysis also confirmed the synthesis of all PUs, with the broad O-H stretch previously observed in the diols being replaced with an N-H stretch appearing at ~3300 cm^{-1} and the C=O urethane stretch at ~1690–1700 cm^{-1} for each of the ten PUs (Figures S52 and S53). An additional C-H stretch between 2800 and 2900 cm^{-1} was observed for PU-1P to PU-5P, confirming the presence of the aliphatic carbons from PDI.

Table 1. Characterization Data of PU Polymers.

Sample	Diol	Diisocyanate	Yield (%)	T_g (°C)	T_d (5%) (°C)	T_d (50%) (°C)	M_W (Da)	M_n (Da)	$Đ$ (M_W/M_n)
PU-1M	1	MDI	23	128.6	235	467	3186	2477	1.3
PU-1P	1	PDI	18	21.0	187	346	65,500	36,700	1.8
PU-2M	2	MDI	98	163.8	205	438	158,000	71,200	2.3
PU-2P	2	PDI	40	−1.7	196	337	48.400	32,600	1.5
PU-3M	3	MDI	40	153.3	256	386	12,460	4645	2.7
PU-3P	3	PDI	31	34.3	201	332	61,000	34,700	1.8
PU-4M	4	MDI	97	183.8	167	372	-	-	-
PU-4P	4	PDI	63	62.2	231	406	10,700	3910	2.7
PU-5M	5	MDI	95	103.8	164	389	-	-	-
PU-5P	5	PDI	86	18.3	202	367	-	-	-

DSC analysis demonstrated that all of the PUs are amorphous, with only one glass transition temperature (T_g) being present in each PU (Figure 1a,b). The T_gs ranged from −1.7 to 183.8 °C, with the T_gs of the PUs synthesized with MDI being greater than those synthesized with PDI. This was to be expected as MDI is a more rigid diisocyanate due to the presence of the phenyl rings and therefore requires a higher temperature for the PU to transition from a glassy hard state to a rubbery state. Conversely, PDI is a more flexible diisocyanate since it contains aliphatic carbons, thereby not needing as much energy to cause a transition in the PU. It was also observed that PU-4M and PU-4P had the highest T_gs compared to the other PUs, which could be due to the large size of the starting diol 4 compared to diols 1–3 and the lack of steric hindrance due to the absence of methoxy substituents, although no noticeable trend was observed between the PUs. TGA analysis investigated the decomposition temperature (T_d) of the PUs at 5% and 50% weight loss (Figure 1c,d). For the PUs synthesized from diols **1–3**, the MDI containing polymers exhibited higher thermal stability at 5% and 50% compared to the PDI containing polymers due to the rigidity and flexibility of the respective starting diisocyanates. At 50% weight loss, it was observed that increasing steric hindrance resulted in decreasing thermal stability for these polymers. For the PUs synthesized from diols **4–5**, the opposite trend was observed, where the PDI containing polymers showed slightly greater thermal stability than the MDI containing polymers, particularly at 5% weight loss. This may be due to the ether linkage present on the starting diol, leading to an increase in flexibility of the overall polymer. It was also observed that PU-3M and PU-1M had the highest thermal stability at 5% and 50% weight loss, respectively, compared to the other PUs, which may also be due to the presence or absence of methoxy groups and the bulkiness of the aromatic rings. The derivative weight change for all PUs typically showed two or three decomposition stages (Figures S55 and S56), indicative of the breakdown of not only the aromatic rings but also the urethane linkages [26].

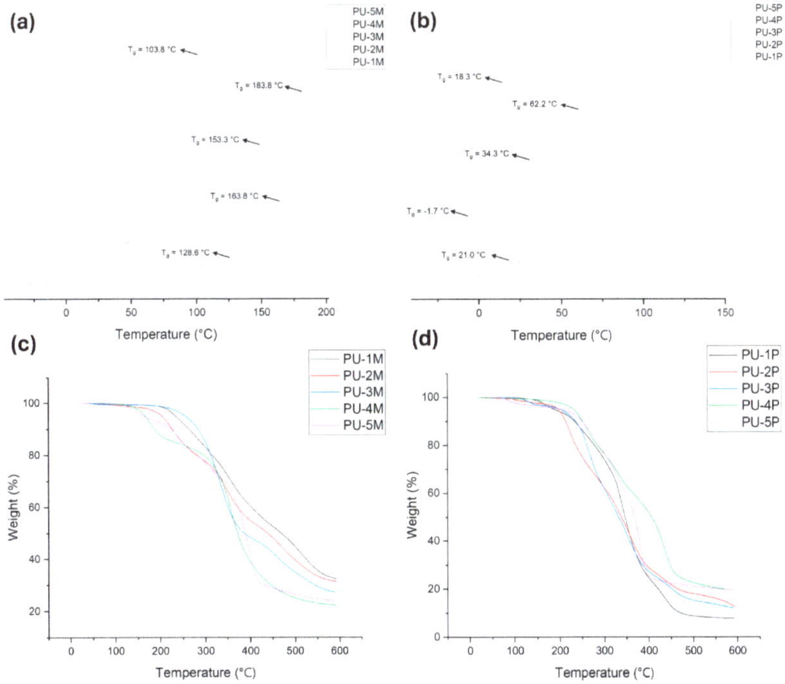

Figure 1. (**a**) DSC traces of PUs synthesized from MDI (exotherm down). (**b**) DSC traces of PUs synthesized from PDI (exotherm down). (**c**) TGA curves of PUs synthesized from MDI. (**d**) TGA curves of PUs synthesized from PDI.

GPC analysis was performed on all PUs except for PU-4M, PU-5M, and PU-5P due to their poor solubility in DMSO. Apart from PU-1M, all PUs had M_W's greater than 10,000 Da. PU-2M was found to have the highest M_W while PU-1M had the lowest M_W. The high M_W of PU-2M could be attributed to having a higher degree of polymerization; however, no obvious trends were observed between PU-2M and the other PUs. All analyzed PUs were also found to have broad dispersities greater than or equal to 1.3, which was to be expected due to the polyaddition reaction taking place between the diol and diisocyanate. Elemental analysis (EA) revealed the carbon and hydrogen content found in the PUs to closely match what was expected for the carbon end groups on the polymer chains within 3% accuracy (Table S1). The only exceptions were PU-1P and PU-2P, where the found carbon content was significantly lower or higher than the calculated content by at least 10%.

3.3. Synthesis and Characterization of PEs

For PE synthesis, diols **1**–**5** underwent an acylation reaction following a previously reported literature procedure using an esterifying agent, anhydrous pyridine as the catalyst, and anhydrous THF as the solvent (Scheme 2b) [26]. SC was selected as the diacid chloride due to the fact that the polymerization reaction can occur at RT as opposed to the higher temperatures required of its carboxylic acid counterpart, SA [38]. After the reaction proceeded for 24 h and the by-product was removed, the resulting polymers were either oil (PE-1, PE-2, and PE-3) or powder (PE-4 and PE-5). Yields ranged from 50 to 79%. The overall properties of the PEs are summarized in Table 2.

Table 2. Characterization Data of PE Polymers.

Sample	Diol	Yield (%)	T_g (°C)	T_d (5%) (°C)	T_d (50%) (°C)
PE-1	**1**	56	-	127	311
PE-2	**2**	58	20.4	138	329
PE-3	**3**	50	−5.1	151	364
PE-4	**4**	79	-	176	395
PE-5	**5**	70	40.4	192	398

In all five PEs, ^{1}H NMR analysis introduced a new broad singlet at ~2.7 ppm representative of the SC protons (Figures S42, S44, S46, S48, and S50) while ^{13}C NMR analysis showed two new signals at ~172 ppm for the C=O ester carbons and ~29 ppm for the aliphatic ester carbons (Figures S43, S45, S47, S49 and S51). FTIR analysis also confirmed the C=O ester stretch at ~1720 cm^{-1} (Figure S54). DSC analysis confirmed the amorphous nature of the PEs, with the T_gs ranging from −5.1 to 40.4 °C (Figure 2a). The only exceptions were PE-1 and PE-4, which did not exhibit a noticeable T_g. This could imply that these PEs exhibit a very weak Tg that is not detectable by DSC, although it may be related to the fact that both PEs were derived from 4-hydroxybenzaldehyde. TGA analysis of the PEs revealed two decomposition stages, which was previously reported to be due to the breakdown of the ester groups and aromatic rings (Figure 2b and Figure S57) [26]. It was also observed that increasing the number of methoxy groups resulted in higher thermal stability at 5% and 50% weight loss. Additional results showed that increasing the bulkiness of the diol also resulted in greater thermal stability as the PEs synthesized from diols **4–5** exhibited higher T_ds than the PEs synthesized from diols **1–3**. EA revealed that the found carbon and hydrogen content were significantly lower than the calculated content (Table S2). This could be attributed to unremoved by-products, particularly pyridine hydrochloride.

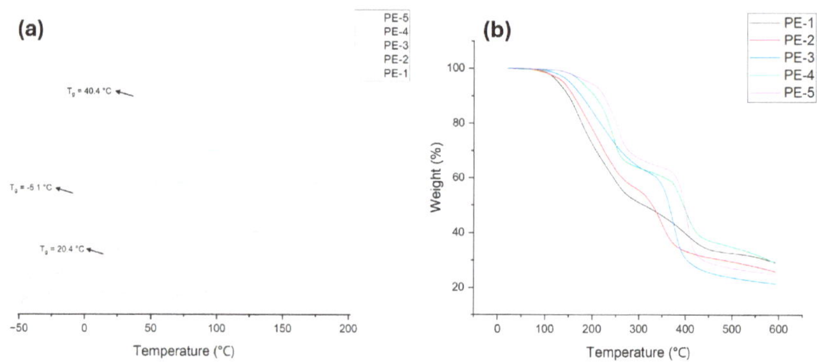

Figure 2. (**a**) DSC traces of PEs. (**b**) TGA curves of PEs.

3.4. Preparation and Characterization of PE Polyols

Of the previously synthesized diols, diol **2** was selected for further investigation in the preparation of PE polyols for their use in PU foams (Scheme 3). The polyols synthesized in this work were compared to commercially available control samples PSA 3000 and PSA 1500, which are PE polyols derived from SA and PDO that have previously been used in the synthesis of flexible PU foams [39,40]. Two polyols were prepared by replacing either 10 wt% or 25 wt% of PDO with diol **2** to make them more sustainable, labeled as polyol **2–10** and polyol **2–25**, respectively. The diol PDO was first added and heated to allow for better solubility of the diacid SA and the biobased diol **2**. After removing the water, titanium catalyst Organotix TC-400 was added to speed up the reaction, and additional PDO was added to maintain the viscosity. Both polyols **2–10** and **2–25** were a dark brown color after reaction completion, whereas PSA 3000 and PSA 1500 were yellow. This brown color is consistent with most lignin-derived materials [37]. AV measurements were calculated following Equation (S1), and the final properties of the polyols are reported in Table 3.

Table 3. PE Polyol Properties.

Sample	Viscosity (cps)	AV # (mg KOH/g)	OH # (mg KOH/g)	M_W (g/mol)
PSA 3000	4335	0.9	39.0	2875
PSA 1500	940	0.5	73.7	1520
2–10	3547	3.2	63.0	1780
2–25	2332	9.2	78.3	1433

The AV number can negatively impact the reactivity of polyols during PU foam synthesis, and therefore it should be as small as possible to avoid competing with the main reaction [41]. The AV numbers of PSA 3000 and PSA 1500 were lowered to 0.9 and 0.5 mg KOH/g, respectively, after 24 h; however, polyols **2–10** and **2–25** were found to have consistent AV numbers by the 70 h mark (Tables S3 and S4). Letting the reaction proceed for longer periods did not result in a decrease in the AV number, and the final AV numbers were determined to be 3.16 mg KOH/g for polyol **2–10** and 9.21 mg KOH/g for polyol **2–25**.

The viscosity is directly related to the M_W of the polyol, where lower viscosity generally results in lower M_W polyols. The M_W is also inversely related to the OH number, as seen in Equation (S2), where the smaller the OH number, the higher the M_W [17]. The final viscosities of PSA 3000 and PSA 1500 at 70 °C were calculated to be 4335 and 940 cps, which is within the expected viscosity range for these polyols. The viscosities of polyols **2–10** and **2–25** were determined to be 3547 and 2332 cps, being closer to that of PSA 3000. PSA 3000 also had the lowest OH number of 39.02 mg KOH/g, while polyols **2–10** and **2–25** were found to be 63.04 and 78.32 mg KOH/g, respectively. These values were closer to the OH number of PSA 1500, which was 73.7 mg KOH/g, resulting in calculated M_W's of 2875 g/mol for PSA 3000, 1520 g/mol for PSA 1500, 1780 g/mol for polyol **2–10**, and 1433 g/mol for polyol **2–25**. These results suggest that the viscosity of the polyols did not have a huge impact on the M_W.

3.5. Preparation and Characterization of PU Foams

Prior to preparing the PU foam plaques, free rises were conducted to determine the reaction kinetics of the foams with the addition of the newly synthesized polyols. Properties analyzed include the cream time, which is the time when the foam starts to grow, the rise time, which is the time when the foam stops growing [42], and the tack-free time, which is the time when the foam surface cures and is no longer sticky to touch [13,43]. As seen in Figure 3, the control foams with PSA 3000 and PSA 1500, labelled PU-3000 and PU-1500, respectively, were bright orange in color due to the addition of the orange dye, while both PU-2-10 and PU-2-25 resulted in light brown foams, even with the orange dye added. This is most likely due to the color of their respective polyols. After 24 h, the foams were cut open to observe their insides (Figure 4). Only PU-1500 was found to exhibit shrinkage after 24 h, while PU-3000, PU-2-10, and PU-2-25 remained intact, suggesting that the foams made with polyols **2–10** and **2–25** exhibited stability that was in line with PU-3000. The kinetic data for all PU foam free rises are shown in Table 4.

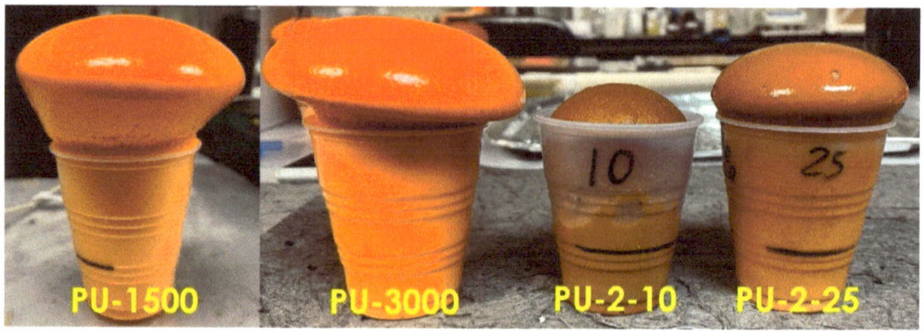

Figure 3. PU foam free rises made from their respective polyols on the day of synthesis.

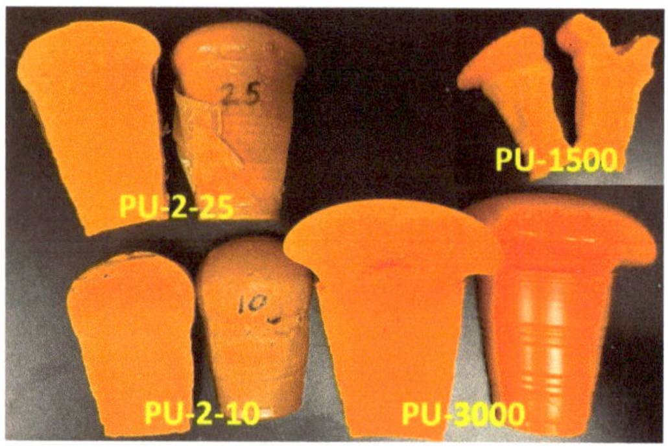

Figure 4. Cut PU foam free rises 24 h after synthesis.

Table 4. Analysis of PU foam free rises.

Sample	Cream Time (s)	Rise Time (min:s)	Tack Free Time (min:s)
PU-3000	15	1:30	3:30
PU-1500	20	1:47	2:55
PU-2-10	16	1:00	0:50
PU-2-25	15	1:30	2:50

The cream times for PU-2-10 and PU-2-25 were closer to that of PU-3000, ranging from 15 to 16 s compared to 20 s for PU-1500. The rise time and tack-free time for PU-2-10 was also faster than the other foam samples. It was found that the rise time of PU-2-25 was more on par with PU-3000 compared to the slower rise time of PU-1500. However, the tack-free time for PU-2-25 was much closer to that of PU-1500, which is faster than the tack-free time for PU-3000. This suggests that adding a smaller amount of diol 2 to the polyol formulation results in PU foams that form faster than the control foams, while adding more of diol 2 slows the kinetics down enough to closer match the controls.

Using this information, PU foam plaques were prepared from the previously mentioned polyols. Figure 5 shows the initial attempt at preparing the foam plaques. Both PU-1500 and PU-3000 appeared as bright orange rectangular foams that filled the mould entirely. PU-2-10 exhibited some deformity as it did not fully form in the heated mould, but it cured at the same rate as both control plaques. PU-2-25 appeared to mostly form the standard rectangular shape, but upon removing it from the heated mould the foam was soft and difficult to remove, making it unsuitable for further testing. The poorly formed foams could be attributed to the fast kinetics previously observed in Table 4, where the foam began to form so quickly that everything did not have time to properly mix. Since PU-2-10 showed the most promise in terms of its appearance, the original formulation was modified to reduce the catalyst loading amount from 0.8 parts per hundred of polyol (pphp) to 0.6 and even 0.4 pphp. As seen in Figure 6, as the catalyst amount decreased, the appearance of the foams significantly improved. Thus, these three foams and the two control foams were further tested for their mechanical properties, which are shown in Table 5.

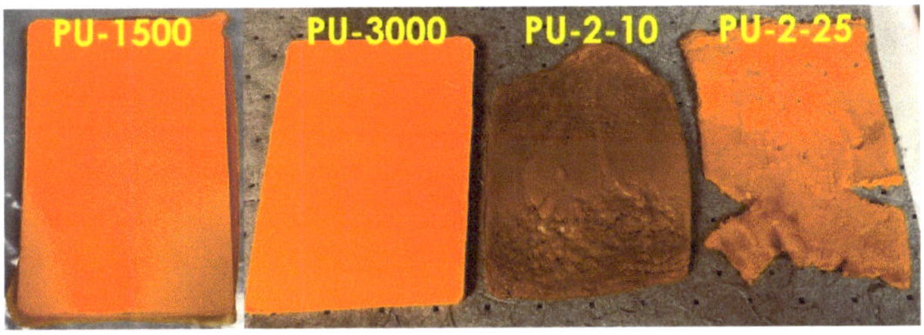

Figure 5. PU foam plaques made using the original formulation.

Figure 6. PU foam plaques of **PU-2-10** with modified catalyst loading.

Table 5. Mechanical properties of PU foams.

Sample	33LV Amount (pphp)	Hardness (Benchtop)	Resiliency (%) (Ball)	Tensile Strength (kg/cm^2)	Elongation (%)	Tear	
						Die-C (N/mm)	Split (N/mm)
PU-3000	0.8	35	33	12.1	399	6.5	2.0
PU-1500	0.8	25	22	10.3	277	7.8	1.5
PU-2-10	0.8	-	-	7.5	103	3.9	-
PU-2-10	0.6	35	13	11.9	119	5.0	0.8
PU-2-10	0.4	35	13	12.4	117	4.8	0.7
PU-3000	0.8	35	33	12.1	399	6.5	2.0

Looking at Table 5, PU-3000 had higher mechanical properties compared to PU-1500 apart from the Die-C tear strength. When comparing PU-2-10 with the different catalyst amounts, the mechanical properties improved with a decrease in the amount of catalyst added. PU-2-10 exhibited similar hardness and tensile strength as PU-3000, whereas PU-1500 had slightly lower values than PU-2-10. Areas where PU-2-10 lacked in mechanical properties compared to both PU-3000 and PU-1500 were in resiliency, elongation, and both Die-C and split tear strength. This is similar to what was observed with the addition of modified lignin in PU foam elastomers, where an increasing amount of lignin content led to a decrease in both the resiliency and elongation of the foams while maintaining other mechanical properties [44].

4. Conclusions

Five lignin-based diols were prepared from phenolic aldehydes to synthesize bio-derived polyurethanes and polyesters. The synthesis of ten thermoplastic PUs and five PEs was confirmed by NMR, EA, and FTIR analysis. DSC analysis of the PUs was consistent with the degree of rigidity or flexibility of the diisocyanate source, whereas DSC analysis of the PEs showed no noticeable T_g for PE-1 and PE-4. TGA analysis confirmed good thermal stability of the PUs and increasing thermal stability in the PEs with increasing bulkiness of the starting diol. GPC analysis also showed relatively high M_Ws for the PUs with the exception of PU-1M. Two PE polyols were prepared with 10 wt% and 25 wt% of diol **2** to produce polyols with higher acid values than either PSA 3000 and PSA 1500 and hydroxyl numbers closely matching that of PSA-1500. PU foams were prepared with polyols PSA 3000, PSA 1500, **2–10**, and **2–25**, showing that PU-2-10 and PU-2-25 had faster reaction kinetics than PU-3000 and PU-1500. PU-2-10 exhibited similar or slightly higher tensile strength and hardness as PU-3000 and PU-1500 foams, but lower tear strength, resiliency, and elongation.

Supplementary Materials: The following supporting information can be downloaded at: https://www.mdpi.com/article/10.3390/polym17050643/s1, Figures S1–S20: ^{1}H and ^{13}C NMR spectra of HEB, 1, HMBD, 2, HEDB, 3, HBA, 4, VA, and 5; Figure S21: FTIR analysis of diols 1–5; Figures S22–S41: ^{1}H and ^{13}C NMR spectra of PU 1M, PU-1P, PU-2M, PU-2P, PU-3M, PU-3P, PU-4M, PU-4P, PU-5M, and PU-5P; Figures S42–S51: ^{1}H and ^{13}C NMR spectra of PE-1, PE-2, PE-3, PE-4, and PE-5; Figure S52: FTIR analysis of PU-1M to PU-5M; Figure S53: FTIR analysis of PU-1P to PU-5P; Figure S54: FTIR analysis of PE-1 to PE-5; Figure S55: Derivative weight change of PU-1M to PU-5M; Figure S56: Derivative weight change of PU-1P to PU-5P; Figure S57: Derivative weight change of PE-1 to PE-5; Table S1: Elemental analysis of PU polymers; Table S2: Elemental analysis of PE polymers; Equation (S1): Determining acid value of polyols using 0.1 N KOH solution; Equation (S2): Determination of molecular weight of polyols; Table S3: Reaction conditions and observations during the synthesis of polyol **2–10**; Table S4: Reaction conditions and observations during the synthesis of polyol **2–25**.

Author Contributions: Conceptualization: R.N.C. and G.G.S.; Methodology: R.N.C. and G.G.S.; Validation: R.N.C. and B.V.F.; Formal analysis: R.N.C. and B.V.F.; Investigation: R.N.C., B.V.F., C.R. and S.R.; Resources: D.A.F. and G.G.S.; Data Curation: R.N.C., B.V.F., C.R. and S.R.; Writing—original draft: R.N.C.; Writing—reviewing and editing: R.N.C., D.A.F. and G.G.S.; Visualization: R.N.C. and C.R.; Supervision: R.N.C., D.A.F. and G.G.S. All authors have read and agreed to the published version of the manuscript.

Funding: This work was supported by the NSERC Discovery Grant program.

Institutional Review Board Statement: Not applicable.

Data Availability Statement: The original contributions presented in this study are included in the article/Supplementary Material. Further inquiries can be directed to the corresponding author.

Acknowledgments: We thank Alex Caschera from McMaster University for running the NMRs of select PUs and EA for all the polymers and Anthony Fasulo for his assistance with preparing the IR, DSC and TGA figures and for running the TGAs for all polymers. R.N.C. was the recipient of a Queen Elizabeth II Graduate Scholarship in Science and Technology, a Toronto Met Graduate Scholarship, and an Ontario Graduate Scholarship.

Conflicts of Interest: The authors declare no conflicts of interest.

References

1. Delidovich, I.; Hausoul, P.J.C.; Deng, L.; Pfützenreuter, R.; Rose, M.; Palkovits, R. Alternative Monomers Based on Lignocellulose and Their Use for Polymer Production. *Chem. Rev.* **2016**, *116*, 1540–1599. [CrossRef] [PubMed]
2. Zhu, Y.; Romain, C.; Williams, C.K. Sustainable Polymers from Renewable Resources. *Nature* **2016**, *540*, 354–362. [CrossRef]
3. Nakajima, H.; Dijkstra, P.; Loos, K. The Recent Developments in Biobased Polymers toward General and Engineering Applications: Polymers That Are Upgraded from Biodegradable Polymers, Analogous to Petroleum-Derived Polymers, and Newly Developed. *Polymers* **2017**, *9*, 523. [CrossRef] [PubMed]

4. Pellis, A.; Malinconico, M.; Guarneri, A.; Gardossi, L. Renewable Polymers and Plastics: Performance beyond the Green. *N. Biotechnol.* **2021**, *60*, 146–158. [CrossRef]
5. Isikgor, F.H.; Becer, C.R. Lignocellulosic Biomass: A Sustainable Platform for the Production of Bio-Based Chemicals and Polymers. *Polym. Chem.* **2015**, *6*, 4497–4559. [CrossRef]
6. Carafa, R.N.; Foucher, D.A.; Sacripante, G.G. Biobased Polymers from Lignocellulosic Sources. *Green. Chem. Lett. Rev.* **2023**, *16*, 2153087. [CrossRef]
7. Rico-García, D.; Ruiz-Rubio, L.; Pérez-Alvarez, L.; Hernández-Olmos, S.L.; Guerrero-Ramírez, G.L.; Vilas-Vilela, J.L. Lignin-Based Hydrogels: Synthesis and Applications. *Polymers* **2020**, *12*, 81. [CrossRef]
8. Fache, M.; Boutevin, B.; Caillol, S. Vanillin Production from Lignin and Its Use as a Renewable Chemical. *ACS Sustain. Chem. Eng.* **2016**, *4*, 35–46. [CrossRef]
9. Rao, S.R.; Ravishankar, G.A. Vanilla Flavour: Production by Conventional and Biotechnological Routes. *J. Sci. Food Agric.* **2000**, *80*, 289–304. [CrossRef]
10. Pandey, M.P.; Kim, C.S. Lignin Depolymerization and Conversion: A Review of Thermochemical Methods. *Chem. Eng. Technol.* **2011**, *34*, 29–41. [CrossRef]
11. Foyer, G.; Chanfi, B.H.; Virieux, D.; David, G.; Caillol, S. Aromatic Dialdehyde Precursors from Lignin Derivatives for the Synthesis of Formaldehyde-Free and High Char Yield Phenolic Resins. *Eur. Polym. J.* **2016**, *77*, 65–74. [CrossRef]
12. Rubens, M.; Van Wesemael, M.; Feghali, E.; Luntadila Lufungula, L.; Blockhuys, F.; Vanbroekhoven, K.; Eevers, W.; Vendamme, R. Exploring the Reactivity of Aliphatic and Phenolic Hydroxyl Groups in Lignin Hydrogenolysis Oil towards Urethane Bond Formation. *Ind. Crops Prod.* **2022**, *180*, 114703. [CrossRef]
13. Furtwengler, P.; Avérous, L. Renewable Polyols for Advanced Polyurethane Foams from Diverse Biomass Resources. *Polym. Chem.* **2018**, *9*, 4258–4287. [CrossRef]
14. Caschera, A.; Calayan, T.; Piccolo, N.; Kakroodi, A.; Robinson, J.J.; Sacripante, G. Degradation of Bioderived Polyurethane Composites by Spectroscopy in ISO20200 Composting Conditions. *Polymers* **2024**, *16*, 2071. [CrossRef] [PubMed]
15. McKenna, S.T.; Hull, T.R. The Fire Toxicity of Polyurethane Foams. *Fire Sci. Rev.* **2016**, *5*, 3. [CrossRef]
16. Sardon, H.; Pascual, A.; Mecerreyes, D.; Taton, D.; Cramail, H.; Hedrick, J.L. Synthesis of Polyurethanes Using Organocatalysis: A Perspective. *Macromolecules* **2015**, *48*, 3153–3165. [CrossRef]
17. Amran, U.A.; Zakaria, S.; Chia, C.H.; Roslan, R.; Jaafar, S.N.S.; Salleh, K.M. Polyols and Rigid Polyurethane Foams Derived from Liquefied Lignocellulosic and Cellulosic Biomass. *Cellulose* **2019**, *26*, 3231–3246. [CrossRef]
18. Palacios-Mateo, C.; van der Meer, Y.; Seide, G. Analysis of the Polyester Clothing Value Chain to Identify Key Intervention Points for Sustainability. *Environ. Sci. Eur.* **2021**, *33*, 2. [CrossRef]
19. Sokolsky-Papkov, M.; Langer, R.; Domb, A.J. Synthesis of Aliphatic Polyesters by Polycondensation Using Inorganic Acid as Catalyst. *Polym. Adv. Technol.* **2011**, *22*, 502–511. [CrossRef] [PubMed]
20. Kim, S.; Chung, H. Synthesis and Characterization of Lignin-Graft-Poly(Ethylene Brassylate): A Biomass-Based Polyester with High Mechanical Properties. *ACS Sustain. Chem. Eng.* **2021**, *9*, 14766–14776. [CrossRef]
21. Zhang, S.; Cheng, Z.; Zeng, S.; Li, G.; Xiong, J.; Ding, L.; Gauthier, M. Synthesis and Characterization of Renewable Polyesters Based on Vanillic Acid. *J. Appl. Polym. Sci.* **2020**, *137*, 4–13. [CrossRef]
22. Jeong, H.; Park, J.; Kim, S.; Lee, J.; Ahn, N.; Roh, H. Preparation and Characterization of Thermoplastic Polyurethanes Using Partially Acetylated Kraft Lignin. *Fibers Polym.* **2013**, *14*, 1082–1093. [CrossRef]
23. Zhang, X.; Jeremic, D.; Kim, Y.; Street, J.; Shmulsky, R. Effects of Surface Functionalization of Lignin on Synthesis and Properties of Rigid Bio-Based Polyurethanes Foams. *Polymers* **2018**, *10*, 706. [CrossRef] [PubMed]
24. Kühnel, I.; Podschun, J.; Saake, B.; Lehnen, R. Synthesis of Lignin Polyols via Oxyalkylation with Propylene Carbonate. *Holzforschung* **2015**, *69*, 531–538. [CrossRef]
25. Kühnel, I.; Saake, B.; Lehnen, R. Oxyalkylation of Lignin with Propylene Carbonate: Influence of Reaction Parameters on the Ensuing Bio-Based Polyols. *Ind. Crops Prod.* **2017**, *101*, 75–83. [CrossRef]
26. Zhao, C.; Huang, C.; Chen, Q.; Ingram, I.D.V.; Zeng, X.; Ren, T.; Xie, H. Sustainable Aromatic Aliphatic Polyesters and Polyurethanes Prepared from Vanillin-Derived Diols via Green Catalysis. *Polymers* **2020**, *12*, 586. [CrossRef]
27. Canceill, J.; Collet, A. ChemInform Abstract: Two-Step Synthesis of D3 and C3h Cryptophanes. *J. Chem. Soc. Chem. Commun.* **1988**, *19*, 582–584. [CrossRef]
28. Mouren, A.; Avérous, L. Aromatic Thermoplastic Polyurethanes Synthesized from Different Potential Sustainable Resources. *Eur. Polym. J.* **2023**, *197*, 112338. [CrossRef]
29. Tarnchompoo, B.; Chitnumsub, P.; Jaruwat, A.; Shaw, P.J.; Vanichtanankul, J.; Poen, S.; Rattanajak, R.; Wongsombat, C.; Tonsomboon, A.; Decharuangsilp, S.; et al. Hybrid Inhibitors of Malarial Dihydrofolate Reductase with Dual Binding Modes That Can Forestall Resistance. *ACS Med. Chem. Lett.* **2018**, *9*, 1235–1240. [CrossRef] [PubMed]
30. Gu, Y.-K.; Han, Y.; Chen, C.-F. Guest-Dependent Complexation of Triptycene-Derived Macrotricyclic Host Containing One Anthracene Moiety with Paraquat Derivatives: Construction of [2]Rotaxanes. *Supramol. Chem.* **2015**, *27*, 357–363. [CrossRef]

31. *ASTM E1899-08;* Standard Test Method for Hydroxyl Groups Using Reaction with p-Toluenesulfonyl Isocyanate (TSI) and Potentiometric Titration with Tetrabutylammonium Hydroxide. ASTM International: West Conshohocken, PA, USA, 2016.
32. *ASTM D412;* Standard Test Methods for Vulcanized Rubber and Thermoplastic Elastomers—Tension. ASTM International: West Conshohocken, PA, USA, 2021.
33. *ASTM D3574;* Standard Test Methods for Flexible Cellular Materials—Slab, Bonded, and Molded Urethane Foams. ASTM International: West Conshohocken, PA, USA, 2017.
34. Sacripante, G.G.; Zhou, K.; Farooque, M. Sustainable Polyester Resins Derived from Rosins. *Macromolecules* **2015**, *48*, 6876–6881. [CrossRef]
35. Fowler, R.G. Microscale Reactions of Vanillin. *J. Chem. Educ.* **1992**, *69*, 43–45. [CrossRef]
36. Song, J.; Xie, C.; Xue, Z.; Wang, T. Strong Oxophilicity of Zr Species in Zr4+-Exchanged Montmorillonite Boosted Meerwein-Ponndorf-Verley Reduction of Renewable Carbonyl Compounds. *ACS Sustain. Chem. Eng.* **2022**, *10*, 12197–12206. [CrossRef]
37. Alinejad, M.; Henry, C.; Nikafshar, S.; Gondaliya, A.; Bagheri, S.; Chen, N.; Singh, S.K.; Hodge, D.B.; Nejad, M. Lignin-Based Polyurethanes: Opportunities for Bio-Based Foams, Elastomers, Coatings and Adhesives. *Polymers* **2019**, *11*, 1202. [CrossRef] [PubMed]
38. Hanson, K.G.; Lin, C.H.; Abu-Omar, M.M. Preparation and Properties of Renewable Polyesters Based on Lignin-Derived Bisphenol. *Polymers* **2021**, *233*, 124202. [CrossRef]
39. Robinson, J.J.; Sacripante, G.G.; Vicol, R.L.; Ghalia, M.A.; Wang, Y.; Mullick, S. Polyurethane Elastomer Compositions, and Processes Thereof. U.S. Patent 10,934,384 B1, 2021.
40. Robinson, J.J.; Ghalia, M.A.; Vicol, R.L.; Sacripante, G.G. Plasticizer Lignin Compositions. U.S. Patent 2022/095182 A1, 23 July 2022.
41. Omisol, C.J.M.; Aguinid, B.J.M.; Abilay, G.Y.; Asequia, D.M.; Tomon, T.R.; Sabulbero, K.X.; Erjeno, D.J.; Osorio, C.K.; Usop, S.; Malaluan, R.; et al. Flexible Polyurethane Foams Modified with Novel Coconut Monoglycerides-Based Polyester Polyols. *ACS Omega* **2024**, *9*, 4497–4512. [CrossRef]
42. Zieglowski, M.; Trosien, S.; Rohrer, J.; Mehlhase, S.; Weber, S.; Bartels, K.; Siegert, G.; Trellenkamp, T.; Albe, K.; Biesalski, M. Reactivity of Isocyanate-Functionalized Lignins: A Key Factor for the Preparation of Lignin-Based Polyurethanes. *Front. Chem.* **2019**, *7*, 562. [CrossRef] [PubMed]
43. Liszkowska, J.; Moraczewski, K.; Borowicz, M.; Paciorek-Sadowska, J.; Czupryński, B.; Isbrandt, M. The Effect of Accelerated Aging Conditions on the Properties of Rigid Polyurethane-Polyisocyanurate Foams Modified by Cinnamon Extract. *Appl. Sci.* **2019**, *9*, 2663. [CrossRef]
44. Robinson, J.J.; Carafa, R.N.; Li, J.; Sacripante, G.G. Polyurethane Elastomer Composition Comprising Lignin. U.S. Patent 17/958266, 30 September 2022.

polymers MDPI

Review

Revolutionizing Dental Polymers: The Versatility and Future Potential of Polyetheretherketone in Restorative Dentistry

Noha Taymour [1], Ahmed Abd El-Fattah [2,3], Sherif Kandil [2], Amal E. Fahmy [4], Naif H. Al-Qahtani [5], Abdulrahman Khaled [2,*], Yousif A. Al-Dulaijan [1,*] and Mohamed Abdel-Hady Gepreel [6]

1 Department of Substitutive Dental Sciences, College of Dentistry, Imam Abdulrahman Bin Faisal University, P.O. Box 1982, Dammam 31441, Saudi Arabia; ntyoussef@iau.edu.sa
2 Department of Materials Science, Institute of Graduate Studies and Research, Alexandria University, El-Shatby, Alexandria 21526, Egypt; aahussein@uob.edu.bh or a_abdelfattah@alexu.edu.eg (A.A.E.-F.); s.kandil@usa.net (S.K.)
3 Department of Chemistry, College of Science, University of Bahrain, Sakhir P.O. Box 32038, Bahrain
4 Department of Dental Materials, Faculty of Dentistry, Alexandria University, Azarita, Alexandria 21526, Egypt; amal_ezzeldin99@yahoo.com
5 College of Dentistry, Imam Abdulrahman Bin Faisal University, P.O. Box 1982, Dammam 31441, Saudi Arabia; 2210000179@iau.edu.sa
6 Department of Materials Science and Engineering, Egypt-Japan University of Science and Technology (E-JUST), New Borg El-Arab City 21934, Egypt; geprell@yahoo.com
* Correspondence: abdulrahman.kh1122@gmail.com (A.K.); yaaldulaijan@iau.edu.sa (Y.A.A.-D.); Tel.: +20-106-182-9400 (A.K.)

Abstract: Polyetheretherketone (PEEK) has emerged as a revolutionary material in modern dentistry because of its unique combination of mechanical strength, biocompatibility, and versatility. This literature review examines the current applications and future potential of PEEK in various dental disciplines. PEEK's favorable properties, including its low specific weight, high strength-to-weight ratio, and ability to be easily machined, have led to its adoption in prosthetics, implantology, and dental esthetic restorations. This material has shown promise for fabricating crowns, bridges, removable partial denture frameworks, and implant components. PEEK's radiolucency and bone-like elastic modulus make it particularly suitable for dental implants and abutments. Additionally, its resistance to degradation and compatibility with various surface treatments enhances its long-term performance in the oral environment. While challenges such as bonding to other dental materials and aesthetic limitations exist, ongoing research is addressing these issues through surface modifications and composite formulations. As the dental field continues to evolve, PEEK's adaptability and biocompatibility position it a key player in the development of next-generation dental materials and techniques, potentially transforming patient care and treatment outcomes in dentistry.

Keywords: Polyetheretherketone (PEEK); implants; prosthesis; thermoplastics; tissue regeneration; bone regeneration

Academic Editors: Vishal Gavande, Vasi Shaikh and Won-Ki Lee

Received: 21 November 2024
Revised: 22 December 2024
Accepted: 28 December 2024
Published: 31 December 2024

Citation: Taymour, N.; Abd El-Fattah, A.; Kandil, S.; Fahmy, A.E.; Al-Qahtani, N.H.; Khaled, A.; Al-Dulaijan, Y.A.; Gepreel, M.A.-H. Revolutionizing Dental Polymers: The Versatility and Future Potential of Polyetheretherketone in Restorative Dentistry. *Polymers* **2025**, *17*, 80. https://doi.org/10.3390/polym17010080

1. Introduction

PEEK (Polyether ether ketone) is a semi-crystalline, high-performance engineering thermoplastic material. It is a special polymer material composed of repeated units of one ketone bond and two ether bonds in the backbone of the polymer, as shown in Figure 1 [1]. PEEK polymers are obtained by step-growth polymerization by the dialkylation of bisphenolate salts. The reaction is conducted at around 300 °C in polar aprotic solvents, such as diphenyl sulfone [2]. PEEK's unique properties enable its use in diverse fields, including biomedical, aerospace, automotive, and electrical industries. It demonstrates exceptional resistance to fatigue, creep, chemical degradation, and wear, making it particularly valuable for high-performance applications. These characteristics have led to PEEK's widespread adoption in structural components, high-temperature environments, and advanced engineering solutions [3]. PEEK's high resistance to oxidative deterioration and its strength-to-weight profile are high enough to compete with metals like aluminum [4]. Regarding thermal properties, PEEK is highly temperature-resistant, with a glass transition temperature of around 143 °C (289 °F) and a melting point of around 343 °C (662 °F) [5]. In addition, PEEK has excellent chemical resistance properties; it is chemically resistant to most organic and inorganic chemicals, as well as to hydrolysis, radiation, and sterilization [5]. PEEK is a highly durable material with excellent mechanical and chemical resistance properties that are retained at high temperatures. Its density is 1320 kg/m^3, Young's modulus is 3.6 GPa, and tensile strength is 90 to 100 MPa [6]. PEEK is readily machinable in the solid state, allowing the creation of complex shapes and designs [7].

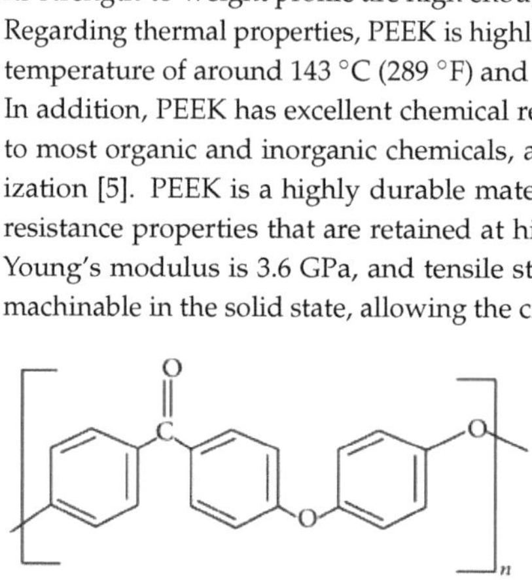

Figure 1. Chemical structure of PEEK [1].

There are several grades of PEEK, including both extremely pure PEEK and PEEK reinforced with glass, carbon fiber, or other technical plastics. In the industry, PEEK comes in four main grades: 30% carbon-filled PEEK, 30% glass-filled PEEK, bearing-grade PEEK, and unfilled PEEK [6]. Different processing techniques are used to enhance PEEK's properties [8,9]. PEEK pellets are generally recommended for extrusion, monofilament, and wire coating operations, while PEEK powder is used for extrusion compounding when fillers are added to enhance PEEK properties, such as expanded graphite (EG) [10]. PEEK can be manufactured using conventional injection molding equipment. To achieve a semi-crystalline component that fully utilizes PEEK's special set of characteristics, the mold temperature is crucial [11]. Compression molding is another process that produces PEEK with excellent mechanical properties; it is recommended for use with PEEK granules/rods. However, the key factors affecting the uniformity of compression-molded PEEK are the mold temperature, molding pressure, and ram speed (or the rate of application of pressure) [12]. Compression-molded PEEK samples showed higher crystallinity in comparison to injection-molded samples; subsequently, they were found to have higher strength compared to injection-molded PEEK samples [13]. Lee DJ et al. [14] have compared the mechanical properties (tensile, fatigue, and creep) of carbon fiber-filled PEEK using injection and compression molding techniques. Specimens with varying initial fiber lengths were created using both methods. Compression-molded specimens showed clear differences in mechanical properties based on fiber length, whereas injection-molded samples displayed consistent tensile and fatigue strengths regardless of the initial fiber length. Further research is needed to optimize the process parameters for specific applications and evaluate the long-term performance and success rates of compression-molded PEEK.

In addition to injection molding and compression molding, two other manufacturing methods have become increasingly significant for PEEK in dental applications. Subtractive manufacturing (CAD/CAM) has emerged as a primary technique for fabricating fixed and removable prosthodontics using PEEK [15]. This method encompasses milling and grinding processes to remove material from a PEEK block, shaping it into the desired form with high precision and accuracy, which are key factors for dental applications [4]. The resultant components exhibit smooth surfaces, often requiring minimal post-processing [16]. These features make subtractive manufacturing particularly suitable for creating fixed and removable prostheses that require precise anatomical fit and functional reliability [17]. 3D printing of PEEK offers several benefits for dental applications. Customization of implants and prostheses to patient-specific anatomy, reduced material waste compared to subtractive methods, ability to create complex geometries that may be challenging with other manufacturing methods, and the potential for faster production times for certain applications [18]. These manufacturing methods have significantly impacted the production of PEEK-based fixed and removable prosthodontics, offering dental professionals more options for creating precise patient-specific solutions [15].

CAD/CAM-milled PEEK remains the superior manufacturing method for dental applications demanding high mechanical strength and precision due to its homogeneous microstructure, which consistently delivers superior tensile, flexural, and fatigue strengths compared to 3D-printed PEEK [19]. While 3D printing of PEEK shows promise in customized and lower-load scenarios, it suffers from porosity and anisotropy unless optimized through parameters like infill rate, temperature, and post-print annealing [20,21]. Both methods offer similar baseline biocompatibility; however, the rougher surfaces of 3D-printed PEEK may promote higher osteoblast proliferation [15]. Surface treatments, such as sandblasting and plasma treatment, can enhance bioactivity for both methods, although uniformity and bacterial adhesion present challenges, especially for 3D-printed surfaces [15,19]. Clinically, milled PEEK is preferred for load-bearing applications like frameworks and implants due to its superior durability, while 3D-printed PEEK excels in customization for non-load-bearing or less demanding restorations, making it ideal for rapid prototyping [19,21]. Optimized parameters can improve the marginal fit of 3D-printed PEEK and rival milled PEEK in some cases. However, variability in printer settings, operator skills, and lack of standardized post-processing protocols limit the reliability of 3D-printed PEEK for clinical dentistry [22]. Advances such as carbon fiber-reinforced 3D-printed PEEK may help narrow the performance gap with milled PEEK, highlighting the need for further research and standardization [23].

PEEK surface modification can be performed either by surface treatment alone or in conjunction with surface coating [24]. Surface treatments, both chemical and physical, can significantly increase PEEK's bioactivity. For instance, PEEK surfaces' chemical and physical characteristics have been modified via sulfonation and sub-millimeter laser machining, improving their bio-interaction [25]. Electron beam irradiation has also been used to modify the surface of PEEK, resulting in increased hydrophilicity and protein adsorption [26]. It has been observed that nitrogen plasma performs particularly well in PEEK implant applications. There has also been an investigation into further surface treatments, such as plasma ion implantation and nanomodification [27,28]. Rather than using physical treatments and composites, surface coatings have been employed to increase PEEK's bioactivity [29,30]. For example, PEEK can be coated with bioactive substances such as hydroxyapatite or titanium to enhance its bioactivity and antibacterial properties [31,32]. The process of creating a nano-level surface topography by coating or mixing PEEK with nanoparticles has recently been the focus of a significant amount of research [33,34]. PEEK's bioactivity can be significantly increased by surface treatment either by itself or in conjunction with surface coating. The use of surface coatings as substitutes for physical treatments and composites has been documented in the literature since the addition of bioactive particles to PEEK has raised questions about how to preserve the material's mechanical properties. Figure 2 shows the different methods of PEEK processing for surface bioactivation.

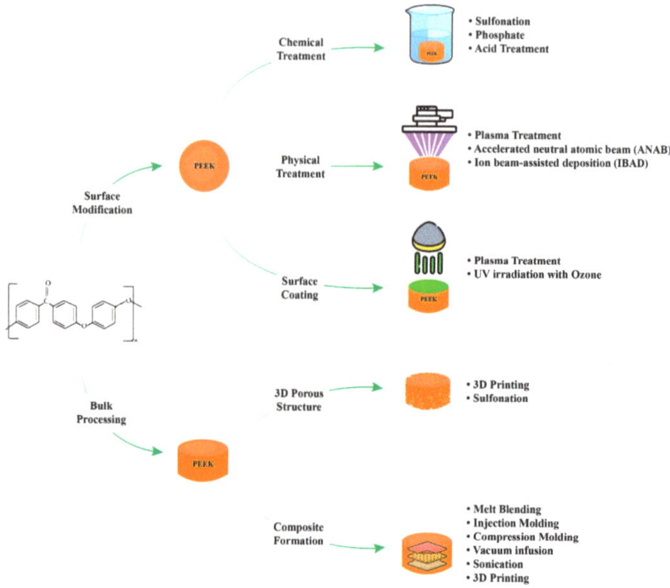

Figure 2. PEEK modifications and coating for enhanced bioactivity.

PEEK's biocompatibility and ability to be blended with fibers and ceramics to enhance its mechanical properties and bioactive properties make it an ideal alternative for advanced medical and dental uses. As shown in Figure 3, PEEK is now utilized in crowns, posts and cores, maxillofacial prostheses, removable and fixed prostheses, dental implants, and implant abutments [35,36].

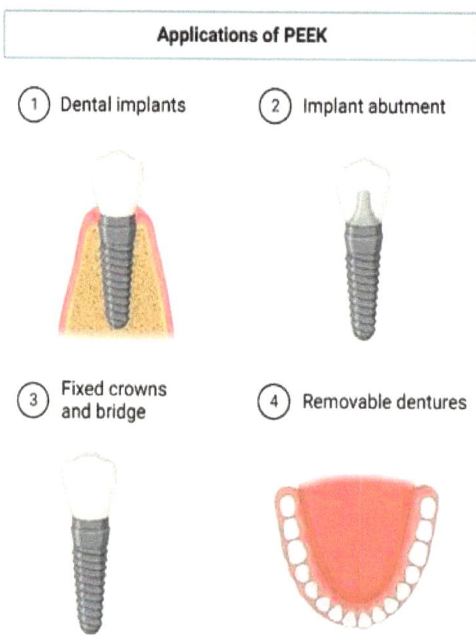

Figure 3. Common applications of PEEK in dentistry.

2. Methodology

2.1. Question Addressed by This Review

What are the different applications and benefits of PEEK in the field of restorative dentistry?

2.2. Literature Research

A narrative exploratory review was conducted. A literature search was carried out utilizing electronic databases such as PubMed (MEDLINE), Scopus, and Web of Science (WoS). Following that, a manual search of the literature was performed, including the reference lists of related and similar studies, to identify any new relevant research.

The main search terms were "PEEK", "dentistry", "implant", "prosthesis" and "abutment". Articles published between 2013 and 2024 were included. Any duplicate entries in the databases were identified and subsequently eliminated using Mendeley Reference Manager software (version 2.128) by Elsevier, based in Amsterdam, Netherlands.

Inclusion criteria encompassed peer-reviewed articles in English-written articles focusing on PEEK applications in the field of restorative dentistry, while exclusion criteria involved non-dental applications and non-English publications. Subtopics within PEEK in dentistry were identified and categorized based on the primary application areas, such as implants, prosthetics, and bone regeneration, as well as material properties, such as mechanical characteristics and biological activity. Papers were organized using a custom classification framework developed in Microsoft Excel, which allowed for efficient categorization and analysis.

2.3. Data Extraction

Data extraction was performed independently by two reviewers using a standardized form to collect information on the study design, sample size, key findings, and conclusions. Any discrepancies in data extraction were resolved through discussion to reach a consensus. The extracted data were then synthesized to identify trends, gaps in research, and potential future directions for PEEK applications in dentistry.

3. Results and Discussion

3.1. Advantages of PEEK over Traditional Materials

PEEK is a desirable material for various applications due to its exceptional mechanical, thermal, and chemical properties. PEEK has a higher strength-to-weight ratio, is more flexible, and is more biocompatible than traditional materials such as metals and ceramics [37].

According to Luo et al. [35], PEEK dental implants are more compatible with the mechanical characteristics of bone than metal and ceramic implants and show less stress shielding. Yongan et al. [38] focused on the improvement schemes for PEEK composites, aiming to enhance their mechanical properties, friction resistance, electrical conductivity, and thermal resistance. Tekin et al. [39] highlighted that PEEK's low elasticity modulus can decrease stress-related issues and make it an effective replacement for current dental materials. The suitability of PEEK as a fixed partial denture framework was examined by Sinha et al. [40], who emphasized its superior mechanical properties and chemical resistance. JianBing et al. [41] mentioned that PEEK has prominent advantages in mechanical strength, thermal stability, and chemical stability. Researchers have suggested that PEEK offers superior properties and versatility compared to traditional materials in dental applications. Detailed comparisons between PEEK and other dental polymers are presented in Table 1.

Table 1. Properties of PEEK compared to those of other polymers.

Material	Specific Gravity (g/cm^3)	Glass Transition Temperature (°C)	Melting Point (°C)	Tensile Strength (MPa)	Young Modulus (GPa)	Operating Temperature (°C)	Elongation at Break (In %)	Ref.
PEEK	1.3	143	340	86–94	3–4	250	-	[4,42]
PEKK [1]	1.27	162	305	105	5.1	300	-	[4]
UHMWPE [2]	0.933	−110	200–220	48	0.69	<100	-	[4]
PLLA [3]	-	51–64	172–189	27.98–49	-	-	5	[43]
DL-PLA [4]	-	50–52	Amorphous	28.9–35	-	-	6	[43]
Tendon Chitosan	-	-		56.48–79.21	-	-	11.3–13.92	[43]

[1] Polyetherketoneketone. [2] Ultra-high molecular weight polyethylene. [3] Poly-l-lactic Acid. [4] Poly(dl-lactic acid).

3.1.1. PEEK vs. Conventional Dental Alloy

PEEK material could be a viable alternative to conventional dental alloys. Tekin et al. [39] highlighted the good mechanical and electrical properties of PEEK, as well as its high biocompatibility, making it suitable for orthopedic and trauma cases. Schwitalla et al. [44] demonstrated that PEEK materials have a modulus of elasticity and strengths that meet the minimum requirements for dental applications. Additionally, Bathala et al. [36] discussed the advantages of PEEK, such as flexibility, radiolucency, thermal resistance, and biocompatibility, which make it a potential alternative to titanium and zirconia for dental implants. Researchers found that PEEK has the potential to be used in various areas of dentistry as a substitute for traditional dental alloys [45–47].

3.1.2. PEEK vs. Dental Ceramics

PEEK and dental ceramics are contrasting materials in modern dentistry, each with distinct properties that influence their clinical applications. PEEK's significant flexibility stands in stark contrast to the rigidity of dental ceramics [6]. This flexibility grants PEEK superior stress distribution and shock absorption capabilities, potentially reducing the risk of fractures and chips that often plague ceramic restorations [48]. Moreover, PEEK's bone-like elasticity may contribute to better temporomandibular joint (TMJ) health by allowing more natural force distribution and adaptive load transfer [49,50]. While ceramics maintain their stronghold in aesthetically critical areas, PEEK's unique combination of flexibility, fracture resistance, and biocompatibility has driven its increased adoption in prosthetics, implant components, and TMJ reconstruction [51,52]. As dentistry continues to evolve, the choice between PEEK and ceramics increasingly depends on balancing the mechanical properties, aesthetics, and patient-specific needs in each clinical scenario. According to Reddy et al. [53], PEEK dental implants better mimic the mechanical characteristics of bone and show less stress shielding. Elawadly et al. [54] evaluated the surface roughness and wettability of different PEEK specimens and suggested that filled PEEK materials exhibit favorable roughness and wettability properties, making them potential substrates for dental implants. Bekhiet et al. [55] investigated the wear and volumetric loss of PEEK against other dental materials and aimed to determine its efficacy as a permanent restoration.

3.2. PEEK in Dental Implantology

PEEK has been explored for various applications in dental implantology due to its unique properties. Since PEEK and bone have more similar mechanical characteristics, PEEK dental implants have been observed to have less stress shielding than titanium dental implants (Table 2) [6,56]. It is a promising material for dental implants due to its biocompatibility, mechanical strength, and resistance to wear and fatigue.

Table 2. PEEK properties compared to bone and other dental materials.

Material	Young's Modulus (GPa)	Tensile Strength (MPa)	Flexural Strength (MPa)	Hardness	Degradability	Ref.
Cancellous Bone	2–4	13–17	12–18	0.46 ± 0.08	Biodegradable	[42,57]
Cortical Bone	27–33	105–115	118–122	0.43 ± 0.13	Biodegradable	[42,57]
PEEK	3–4	86–94	110–120	85–109	Nonbiodegradable	[42,57]
20 wt.% CFR-PEEK	18–20	125–131	160–168	-	Nonbiodegradable	[42]
30 wt.% CFR-PEEK	22–28	151–157	163–173	-	Nonbiodegradable	[42]
GFR-PEEK	10–12	115–158	198–228	-	Nonbiodegradable	[42]
Zirconia	200–10	320–340	240–260	-	Nonbiodegradable	[42]
Titanium alloy	110–119	862–1200	-	337–357	Nonbiodegradable	[57]

PEEK can be used as an alternative to traditional metal and ceramic materials for implant abutments [58]. PEEK abutments have been found to have a lower modulus of elasticity than titanium abutments, which can reduce the risk of implant failure due to stress shielding [6]. In addition, PEEK material radiolucency prevents interference with X-rays or other imaging techniques, which allows for better visualization of dental structures and implants during after-implantation diagnosis and treatment planning [59].

Nevertheless, PEEK is considered bioinert and does not readily bond with bone tissue, hindering its full integration with the surrounding bone following implantation. This hydrophobic nature arises from its aromatic ring and polyester functional groups, which prevent osseointegration and result in its separation from the bone [24,60]. The goal of recent research has been to increase the nanoscale bioactivity of PEEK implants in order to promote bone remodeling and osseointegration [24,61]. Comparing PEEK to bioglass-based PEEK nanocomposites, Taymour et al. [6] reported that PEEK with 20 wt.% forsterite in the nanocomposite formulation was the most efficient in PEEK's bioactivation. This particular formulation greatly increased the material's microhardness, flexural strength, and elastic modulus, in addition to enhancing PEEK's bioactivation. This particular formulation greatly increased the material's microhardness, flexural strength, and elastic modulus, in addition to enhancing PEEK's bioactivation. Additionally, as illustrated in Figure 4 [6], both PEEK nanocomposites loaded with bioglass and those with forsterite nanofillers showed the potential to precipitate phosphate and calcium bone minerals on their surfaces during in vitro bioactivity evaluation using biomimetic simulated bodily fluid. Chayanun et al. [25] studied how sulfonation and sub-millimeter laser machining could improve the bioactivity of PEEK surfaces. The outcomes demonstrated that PEEK surfaces' bioactivity was enhanced by the combination of these two methods. The classification of PEEK modifications and coatings for enhanced osseointegration is discussed in greater detail in the next section. However, further research is required to fully understand the potential of PEEK in implantology and improve its bioactivity without compromising its mechanical properties.

PEEK Modifications and Coatings for Enhanced Osseointegration

One of the major drawbacks of PEEK in dental implants is its bioinertness, as discussed above [6,27]. Surface modifications and blends can be applied to PEEK implants to enhance osseointegration. Three primary categories can be used to group these modifications as follows:

1. Bioactive agent surface functionalization: Bioactive agents can be introduced to the PEEK implant surface chemically or physically, as shown in Figure 2. Bioactive agents such as growth factors or peptides can promote cell adhesion, proliferation, and differentiation, leading to improved osseointegration [28]. Cyclic peptides, which have unique characteristics that promote the development of the endothelium, were bound to the PEEK surface by Young et al. [62] using ammonia plasma treatment.

Furthermore, these conjugated cyclic peptides on the PEEK surface exhibit anticoagulant qualities by preventing platelet adhesion and activation, which may prevent blood clot formation. An et al. [63] created a dialdehyde cross-linked hyaluronic acid hydrogel coating on the surface of PEEK that was loaded with nerve growth factor (NGF) and platelet-rich plasma (PRP). These findings demonstrate that the hybrid hydrogel coating on the PEEK surface was capable of continuously releasing growth factors and exhibited good hydrophilicity. The hybrid hydrogel coating exhibited strong cell adhesion and facilitated the differentiation of MC3T3-E1 cells and angiogenesis of human umbilical vein endothelial cells (HUVECs), according to in vitro cell tests. Nevertheless, NGF did not enhance the hybrid hydrogel's capacity for cell growth. Excellent osteogenic and angiogenic properties were also demonstrated by PEEK samples coated with the hybrid hydrogel in a rat tibial defect model. A comparatively green technique for grafting heparin onto PEEK through the thiol-ol reaction was developed by Goh et al. [64] by combining ozone and UV irradiation without the use of any chemical reagents or organic solvents. Instead, they used biocompatible cysteine, an amino acid, to thiolate heparin. In contrast to the rapid release of loaded bone morphogenetic protein-2 (BMP-2) from pure PEEK, heparin grafting on PEEK effectively immobilized BMP-2. Comparing heparin grafting to pure PEEK, the bioactivity of PEEK was improved in terms of MG-63 proliferation and osteogenic differentiation. It was only after heparin grafting that BMP-2 loading onto PEEK could have an impact on more specialized osteogenic activities such as alkaline phosphatase (ALP) activity and calcium deposition in MG-63.

2. Incorporation of PEEK with bioactive materials: Bioactive materials can be applied as surface coatings or incorporated into the PEEK structure as composites, which provide a favorable environment for cell attachment, proliferation, and differentiation, enhancing osseointegration [61]. Bioactive particles, such as hydroxyapatite (HA), strontium-containing hydroxyapatite, titanium dioxide (TiO2), tricalcium phosphate (TCP), bioactive glass, forsterite, and graphene, have been compounded with PEEK to increase the bioactivity and bone osseointegration of PEEK implants [6,65,66]. Silicate-based bioceramic-reinforced PEEK nanocomposites have been developed to improve the mechanical properties, bioactivity, and bone osseointegration of PEEK. PEEK/Bioceramics nanocomposites were fabricated using melt blending and compression molding techniques. The uniform dispersion of nanofiller particles within the PEEK matrix is crucial for improving the mechanical properties and bioactivity [9]. While surface treatments can offer some benefits in terms of improving the surface characteristics of PEEK, such as wettability and bioactivity, PEEK-based composites are designed to provide a comprehensive solution that addresses both mechanical strength and bioactive properties. These composites have shown promise in enhancing the performance of dental implants by promoting osseointegration and ensuring long-term implant stability [67]. PEEK composites offer a consistent distribution of bioactive fillers throughout the material, ensuring uniform bioactivity and mechanical performance across the entire implant, as shown in Figure 4. Surface treatments, on the other hand, may vary in effectiveness and may not provide uniform coverage or bioactivity throughout the PEEK surface. Researchers are tailoring the composition and content of bioactive fillers to optimize these properties. Surface treatments, on the other hand, may not provide the same degree of control over material properties and bioactivity [68]. In addition, PEEK-based composites typically incorporate reinforcing materials, such as hydroxyapatite (HA), glass fibers, or other bioactive fillers. These fillers can significantly enhance the mechanical properties of PEEK, including its tensile strength, elastic modulus, and microhardness. This improved mechanical

strength can better withstand the stresses and loading conditions experienced in the oral environment, contributing to implant stability and longevity [69]. The mechanical properties of PEEK are dependent on the degree of crystallinity since higher crystallinity leads to higher tensile and flexural strengths. The degree of crystallinity is a positive factor in the mechanical properties of PEEK [70]. Pérez-Martín et al. [71] showed that carbon fiber-reinforced PEEK has improved mechanical properties with a higher crystallinity. However, Taymour et al. [6] found that by increasing the number of nanofillers by more than 20% by weight, despite the increased crystallinity, the mechanical properties were reduced, as shown in Table 3. The decline in the mechanical properties at higher filler content might be due to the fact that the higher amounts of nanofillers could restrict the mobility of the polymer chains, decreasing the degree of crystallinity, crystallite size, and crystallization growth rate, as shown in Figures 5 and 6 and Table 3 [72].

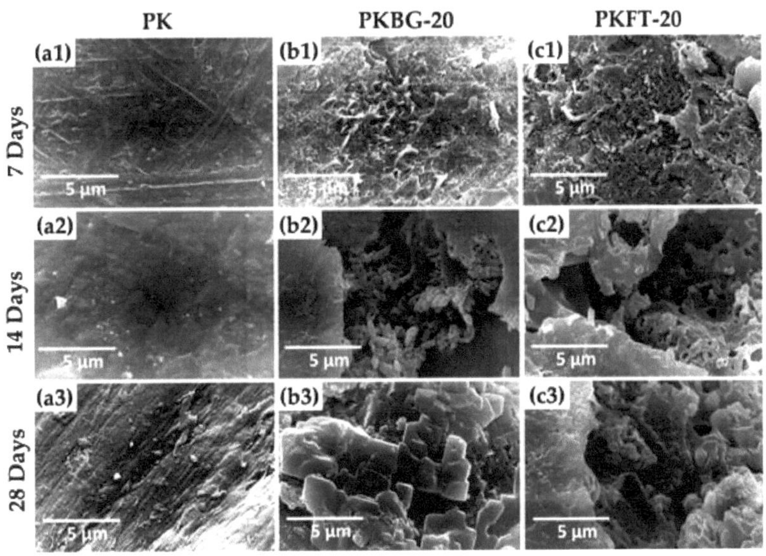

Figure 4. SEM images of (**a1–a3**) pure PEEK and (**b1–b3**) PEEK containing 20 wt.% 45S5 bioglass nanofillers, and (**c1–c3**) PEEK containing 20 wt.% forsterite (Mg_2SiO_4) nanofillers, demonstrating the apatite-formation capability following 7, 14, and 28 days of submersion in SBF. Pure PEEK (**a1–a3**) showed no changes on its surface. PEEK nanocomposites (**b1–b3**,**c1–c3**), however, encourage the production of apatite after immersion in SBF at all time points [6].

Table 3. Mean compression elastic modulus and flexural strength of PEEK nanocomposites with respect to the proportions of bioglass (45S5 bioglass) and forsterite (Mg_2SiO_4) nanofillers. PKBG 10, PKBG 20, and PKBG 30 are PEEK containing 10, 20, and 30 wt. %, bioglass nanofillers, respectively, and PKFT 10, PKFT 20, PKFT 30 is PEEK containing 10, 20, 30 wt.% Forsterite nanofillers, respectively [6].

PEEK Nanocomposites	Elastic Modulus (MPa) Mean ± SD	Flexural Strength (GPa) Mean ± SD
Pure PEEK	3.8 ± 0.13	143.63 ± 3.26
PKBG 10	4.19 ± 0.13	149.43 ± 6.11
PKBG 20	4.63 ± 0.12	151.3 ± 6.69
PKBG 30	3.64 ± 0.14	121.2 ± 6.4
PKFT 10	4.4 ± 0.14	166.33 ± 9.2
PKFT 20	6.02 ± 0.11	200.19 ± 4.12
PKFT 30	4.98 ± 0.17	188.32 ± 4.13

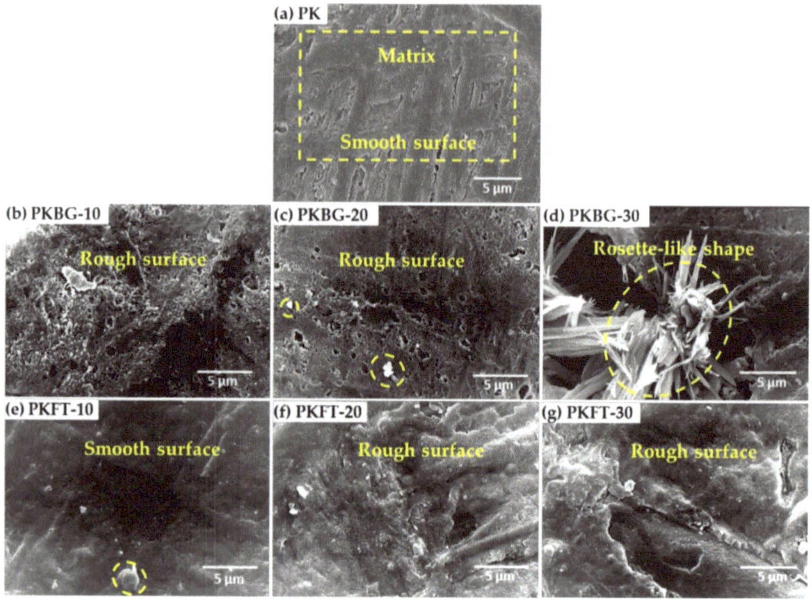

Figure 5. SEM images of pristine PEEK and PEEK loaded with 10,20,30 wt.% BG (45S5 bioglass) nanofillers and PEEK loaded with 10,20,30 wt.% FT (Mg_2SiO_4) nanofillers showing the morphologies of the surface displaying information about the rough and smooth areas: (**a**) PEEK's smooth surface, (**b**,**c**) PKBG-10 and PKBG-20 rough surfaces, (**d**) rough bioglass nanoparticle aggregation on PKBG-30, (**e**) PKFT-10's smooth surface, and (**f**,**g**) PKFT-20 and PKFT-30 low roughness with forsterite nanoparticles dispersed arbitrarily throughout the PEEK matrix. The average particle sizes of the BG and FT nanofillers are 40 ± 4 nm and 30 ± 5 nm, respectively [6].

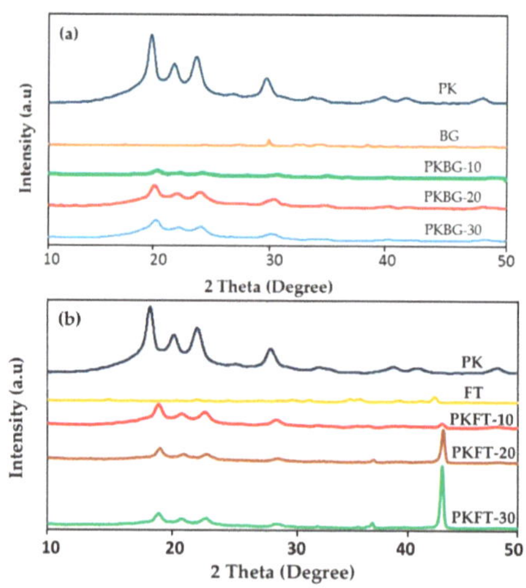

Figure 6. XRD patterns of (**a**) nanocomposites containing Bioglass (BG) nanoparticles and (**b**) nanocomposites including Forsterite (FT) nanoparticles, demonstrating how the addition of these particles affects the crystallinity of PEEK [6].

3. Construction of three-dimensional porous structures: Three-dimensional porous structures can be created on the PEEK surface to promote cell infiltration and vascularization, facilitating osseointegration. These structures can be fabricated using

techniques such as 3D printing or surface modification methods, as shown in Figure 2. Surface porous PEEK has been shown to promote the proliferation, differentiation, and mineralization of osteoblasts, leading to improved osseointegration [67,73,74]. A study conducted by Wei et al. [75] used magnesium surface-activated 3D-printed porous PEEK scaffolds to enhance the osseointegration capacity of PEEK materials. The surface was coated with polydopamine (PDA) chelated with magnesium ions (Mg^{2+}). Following surface modification, bioactive Mg^{2+} was released, and the hydrophilicity of the PEEK scaffolds was greatly increased. The results showed that the customized three-dimensional porous structure facilitated bone ingrowth within the PEEK scaffolds, and the released Mg^{2+} accelerated early bone ingrowth by promoting early angiogenesis during the coating degradation process. Xu et al. [76] developed a Dex/Mino liposome-modified PEEK surface; this surface modification presented favorable stability, cytocompatibility, and improved osseointegration compared to bare PEEK, giving a potential as an orthopedic/dental implant material for clinical application. A recent study proposed a novel surface modification method to obtain a three-dimensional (3D) hierarchical porous structure on the PEEK surface, which showed a boosted osseointegration potential. The hierarchical porous architecture was designed to mimic the natural bone structure and enhance the interaction between the implant and surrounding bone tissue, promoting better osseointegration. More contact areas with the host bone and tissue ingrowth are made possible by the hierarchical topological structure, which results in better mechanical interlocking with the newly created bone tissue [77].

The clinical outcomes and success rates of PEEK dental implants have been evaluated in several studies [78]. A study evaluated the clinical survival and success rate of PEEK composite dental implants using immediate and delayed loading protocols, and the results showed that PEEK composite dental implants had a high clinical survival and success rate (64–76%) [79]. Another study conducted by Ayyadanveettil et al. [30] showed that zirconia and PEEK abutments exhibited the same survival rate with similar biological and esthetic outcomes at the 5-year evaluation. However, there is still a lack of evidence regarding dental implant materials. A longer follow-up period, greater number of patients, and different tooth types (variation in occlusal forces) are required for more accurate results.

3.3. PEEK-Based Dental Prosthetics

PEEK has also been explored for various applications in prosthodontics, including crowns, bridges, and dentures [36] (Figure 7). According to a narrative review of CAD-CAM PEEK dental prostheses, PEEK exhibits excellent mechanical characteristics and strong bonding with veneering composite materials, making it an attractive material for fixed dental prostheses. Similar to zirconia and lithium disilicate crowns, PEEK prostheses have been shown to have high fracture resistance and capacity to tolerate occlusal stresses in the molar region. Additionally, PEEK could be considered a material that works effectively for denture bases because it resists fracture and notch concentration [80]. A systematic review of PEEK-fixed partial dentures found that PEEK frameworks showed better esthetics compared to metal frameworks [81]. PEEK biomaterial has been investigated for a number of clinical dentistry applications, such as implant-supported prostheses, fixed dental prostheses, and removable dental prostheses, according to a literature review of PEEK biomaterial in prosthodontics [82]. Fixed Dental Prostheses (FDPs) made from pre-pressed PEEK blanks utilizing CAD-CAM milling showed greater fracture loads. (2354 N) and less deformation than those that were granularly pressed (1738 N). This implies that employing pre-pressed PEEK and CAD-CAM technologies can produce FDPs that are more resilient [83]. PEEK has been shown in other in vitro investigations to be an excellent substitute material for permanent dental prostheses and single crowns. Its potential for load-bearing areas was verified by three-unit PEEK frameworks, which showed FDP connector fracture at 1383 N and deformation at 1200 N [84]. The fracture resistance of CAD-CAM-milled PEEK FDPs is substantially higher than that of zirconia (981–1331 N), alumina (851 N), and lithium disilicate (950 N) when placed against other dental materials [80,83]. This suggests that PEEK (and its composites) may offer superior durability in certain clinical scenarios. When compared to materials like PMMA, composite resin paste, and fiber-reinforced composites, PEEK demonstrated the maximum load-bearing capacity in an in vitro analysis comparing several inlay-retained FDPs [85].

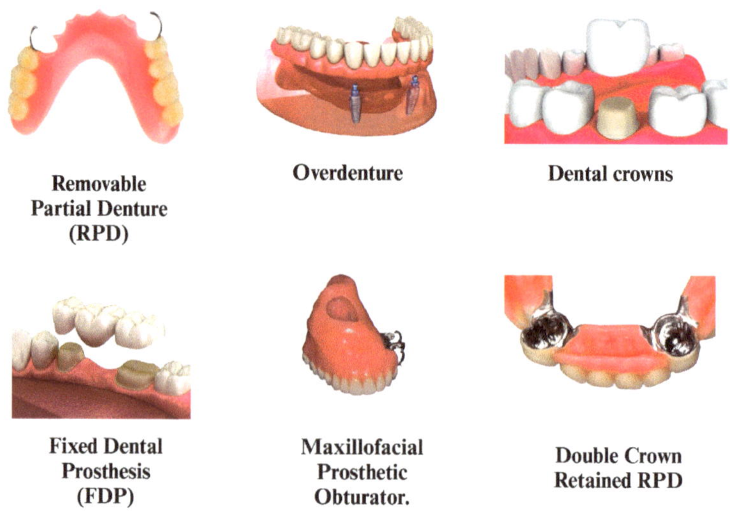

Figure 7. PEEK-based prosthetic restorations.

PEEK-based dental prostheses offer a viable alternative to traditional materials owing to their superior biocompatibility, patient acceptance, and advantageous mechanical characteristics. However, further investigation is required to assess their long-term performance and success rates. From a biocompatibility point of view, PEEK exhibits superior biocompatibility, making it suitable for dental applications. Its insolubility in water makes it ideal for use in patients with allergies. PEEK has been found to be non-mutagenic and non-cytotoxic, ensuring its safety for intraoral use [86]. In terms of patient acceptance, PEEK-based dental prostheses have shown good patient acceptance due to their esthetics and metal-free nature. Implant restorations are frequently chosen by patients who have lost their teeth, and PEEK has been suggested for a variety of implant-supported prostheses [8,86]. Regarding mechanical properties, PEEK-based dental prostheses have been found to have high fracture resistance and load-bearing capacity, comparable to other fixed dental prosthesis materials [35]. In regard to bonding issues, bonding between PEEK and resin-based luting materials, including Panavia V5, has been achieved, ensuring the stability and longevity of PEEK-based dental prostheses [87–90]. PEEK dental prosthesis' long-term survival cannot be confirmed due to the lack of evidence [81]. However, some studies have reported high clinical survival and success rates of PEEK-based dental prostheses for up to 5–8 years [81,91]. To assess the long-term effectiveness and success rates of dental prostheses made from PEEK, additional studies are required [92]. PEEK-based dental prostheses require regular maintenance, including cleaning and polishing, to prevent plaque buildup and staining. PEEK is a highly durable material that can withstand mechanical and physical stress, which makes it a perfect material for dental implants and devices that require long-term durability. However, regular maintenance is still necessary to ensure the longevity of PEEK-based dental prostheses [86]. PEEK-based dental prostheses can be repaired using composite resin materials that bond well with PEEK. However, the repairability of PEEK-based dental prostheses may depend on the extent of damage and location of the prosthesis. PEEK-based dental prostheses may need to be replaced over time due to wear and tear or damage. Replacement may be necessary if the prosthesis becomes loose or damaged beyond repair [93].

A one year single-arm pilot study evaluated the clinical acceptability of PEEK-fixed dental prostheses in partially edentulous patients. The study found that PEEK-fixed dental prostheses were clinically acceptable, but further research is needed to evaluate their long-term survival and success rates [94]. Another study evaluated the survival rates and treatment success of full-arch implant-supported PEEK prostheses. The study found that the dental implant survival rate was 99% for up to 77 months (6 years and 5 months) and the PEEK prosthesis survival rate was 100% for up to 56 months (4 years and 8 months). Bone loss after an average of 54 months (4 years and 6 months) was minimal [92]. However, systematic reviews of clinical studies have found inadequate data to determine the long-term survival of PEEK dental prostheses. Another interesting study evaluated the clinical results of edentulous patients receiving implant-supported fixed complete dentures with PEEK CAD-CAM frameworks compared to titanium frameworks. The study found that PEEK CAD-CAM frameworks were a viable alternative to titanium frameworks, with a high clinical success rate (100%) [95,96].

In terms of practical applications, a more recent systematic exploration by Mishra et al. [97] highlighted that complete-arch implant-supported fixed dental prostheses fabricated with PEEK exhibited a cumulative survival rate of 97.3% over a follow-up period ranging from 1 to 6 years, indicating promising initial outcomes. Their analysis was informed by 12 clinical studies, with a total of 119 prostheses assessed, underscoring feasibility in clinical settings. While the performance metrics appear encouraging for the short-term, the author's cautionary note about adhesion issues indicates an area requiring additional rigorous inquiry. A noteworthy consideration highlighted by Qin et al. [98] is the treatment of the PEEK surface, which is intrinsically hydrophobic. This characteristic has obstructed its bonding efficacy with dental resin composites, presenting a barrier to its widespread use despite the favorable physical and mechanical properties identified in the literature. Innovative surface modifications are posited as crucial pathways toward enhancing the clinical applicability of PEEK-based systems, aligning with the contemporary focus on improving biomaterials in dentistry. Alexakou et al. [99] reported that despite the promising direction of PEEK as a dental restorative material, the clinical translation remains limited by the scarcity of definitive longitudinal studies that track patient outcomes over an extended timeframe.

PEEK's surface characteristics play a significant role in its polishability and subsequent bacterial adhesion, which is critical for prosthetic applications. It demonstrates good polishing properties, which are important for reducing bacterial adhesion and plaque accumulation. PEEK surfaces can be polished to achieve a roughness (Ra) below 0.13 μm. This level of smoothness is considered favorable for reducing bacterial adhesion [100]. Glazing of PEEK surfaces has a positive effect on surface roughness, regardless of artificial aging or staining protocols [101]. Surface irregularities such as flaws and defects promote bacterial attachment by providing protection against shear forces [100]. Pressed PEEK showed higher bacterial adhesion compared to milled, injection-molded, or 3D-printed PEEK. Horizontally, 3D-printed PEEK exhibited lower bacterial adhesion compared to vertically printed specimens. Surface roughness below 10 μm or additional polishing appears to be essential for reducing bacterial adhesion on PEEK surfaces [102] (Figure 8).

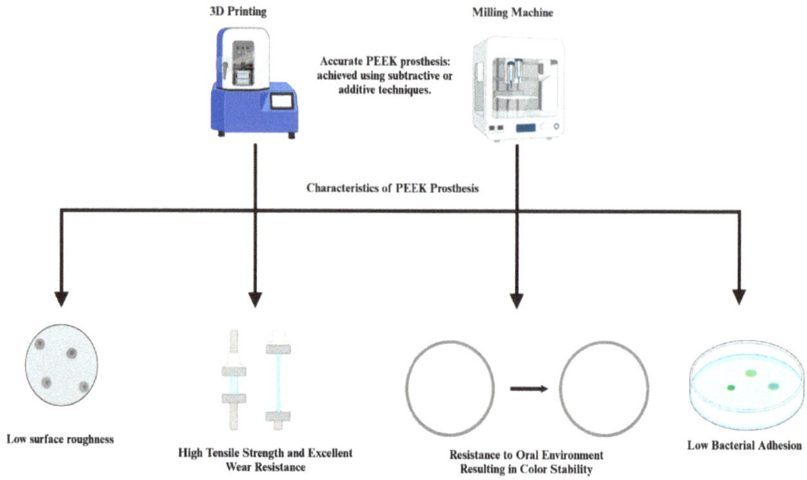

Figure 8. Significant material properties of 3D printed vs. CAD/CAM-milled PEEK prostheses.

3.4. PEEK in Bone Regeneration

3.4.1. PEEK in Alveolar Ridge Preservation and Augmentation

Alveolar ridge augmentation provides sufficient and precise regenerated bone tissue for subsequent dental implant placement [103]. A comparative study based on 3D printing technology found that customized alveolar bone augmentation using PEEK scaffolds is a viable alternative to titanium scaffolds [104–106]. PEEK has been compared to other materials, including titanium mesh, for use as a bone replacement material. PEEK has a higher cost compared to titanium mesh. However, the cost of the combination of titanium mesh and bone cement is reportedly similar to that of PEEK if surgical time and other factors are considered [107]. PEEK exhibits good biocompatibility and radiolucency, making it a potential material for use in cranioplasty and other medical applications. Titanium mesh is commonly used for cranioplasty, but PEEK has been found to have a lower complication rate and implant failure rate compared to titanium mesh [108]. PEEK has a lower elastic modulus than titanium, making it more suitable for craniofacial and orbital floor defects [109]. In comparison to titanium grade 2 and titanium grade 5, PEEK exhibits greater bacterial adhesion [110,111].

3.4.2. Guided Bone Regeneration with PEEK

In order to heal defects in the bone and rebuild the alveolar ridge, PEEK membranes have been examined for use in guided bone regeneration (GBR). A membrane is utilized in GBR to separate soft tissues and boost bone regeneration. The success of GBR is largely dependent on the implementation of barrier membranes. PEEK membranes have been proposed for GBR due to their antibacterial and bone-binding capabilities. However, the economic cost, clinical feasibility, and processing complexity of PEEK modification remain to be evaluated [112–114]. PEEK can be modified to possess antibacterial properties that can reduce the risk of infection and promote bone regeneration. The antibacterial properties of PEEK can be achieved through surface modification or blending modification [115–117]. PEEK has been found to have good bone-binding capabilities, making it a potential material for GBR. PEEK can be modified to enhance its hydrophilicity and protein adsorption capacity, thereby promoting osteogenesis [27,118,119]. PEEK offers a stable environment for bone regeneration and can be utilized as a tailored barrier membrane for GBR. PEEK membranes have been found to have good integration behavior and low immune response, making them a suitable material for GBR. Severe atrophic alveolar bones can be effectively supported using customized PEEK membranes [112,114,120]. However, the practical use of PEEK is restricted due to its poor capacity for osteogenesis and osseointegration. The clinical viability, economic cost, and processing complexity of PEEK modification remain to be evaluated. The antibacterial and bone-binding capabilities of PEEK can be enhanced through surface modification or blending modification. Further research is needed to evaluate the clinical feasibility and long-term performance of PEEK membranes for GBR compared to other materials [112].

3.4.3. PEEK-Based Scaffolds for Tissue Engineering

Cells, scaffolds, and growth factors are used in tissue engineering to replace or regenerate diseased or damaged tissues. The mechanical, biodegradable, and biocompatible characteristics of polymeric scaffolds have made them popular for tissue engineering applications. PEEK scaffolds have been found to have good mechanical properties, including high fracture resistance and load-bearing capacity, making them suitable materials for tissue engineering [75,121,122].

Because of their unique features, such as their high mechanical strength, radiolucency, and good biocompatibility, tissue engineering has made use of PEEK scaffolds. Tissue healing is facilitated by scaffolds, which enhance the differentiation of cells and proliferation. PEEK scaffolds have been found to promote osteogenic bone regeneration and delay adjacent bone loss. PEEK scaffolds have also been used in bone tissue engineering to provide a 3-dimensional environment for cell seeding and proliferation, as well as filling bone defects [11,121,123]. The integration of stem cells and growth factors into PEEK scaffolds has been explored in regenerative medicine and tissue engineering [124]. Scaffolds made of PEEK may contain growth factors for extended release, which promotes osteogenesis. However, mismatched release profiles, in which the growth factor release is frequently unsteady, are typically linked to difficulties with growth factor-enriched scaffolds [125]. Stem cells can be seeded onto PEEK scaffolds to promote bone regeneration and tissue healing. Roskeis et al. [126] used a 3D-printed scaffold embedded with mesenchymal stem cells in an effort to increase PEEK bioactivity for craniofacial repair. Computer-aided design software was used to create customized PEEK scaffolds with a trabecular microstructure, which were subsequently printed using selective laser sintering (SLS). The scaffold structure was evaluated using micro-CT, and SEM was used to evaluate scaffold morphology with and without mesenchymal stem cells (MSCs). The study found that PEEK scaffolds maintained the viability of both adipose-derived stem cells (ADSCs) and bone marrow-derived mesenchymal stem cells (BMSCs); however, ADSCs demonstrated higher osteodifferentiation than BMSCs. Another study evaluated the influence of porous silk scaffolds on the delivery of growth factors and stem cells to enhance bone regeneration. PEEK scaffolds can also be used to seed cells that have been altered to express osteoinductive growth factors, producing a comparable result. Gene therapy carried out by viral or nonviral transduction is usually required to modify the cells [127]. When evaluating how human mesenchymal stem cells (hMSCs) function following their seeding on extremely porous trabecular titanium (T-Ti) or PEEK, the study showed that when it comes to the manner in which human adipose stem cells differentiate into cells that generate bone in 3D culture, osteogenic media outperform growth factors. While hMSCs cultured on the T-Ti scaffold demonstrated uniform colonization of the scaffold, those cultured on PEEK in osteogenic media showed a decrease in cellular density [128].

4. Clinical Outcomes and Success Rate of PEEK Materials

The clinical outcomes and success rates of PEEK materials are promising, with high survival rates reported for PEEK implants, PEEK crowns, PEEK abutments, and PEEK bridges in several studies. However, more long-term studies are needed to ascertain the long-term survival and clinical acceptability of PEEK materials.

Siewert et al. [129] reported positive long-term survival rates and patient satisfaction with PEEK implant-supported full-arch prostheses. Montero et al. [130] found that PEEK rehabilitation significantly improves bite force, occlusal arrangement, masticatory function, patient satisfaction, and quality of life. PEEK was shown by Kwan et al. [131] to be a suitable temporary fixed dental prosthesis during implant treatment, with good results and limited complications. However, Gowda et al. [79] highlighted a higher failure rate for immediately loaded PEEK implants compared to those loaded using an early loading protocol. Most research suggests that PEEK dental materials can offer favorable clinical outcomes and patient satisfaction; however, careful consideration of loading protocols may be necessary for optimal success rates.

Wang et al. conducted a comparative study of PEEK and titanium overdentures. The 5-year survival rates were comparable, with those of PEEK at 93.1% and titanium at 93.5%. PEEK showed slightly better outcomes in terms of peri-implant health: bleeding on probing: PEEK 13.8% vs. titanium 16.1%, soft tissue inflammation: PEEK 3.4% vs. titanium 3.2%, vertical bone loss: PEEK 0.70 mm vs. titanium 0.96 mm [8].

Tasopoulos et al. reported a case study of a 47-year-old patient who received a two-piece PEEK maxillary obturator. After one year of follow-up, no complications were reported, suggesting promising outcomes for PEEK in maxillofacial applications [132].

While these studies demonstrate promising outcomes for PEEK in dental applications, it is important to note that long-term data are still limited. A systematic review protocol conducted in (2024) aims to comprehensively assess the clinical efficacy of PEEK in the all-on-four concept, which may provide more definitive evidence in the future [133]. In addition, Khurshid et al. reported that the majority of the evidence regarding the outcomes of PEEK dental prostheses is obtained from case reports and non-randomized observational studies [81].

Although long-term data on PEEK prosthetics in clinical use are still growing, early results from follow-ups highlight their durability, reduced failure rates, and lower complications, such as mechanical fractures or material wear, compared to traditional prosthetic materials.

5. Future Directions and Emerging Trends

One of the emerging trends is the customization of PEEK dental prosthetics. Advanced technologies like 3D printing and CAD/CAM (computer-aided design and computer-aided manufacturing) are being used to create patient-specific dental implants, crowns, and bridges. This approach can enhance the fit and aesthetics of dental prostheses [15,134,135]. 3D printing technology has been explored for the fabrication of dental prostheses, including crowns, bridges, and dentures [136]. 3D printing technology has also been explored for the fabrication of customized periodontal scaffolds for the regeneration of oral soft tissues [137]. The integration of PEEK-based dental materials into digital workflow and teledentistry could streamline treatment planning and delivery, enabling more efficient and patient-centered care. Interest in the development of hybrid materials that combine PEEK with other substances like bioceramics or bioactive agents to leverage PEEK's mechanical properties while enhancing its biological interactions is growing rapidly. Moreover, advanced techniques such as plasma treatment and coating with bioactive materials can potentially enhance PEEK's bioactivity. Continued research on implant design, including macrostructure and microstructure modifications, can lead to more successful and durable PEEK dental implants [42]. These improvements may enhance stability and reduce the risk of implant failure. The use of PEEK-based dental materials and 3D printing technology in dentistry raises regulatory considerations. The safety and efficacy of PEEK-based dental materials and 3D printing technology need to be evaluated through clinical trials and regulatory approval processes.

6. Conclusions

PEEK is a highly promising material in the field of dentistry, with a wide range of applications across various specialties. Key findings from the existing literature highlight PEEK's exceptional mechanical properties and its versatility in removable and fixed dental prostheses. Researchers have also been actively exploring ways to enhance the bioactivity of PEEK implants, opening up new avenues for implantology. PEEK's notable qualities, including mechanical strength, biocompatibility, chemical stability, and radiolucency, position it as an ideal material for removable dental prostheses. Furthermore, its applications extend to implantology, removable denture frameworks, fixed partial dentures, and orthodontic wires. Moreover, PEEK's unique properties have paved the way for groundbreaking developments in tissue engineering and regenerative medicine, particularly in areas such as bone regeneration and the creation of customized periodontal scaffolds. By leveraging 3D printing technology, dental professionals have gained greater flexibility in designing and fabricating dental prostheses and scaffolds, thereby ushering in a new era of personalized dentistry. However, it is imperative to stress that clinical trials and regulatory approval procedures should be undertaken to thoroughly assess the safety and effectiveness of PEEK-based dental materials and 3D printing technologies. This step is vital to ensure that these innovations meet the highest standards and adhere to ethical considerations.

In summary, PEEK-based dental materials hold significant promise for the future of dentistry, offering a unique combination of properties and the potential for personalized treatment approaches. However, ongoing research is essential to assess their clinical feasibility and long-term performance and navigate the complex landscape of regulatory and ethical considerations, ultimately working toward improved patient outcomes in the field of dentistry.

Author Contributions: Conceptualization, N.T. and M.A.-H.G.; methodology, N.T.; software, A.A.E.-F.; validation, M.A.-H.G., A.K. and S.K.; formal analysis, A.E.F.; investigation, A.E.F.; resources, N.T.; data curation, A.K.; writing—original draft preparation, N.T.; writing—review and editing, N.T. and A.K.; visualization N.H.A.-Q. and M.A.-H.G.; supervision, Y.A.A.-D.; project administration, A.K.; funding acquisition, Y.A.A.-D. All authors have read and agreed to the published version of the manuscript.

Funding: This research received no external funding.

Institutional Review Board Statement: Not applicable.

Data Availability Statement: All data analyzed in this study are included in this published article.

Conflicts of Interest: The authors declare no conflicts of interest.

References

1. Panayotov, I.V.; Orti, V.; Cuisinier, F.; Yachouh, J. Polyetheretherketone (PEEK) for Medical Applications. *J. Mater. Sci. Mater. Med.* **2016**, *27*, 118. [CrossRef]
2. Behera, A. Advanced Plastic Materials. In *Advanced Materials*; Springer International Publishing: Cham, Switzerland, 2022; pp. 469–506.
3. Russel, M.E.; Madhu, S. A Review on The Tribological Behavior of Polyether Ether Ketone (PEEK). *Ann. Rom. Soc. Cell Biol.* **2021**, *25*, 20312–20321.
4. Zol, S.M.; Alauddin, M.S.; Said, Z.; Mohd Ghazali, M.I.; Hao-Ern, L.; Mohd Farid, D.A.; Zahari, N.A.H.; Al-Khadim, A.H.A.; Abdul Aziz, A.H. Description of Poly(Aryl-Ether-Ketone) Materials (PAEKs), Polyetheretherketone (PEEK) and Polyetherketoneketone (PEKK) for Application as a Dental Material: A Materials Science Review. *Polymers* **2023**, *15*, 2170. [CrossRef] [PubMed]

5. Abd El-Fattah, A.; Youssef, H.; Gepreel, M.A.H.; Abbas, R.; Kandil, S. Surface Morphology and Mechanical Properties of Polyether Ether Ketone (PEEK) Nanocomposites Reinforced by Nano-Sized Silica (SiO_2) for Prosthodontics and Restorative Dentistry. *Polymers* **2021**, *13*, 3006. [CrossRef]
6. Taymour, N.; Fahmy, A.E.; Gepreel, M.A.H.; Kandil, S.; El-Fattah, A.A. Improved Mechanical Properties and Bioactivity of Silicate Based Bioceramics Reinforced Poly(Ether-Ether-Ketone) Nanocomposites for Prosthetic Dental Implantology. *Polymers* **2022**, *14*, 1632. [CrossRef]
7. Nikonovich, M.; Costa, J.F.S.; Fonseca, A.C.; Ramalho, A.; Emami, N. Structural, Thermal, and Mechanical Characterisation of PEEK-Based Composites in Cryogenic Temperature. *Polym. Test.* **2023**, *125*, 108139. [CrossRef]
8. Wang, J.; Wu, P.; Liu, H.L.; Zhang, L.; Liu, L.P.; Ma, C.F.; Chen, J.H. Polyetheretherketone versus Titanium CAD-CAM Framework for Implant-Supported Fixed Complete Dentures: A Retrospective Study with up to 5-Year Follow-Up. *J. Prosthodont. Res.* **2022**, *66*, 279–287. [CrossRef] [PubMed]
9. Almasi, D.; Iqbal, N.; Sadeghi, M.; Sudin, I.; Abdul Kadir, M.R.; Kamarul, T. Preparation Methods for Improving PEEK's Bioactivity for Orthopedic and Dental Application: A Review. *Int. J. Biomater.* **2016**, *2016*, 8202653. [CrossRef] [PubMed]
10. Hanemann, T.; Klein, A.; Baumgärtner, S.; Jung, J.; Wilhelm, D.; Antusch, S. Material Extrusion 3D Printing of PEEK-Based Composites. *Polymers* **2023**, *15*, 3412. [CrossRef]
11. Zheng, J.; Zhao, H.; Dong, E.; Kang, J.; Liu, C.; Sun, C.; Li, D.; Wang, L. Additively-Manufactured PEEK/HA Porous Scaffolds with Highly-Controllable Mechanical Properties and Excellent Biocompatibility. *Mater. Sci. Eng. C* **2021**, *128*, 112333. [CrossRef]
12. Zhang, X.H.; Jiao, M.X.; Wang, X.; Li, B.L.; Zhang, F.; Li, Y.B.; Zhao, J.N.; Jin, H.H.; Yang, Y. Preheat Compression Molding for Polyetherketoneketone: Effect of Molecular Mobility. *Chin. J. Polym. Sci.* **2022**, *40*, 175–184. [CrossRef]
13. Li, T.; Song, Z.; Yang, X.; Du, J. Influence of Processing Parameters on the Mechanical Properties of Peek Plates by Hot Compression Molding. *Materials* **2022**, *16*, 36. [CrossRef] [PubMed]
14. Lee, D.J. Comparison of Mechanical Properties of Compression and Injection Molded PEEK/Carbon Fiber Reinforced Composites. *Key Eng. Mater.* **2006**, *306–308*, 751–756. [CrossRef]
15. Chuchulska, B.; Dimitrova, M.; Vlahova, A.; Hristov, I.; Tomova, Z.; Kazakova, R. Comparative Analysis of the Mechanical Properties and Biocompatibility between CAD/CAM and Conventional Polymers Applied in Prosthetic Dentistry. *Polymers* **2024**, *16*, 877. [CrossRef] [PubMed]
16. Alqutaibi, A.Y.; Alghauli, M.A.; Aljohani, M.H.A.; Zafar, M.S. Advanced Additive Manufacturing in Implant Dentistry: 3D Printing Technologies, Printable Materials, Current Applications and Future Requirements. *Bioprinting* **2024**, *42*, e00356. [CrossRef]
17. Vecsei, B.; Czigola, A.; Róth, I.; Hermann, P.; Borbély, J. Digital Impression Systems, CAD/CAM, and STL File. In *Guided Endodontics*; Springer International Publishing: Cham, Switzerland, 2021; pp. 27–63.
18. Jockusch, J.; Özcan, M. Additive Manufacturing of Dental Polymers: An Overview on Processes, Materials and Applications. *Dent. Mater. J.* **2020**, *39*, 345–354. [CrossRef]
19. Limaye, N.; Veschini, L.; Coward, T. Assessing Biocompatibility & Mechanical Testing of 3D-Printed PEEK versus Milled PEEK. *Heliyon* **2022**, *8*, e12314. [CrossRef] [PubMed]
20. Prechtel, A.; Reymus, M.; Edelhoff, D.; Hickel, R.; Stawarczyk, B. Comparison of Various 3D Printed and Milled PAEK Materials: Effect of Printing Direction and Artificial Aging on Martens Parameters. *Dent. Mater.* **2020**, *36*, 197–209. [CrossRef]
21. Moby, V.; Dupagne, L.; Fouquet, V.; Attal, J.P.; François, P.; Dursun, E. Mechanical Properties of Fused Deposition Modeling of Polyetheretherketone (PEEK) and Interest for Dental Restorations: A Systematic Review. *Materials* **2022**, *15*, 6801. [CrossRef] [PubMed]
22. Al Wadei, M.H.D.; Sayed, M.E.; Jain, S.; Aggarwal, A.; Alqarni, H.; Gupta, S.G.; Alqahtani, S.M.; Alahmari, N.M.; Alshehri, A.H.; Jain, M.; et al. Marginal Adaptation and Internal Fit of 3D-Printed Provisional Crowns and Fixed Dental Prosthesis Resins Compared to CAD/CAM-Milled and Conventional Provisional Resins: A Systematic Review and Meta-Analysis. *Coatings* **2022**, *12*, 1777. [CrossRef]
23. Han, X.; Yang, D.; Yang, C.; Spintzyk, S.; Scheideler, L.; Li, P.; Li, D.; Geis-Gerstorfer, J.; Rupp, F. Carbon Fiber Reinforced PEEK Composites Based on 3D-Printing Technology for Orthopedic and Dental Applications. *J. Clin. Med.* **2019**, *8*, 240. [CrossRef]
24. Duseja, S.; Shah, D.; Vaishnav, K.; Duseja, S.; Kumar, S.; Desai, K. Accuracy of Different Elastomeric Impression Materials Using Dual Arch Impression Trays: An In Vitro Study. *Adv. Hum. Biol.* **2014**, *4*, 3–8.
25. Chayanun, S.; Chanamuangkon, T.; Boonsuth, B.; Boccaccini, A.R.; Lohwongwatana, B. Enhancing PEEK Surface Bioactivity: Investigating the Effects of Combining Sulfonation with Sub-Millimeter Laser Machining. *Mater. Today Bio* **2023**, *22*, 100754. [CrossRef]
26. Martin, A.; Azhagarasan, N.S.; Ravichandran, M.; Ramakrishnan, H.; Jaya Krishnakumar, S.; Mahadevan, V. Evaluation of the Bioactivity of Surface Modified Polyetheretherketone (PEEK) as an Implant Material: An: In Vitro: Study. *Contemp. Clin. Dent.* **2020**, *11*, 356–366. [CrossRef]
27. Yu, D.; Lei, X.; Zhu, H. Modification of Polyetheretherketone (PEEK) Physical Features to Improve Osteointegration. *J. Zhejiang Univ. Sci. B* **2022**, *23*, 189–203. [CrossRef]

28. Dondani, J.R.; Iyer, J.; Tran, S.D. Surface Treatments of PEEK for Osseointegration to Bone. *Biomolecules* **2023**, *13*, 464. [CrossRef]
29. Senra, M.R.; Marques, M.D.; Monteiro, S.N. Poly (Ether-Ether-Ketone) for Biomedical Applications: From Enhancing Bioactivity to Reinforced-Bioactive Composites—An Overview. *Polymers* **2023**, *15*, 373. [CrossRef] [PubMed]
30. Ayyadanveettil, P.; Thavakkara, V.; Latha, N.; Pavanan, M.; Saraswathy, A.; Kuruniyan, M.S. Randomized Clinical Trial of Zirconia and Polyetheretherketone Implant Abutments for Single-Tooth Implant Restorations: A 5-Year Evaluation. *J. Prosthet. Dent.* **2022**, *128*, 1275–1281. [CrossRef] [PubMed]
31. Al-Noaman, A.; Rawlinson, S.C. Polyether Ether Ketone Coated with Nanohydroxyapatite/Graphene Oxide Composite Promotes Bioactivity and Antibacterial Activity at the Surface of the Material. *Eur. J. Oral Sci.* **2023**, *131*, e12946. [CrossRef]
32. Han, X.; Gao, W.; Zhou, Z.; Yang, S.; Wang, J.; Shi, R.; Li, Y.; Jiao, J.; Qi, Y.; Zhao, J. Application of Biomolecules Modification Strategies on PEEK and Its Composites for Osteogenesis and Antibacterial Properties. *Colloids Surf. B Biointerfaces* **2022**, *215*, 112492. [CrossRef]
33. Thakur, A.; Kumar, A.; Kaya, S.; Marzouki, R.; Zhang, F.; Guo, L. Recent Advancements in Surface Modification, Characterization and Functionalization for Enhancing the Biocompatibility and Corrosion Resistance of Biomedical Implants. *Coatings* **2022**, *12*, 1459. [CrossRef]
34. Palekar, G.S.; Uppoor, A.S.; Nayak, S.U. Nano-Scale Surface Modification of Dental Implants–An Emerging Boon for Osseointegration and Biofilm Control. *Acta Marisiensis-Ser. Medica* **2022**, *68*, 154–158.
35. Luo, C.; Liu, Y.; Peng, B.; Chen, M.; Liu, Z.; Li, Z.; Kuang, H.; Gong, B.; Li, Z.; Sun, H. PEEK for Oral Applications: Recent Advances in Mechanical and Adhesive Properties. *Polymers* **2023**, *15*, 386. [CrossRef]
36. Bathala, L.; Majeti, V.; Rachuri, N.; Singh, N.; Gedela, S. The Role of Polyether Ether Ketone (Peek) in Dentistry—A Review. *J. Med. Life* **2019**, *12*, 5–9. [CrossRef]
37. Akay, C.; Ersöz, M.B. PEEK in Dentistry, Properties and Application Areas. *Int. Dent. Res.* **2020**, *10*, 60–65. [CrossRef]
38. Niu, Y.; Zheng, S.; Song, P.; Zhang, X.; Wang, C. Mechanical and Thermal Properties of PEEK Composites by Incorporating Inorganic Particles Modified Phosphates. *Compos. B Eng.* **2021**, *212*, 108715. [CrossRef]
39. Tekin, S.; Cangül, S.; Adıgüzel, Ö.; Değer, Y. Areas for Use of PEEK Material in Dentistry. *Int. Dent. Res.* **2018**, *8*, 84–92. [CrossRef]
40. Sinha, N.; Gupta, N.; Reddy, K.M.; Shastry, Y.M. Versatility of PEEK as a Fixed Partial Denture Framework. *J. Indian Prosthodont. Soc.* **2017**, *17*, 80–83. [CrossRef]
41. Chen, J. PEEK and PES. In *Reference Module in Materials Science and Materials Engineering*; Elsevier: Amsterdam, The Netherlands, 2018.
42. Chen, M.; Luo, C.; Yuan, Y.; Zhou, H.; Li, Z.; Wang, Q.; Gong, B.; Li, Z.; Sun, H. Modification of PEEK for Implants: Strategies to Improve Mechanical, Antibacterial, and Osteogenic Properties. *Rev. Adv. Mater. Sci.* **2024**, *63*, 20240025. [CrossRef]
43. Kadambi, P.; Luniya, P.; Dhatrak, P. Current Advancements in Polymer/Polymer Matrix Composites for Dental Implants: A Systematic Review. *Mater. Today Proc.* **2021**, *46*, 740–745. [CrossRef]
44. Schwitalla, A.D.; Spintig, T.; Kallage, I.; Müller, W.-D. Flexural Behavior of PEEK Materials for Dental Application. *Dent. Mater.* **2015**, *31*, 1377–1384. [CrossRef]
45. Bertotti, K.; Mwenge-Wambel, J.; Sireix, C.; Hüe, O.; Jeannin, C.; Grosgogeat, B. Accurate Analysis of Titanium and PolyEtherEther-Ketone Materials as an Alternative to Cobalt-Chrome Framework in Removable Partial Denture: A Systematic Review. *Dent. Mater.* **2024**, *40*, 1854–1861. [CrossRef]
46. Al Mortadi, N.A.; Khasawneh, L.; Alzoubi, K.H. Manufacturing of PEEK Orthodontic Baseplate and 3D-Printed Alloy Components from an Intraoral Scan. *Medicine* **2024**, *103*, E38004. [CrossRef] [PubMed]
47. Silva, L.S.; Bento, V.; Brunetto, J.; Pesqueira, A. Polyetheretherketone Materials for Removable Partial Denture Frameworks: An Integrative Review. *Gen. Dent.* **2023**, *71*, 58–62. [PubMed]
48. Huang, Z.L.; Shi, J.Y.; Zhang, X.; Gu, Y.X.; Lai, H.C. The Influence of the Shock-Absorbing Restorative Materials on the Stress Distributions of Short Dental Implant Rehabilitations. *Eur. Rev. Med. Pharmacol. Sci.* **2021**, *25*, 24–34. [CrossRef]
49. Shash, Y.H.; El-Wakad, M.T.; Eldosoky, M.A.A.; Dohiem, M.M. Evaluation of Stress and Strain on Mandible Caused Using "All-on-Four" System from PEEK in Hybrid Prosthesis: Finite-Element Analysis. *Odontology* **2023**, *111*, 618–629. [CrossRef]
50. Zhu, Y.; Zheng, F.; Gong, Y.; Zhu, J.; Yin, D.; Liu, Y. Effect of Occlusal Contact on TMJ Loading during Occlusion: An in Silico Study. *Comput. Biol. Med.* **2024**, *178*, 108725. [CrossRef]
51. Emam, M.; Arafa, A.M. Stress Distribution and Fracture Resistance of Green Reprocessed Polyetheretherketone (PEEK) Single Implant Crown Restorations Compared to Unreprocessed PEEK and Zirconia: An in-Vitro Study. *BMC Oral Health* **2023**, *23*, 275. [CrossRef]
52. Genovesi, W.; Comenale, I.C.; Genovesi Filho, W.; Veloso Fernandes, M. Biomechanical Comparative Analysis of Temporomandibular Joint, Glenoid Fossa and Head of the Condyle of Conventional Models Prothesis with New PEEK Design. *J. Oral Biol. Craniofac. Res.* **2022**, *12*, 529–541. [CrossRef]

53. Reddy, K.U.K.; Seth, A.; Vuppuluri, A.; Verma, P.C.; Narala, S.K.R.; Babu, P.J.; Saravanan, P. Exploring the Bio-Mechanical Behavior of PEEK and CFR-PEEK Materials for Dental Implant Applications Using Finite Element Analysis. *J. Prosthodont. Res.* **2024**. [CrossRef]
54. Elawadly, T.; Radi, I.A.W.; El Khadem, A.; Osman, R.B. Can PEEK Be an Implant Material? Evaluation of Surface Topography and Wettability of Filled Versus Unfilled PEEK With Different Surface Roughness. *J. Oral Implantol.* **2017**, *43*, 456–461. [CrossRef]
55. Bekhiet, A.; Katamish, H.; Salah, T.; Samhan, T. The Effect of Two Body Wear on CAD/CAM PEEK against Monolithic Zirconia, Lithium Disilicate, and Tooth Enamel; "An In Vitro Study". *Egypt. Dent. J.* **2022**, *68*, 2691–2700. [CrossRef]
56. Najeeb, S.; Zafar, M.S.; Khurshid, Z.; Siddiqui, F. Applications of Polyetheretherketone (PEEK) in Oral Implantology and Prosthodontics. *J. Prosthodont. Res.* **2016**, *60*, 12–19. [CrossRef] [PubMed]
57. Mehrotra, D.; Kumar, S.; Mehrotra, P.; Khanna, R.; Khanna, V.; Eggbeer, D.; Evans, P. Patient Specific Total Temporomandibular Joint Reconstruction: A Review of Biomaterial, Designs, Fabrication and Outcomes. *J. Oral Biol. Craniofac. Res.* **2021**, *11*, 334–343. [CrossRef] [PubMed]
58. Mishra, S.; Chowdhary, R. PEEK Materials as an Alternative to Titanium in Dental Implants: A Systematic Review. *Clin. Implant. Dent. Relat. Res.* **2019**, *21*, 208–222. [CrossRef]
59. Nagarajan, A.; Perumalsamy, R.; Thyagarajan, R.; Namasivayam, A. Diagnostic Imaging for Dental Implant Therapy. *J. Clin. Imaging Sci.* **2014**, *4*, 4. [CrossRef]
60. Ma, R.; Guo, D. Evaluating the Bioactivity of a Hydroxyapatite-Incorporated Polyetheretherketone Biocomposite. *J. Orthop. Surg. Res.* **2019**, *14*, 32. [CrossRef]
61. Zheng, Z.; Liu, P.; Zhang, X.; Xin, J.; Wang, Y.; Zou, X.; Mei, X.; Zhang, S.; Zhang, S. Strategies to Improve Bioactive and Antibacterial Properties of Polyetheretherketone (PEEK) for Use as Orthopedic Implants. *Mater. Today Bio* **2022**, *16*, 100402. [CrossRef] [PubMed]
62. Young, E.R.; Martin, C.; Ribaudo, J.G.; Xia, X.; Moritz, W.R.; Madira, S.; Zayed, M.A.; Sacks, J.M.; Li, X. Surface Modification of PEEKs with Cyclic Peptides to Support Endothelialization and Antithrombogenicity. *Mater. Today Commun.* **2024**, *39*, 108664. [CrossRef]
63. An, J.; Shi, X.; Zhang, J.; Qi, L.; Xue, W.; Nie, X.; Yun, Z.; Zhang, P.; Liu, Q. Dual Aldehyde Cross-Linked Hyaluronic Acid Hydrogels Loaded with PRP and NGF Biofunctionalized PEEK Interfaces to Enhance Osteogenesis and Vascularization. *Mater. Today Bio* **2024**, *24*, 100928. [CrossRef]
64. Goh, M.; Min, K.; Kim, Y.H.; Tae, G. Chemically Heparinized PEEK via a Green Method to Immobilize Bone Morphogenetic Protein-2 (BMP-2) for Enhanced Osteogenic Activity. *RSC Adv.* **2024**, *14*, 1866–1874. [CrossRef] [PubMed]
65. Khallaf, R.M.; Emam, A.N.; Mostafa, A.A.; Nassif, M.S.; Hussein, T.S. Strength and Bioactivity of PEEK Composites Containing Multiwalled Carbon Nanotubes and Bioactive Glass. *J. Mech. Behav. Biomed. Mater.* **2023**, *144*, 105964. [CrossRef]
66. Jiang, N.; Tan, P.; He, M.; Zhang, J.; Sun, D.; Zhu, S. Graphene Reinforced Polyether Ether Ketone Nanocomposites for Bone Repair Applications. *Polym. Test.* **2021**, *100*, 107276. [CrossRef]
67. Rahmitasari, F.; Ishida, Y.; Kurahashi, K.; Matsuda, T.; Watanabe, M.; Ichikawa, T. PEEK with Reinforced Materials and Modifications for Dental Implant Applications. *Dent. J.* **2017**, *5*, 35. [CrossRef]
68. Pidhatika, B.; Widyaya, V.T.; Nalam, P.C.; Swasono, Y.A.; Ardhani, R. Surface Modifications of High-Performance Polymer Polyetheretherketone (PEEK) to Improve Its Biological Performance in Dentistry. *Polymers* **2022**, *14*, 5526. [CrossRef]
69. Oladapo, B.I.; Zahedi, S.A. Improving Bioactivity and Strength of PEEK Composite Polymer for Bone Application. *Mater. Chem. Phys.* **2021**, *266*, 124485. [CrossRef]
70. Yu, X.; Yao, S.; Chen, C.; Wang, J.; Li, Y.; Wang, Y.; Khademhosseini, A.; Wan, J.; Wu, Q. Preparation of Poly(Ether-Ether-Ketone)/Nanohydroxyapatite Composites with Improved Mechanical Performance and Biointerfacial Affinity. *ACS Omega* **2020**, *5*, 29398–29406. [CrossRef]
71. Pérez-Martín, H.; Buchalik-Bopp, S.; Guettler, B.E.; Mackenzie, P.; Baidak, A.; Ó Brádaigh, C.M.; Ray, D. Effect of Crystallinity and Morphology on the Mechanical Properties of CF/PEKK Composites Manufactured under Compression Moulding and Automated Tape Placement. *Mater. Today Commun.* **2023**, *36*, 106442. [CrossRef]
72. Alvaredo, Á.; Martín, M.I.; Castell, P.; de Villoria, R.; Fernández-Blázquez, J.P. Non-Isothermal Crystallization Behavior of PEEK/Graphene Nanoplatelets Composites from Melt and Glass States. *Polymers* **2019**, *11*, 124. [CrossRef]
73. Blanch-Martínez, N.; Arias-Herrera, S.; Martínez-González, A. Behavior of Polyether-Ether-Ketone (PEEK) in Prostheses on Dental Implants. A Review. *J. Clin. Exp. Dent.* **2021**, *13*, e520–e526. [CrossRef] [PubMed]
74. Ma, Z.; Zhao, X.; Zhao, J.; Zhao, Z.; Wang, Q.; Zhang, C. Biologically Modified Polyether Ether Ketone as Dental Implant Material. *Front. Bioeng. Biotechnol.* **2020**, *8*, 620537. [CrossRef] [PubMed]
75. Wei, X.; Zhou, W.; Tang, Z.; Wu, H.; Liu, Y.; Dong, H.; Wang, N.; Huang, H.; Bao, S.; Shi, L.; et al. Magnesium Surface-Activated 3D Printed Porous PEEK Scaffolds for in Vivo Osseointegration by Promoting Angiogenesis and Osteogenesis. *Bioact. Mater.* **2023**, *20*, 16–28. [CrossRef]

76. Xu, X.; Li, Y.; Wang, L.; Li, Y.; Pan, J.; Fu, X.; Luo, Z.; Sui, Y.; Zhang, S.; Wang, L.; et al. Triple-Functional Polyetheretherketone Surface with Enhanced Bacteriostasis and Anti-Inflammatory and Osseointegrative Properties for Implant Application. *Biomaterials* **2019**, *212*, 98–114. [CrossRef] [PubMed]
77. Schwitalla, A.; Müller, W.D. PEEK Dental Implants: A Review of the Literature. *J. Oral Implantol.* **2013**, *39*, 743–749. [CrossRef] [PubMed]
78. Chen, Z.; Chen, Y.; Wang, Y.; Deng, J.; Wang, X.; Wang, Q.; Liu, Y.; Ding, J.; Yu, L. Polyetheretherketone Implants with Hierarchical Porous Structure for Boosted Osseointegration. *Biomater. Res.* **2023**, *27*, 61. [CrossRef]
79. Gowda, E.M.; Iyer, S.R.; Verma, K.; Murali Mohan, S. Evaluation of PEEK Composite Dental Implants: A Comparison of Two Different Loading Protocols. *J. Dent. Res. Rep.* **2018**, *1*, 1–5. [CrossRef]
80. Papathanasiou, I.; Kamposiora, P.; Papavasiliou, G.; Ferrari, M. The Use of PEEK in Digital Prosthodontics: A Narrative Review. *BMC Oral Health* **2020**, *20*, 217. [CrossRef]
81. Khurshid, Z.; Nedumgottil, B.M.; Ali, R.M.M.; Bencharit, S.; Najeeb, S. Insufficient Evidence to Ascertain the Long-Term Survival of PEEK Dental Prostheses: A Systematic Review of Clinical Studies. *Polymers* **2022**, *14*, 2441. [CrossRef] [PubMed]
82. Suphangul, S.; Rokaya, D.; Kanchanasobhana, C.; Rungsiyakull, P.; Chaijareenont, P. PEEK Biomaterial in Long-Term Provisional Implant Restorations: A Review. *J. Funct. Biomater.* **2022**, *13*, 33. [CrossRef]
83. Stawarczyk, B.; Eichberger, M.; Uhrenbacher, J.; Wimmer, T.; Edelhoff, D.; Schmidlin, P.R. Three-Unit Reinforced Polyetheretherketone Composite FDPs: Influence of Fabrication Method on Load-Bearing Capacity and Failure Types. *Dent. Mater. J.* **2015**, *34*, 7–12. [CrossRef]
84. Stawarczyk, B.; Beuer, F.; Wimmer, T.; Jahn, D.; Sener, B.; Roos, M.; Schmidlin, P.R. Polyetheretherketone—A Suitable Material for Fixed Dental Prostheses? *J. Biomed. Mater. Res. B Appl. Biomater.* **2013**, *101*, 1209–1216. [CrossRef]
85. Cekic-Nagas, I.; Egilmez, F.; Ergun, G.; Vallittu, P.K.; Lassila, L.V.J. Load-Bearing Capacity of Novel Resin-Based Fixed Dental Prosthesis Materials. *Dent. Mater. J.* **2018**, *37*, 49–58. [CrossRef] [PubMed]
86. Wang, B.; Huang, M.; Dang, P.; Xie, J.; Zhang, X.; Yan, X. PEEK in Fixed Dental Prostheses: Application and Adhesion Improvement. *Polymers* **2022**, *14*, 2323. [CrossRef]
87. Tsuka, H.; Morita, K.; Kato, K.; Kawano, H.; Abekura, H.; Tsuga, K. Evaluation of Shear Bond Strength between PEEK and Resin-Based Luting Material. *J. Oral Biosci.* **2017**, *59*, 231–236. [CrossRef]
88. Pourkhalili, H.; Asli, H.N.; Toreihi, N.; Falahchai, M. Shear Bond Strength of Different Luting Agents to Polyether Ether Ketone. *Dent. Res. J.* **2022**, *19*, 45.
89. Soares Machado, P.; Cadore Rodrigues, A.C.; Chaves, E.T.; Susin, A.H.; Valandro, L.F.; Pereira, G.K.R.; Rippe, M.P. Surface Treatments and Adhesives Used to Increase the Bond Strength Between Polyetheretherketone and Resin-Based Dental Materials: A Scoping Review. *J. Adhes. Dent.* **2022**, *24*, 233–245. [CrossRef]
90. Hata, K.; Komagata, Y.; Nagamatsu, Y.; Masaki, C.; Hosokawa, R.; Ikeda, H. Bond Strength of Sandblasted PEEK with Dental Methyl Methacrylate-Based Cement or Composite-Based Resin Cement. *Polymers* **2023**, *15*, 1830. [CrossRef] [PubMed]
91. Tasopoulos, T.; Pachiou, A.; Kouveliotis, G.; Karaiskou, G.; Ottenga, M.; Zoidis, P. An 8-Year Clinical Outcome of Posterior Inlay Retained Resin Bonded Fixed Dental Prosthesis Utilizing High Performance Polymer Materials: A Clinical Report. *J. Prosthodont.* **2021**, *30*, 19–23. [CrossRef]
92. Siewert, B.; De Alarcón, P. Five-Year Clinical Retrospective Study: Survival and Treatment Success of Full-Arch Implant-Supported PEEK Prostheses. *Eur. J. Dent. Implantol.* **2019**, *15*, 60–64.
93. Zoidis, P.; Papathanasiou, I. Modified PEEK Resin-Bonded Fixed Dental Prosthesis as an Interim Restoration after Implant Placement. *J. Prosthet. Dent.* **2016**, *116*, 637–641. [CrossRef]
94. Raj, D.A.; Chander, N.G.; Reddy, J.R.; Balasubramaniam, M. Clinical Acceptability of PEEK Fixed Dental Prosthesis in Partially Edentulous Patient—A One Year Single Arm Pilot Study. *J. Oral Biol. Craniofac. Res.* **2020**, *10*, 523–528. [CrossRef] [PubMed]
95. Cevik, P.; Schimmel, M.; Yilmaz, B. New Generation CAD-CAM Materials for Implant-Supported Definitive Frameworks Fabricated by Using Subtractive Technologies. *Biomed. Res. Int.* **2022**, *2022*, 3074182. [CrossRef] [PubMed]
96. Gama, L.T.; Bezerra, A.P.; Schimmel, M.; Garcia, R.C.; de Luca Canto, G.; Gonçalves, T.M. Clinical Performance of Polymer Frameworks in Dental Prostheses: A Systematic Review. *J. Prosthet. Dent.* **2022**, *131*, 579–590. [CrossRef] [PubMed]
97. Mishra, S.K.; Bhasmey, S.R.; Chowdhary, R. Complete-Arch Implant-Supported Fixed Dental Prostheses Fabricated with PEEK and PEKK Framework: A Systematic Review. *Evid.-Based Dent.* **2023**, *24*, 193. [CrossRef]
98. Qin, L.; Yao, S.; Zhao, J.; Zhou, C.; Oates, T.W.; Weir, M.D.; Wu, J.; Xu, H.H.K. Review on Development and Dental Applications of Polyetheretherketone-Based Biomaterials and Restorations. *Materials* **2021**, *14*, 408. [CrossRef]
99. Alexakou, E.; Damanaki, M.; Zoidis, P.; Bakiri, E.; Mouzis, N.; Smidt, G.; Kourtis, S. PEEK High Performance Polymers: A Review of Properties and Clinical Applications in Prosthodontics and Restorative Dentistry. *Eur. J. Prosthodont. Restor. Dent.* **2019**, *27*, 113–121. [CrossRef]
100. Liu, Y.; Fang, M.; Zhao, R.; Liu, H.; Li, K.; Tian, M.; Niu, L.; Xie, R.; Bai, S. Clinical Applications of Polyetheretherketone in Removable Dental Prostheses: Accuracy, Characteristics, and Performance. *Polymers* **2022**, *14*, 4615. [CrossRef] [PubMed]

101. Porojan, L.; Toma, F.R.; Vasiliu, R.D.; Topală-Ionel, F.; Porojan, S.D.; Matichescu, A. Optical Properties and Color Stability of Dental PEEK Related to Artificial Ageing and Staining. *Polymers* **2021**, *13*, 4102. [CrossRef]
102. Bächle, J.; Merle, C.; Hahnel, S.; Rosentritt, M. Bacterial Adhesion on Dental Polymers as a Function of Manufacturing Techniques. *Materials* **2023**, *16*, 2373. [CrossRef]
103. Li, L.; Gao, H.; Wang, C.; Ji, P.; Huang, Y.; Wang, C. Assessment of Customized Alveolar Bone Augmentation Using Titanium Scaffolds vs Polyetheretherketone (PEEK) Scaffolds: A Comparative Study Based on 3D Printing Technology. *ACS Biomater. Sci. Eng.* **2022**, *8*, 2028–2039. [CrossRef]
104. Asa'ad, F.; Pagni, G.; Pilipchuk, S.P.; Giannì, A.B.; Giannobile, W.V.; Rasperini, G. 3D-Printed Scaffolds and Biomaterials: Review of Alveolar Bone Augmentation and Periodontal Regeneration Applications. *Int. J. Dent.* **2016**, *2016*, 1239842. [CrossRef]
105. Abdelaziz, A.G.; Nageh, H.; Abdo, S.M.; Abdalla, M.S.; Amer, A.A.; Abdal-hay, A.; Barhoum, A. A Review of 3D Polymeric Scaffolds for Bone Tissue Engineering: Principles, Fabrication Techniques, Immunomodulatory Roles, and Challenges. *Bioengineering* **2023**, *10*, 204. [CrossRef]
106. Wu, D.T.; Munguia-Lopez, J.G.; Cho, Y.W.; Ma, X.; Song, V.; Zhu, Z.; Tran, S.D. Polymeric Scaffolds for Dental, Oral, and Craniofacial Regenerative Medicine. *Molecules* **2021**, *26*, 7043. [CrossRef] [PubMed]
107. Sarfraz, S.; Mäntynen, P.H.; Laurila, M.; Rossi, S.; Leikola, J.; Kaakinen, M.; Suojanen, J.; Reunanen, J. Comparison of Titanium and PEEK Medical Plastic Implant Materials for Their Bacterial Biofilm Formation Properties. *Polymers* **2022**, *14*, 3862. [CrossRef] [PubMed]
108. Thien, A.; King, N.K.; Ang, B.T.; Wang, E.; Ng, I. Comparison of Polyetheretherketone and Titanium Cranioplasty after Decompressive Craniectomy. *World Neurosurg.* **2015**, *83*, 176–180. [CrossRef]
109. Yang, J.; Sun, T.; Yuan, Y.; Li, X.; Yu, H.; Guan, J. Evaluation of Titanium Cranioplasty and Polyetheretherketone Cranioplasty after Decompressive Craniectomy for Traumatic Brain Injury: A Prospective, Multicenter, Non-Randomized Controlled Trial. *Medicine* **2020**, *99*, e21251. [CrossRef]
110. D'Ercole, S.; Cellini, L.; Pilato, S.; Di Lodovico, S.; Iezzi, G.; Piattelli, A.; Petrini, M. Material Characterization and Streptococcus Oralis Adhesion on Polyetheretherketone (PEEK) and Titanium Surfaces Used in Implantology. *J. Mater. Sci. Mater. Med.* **2020**, *31*, 84. [CrossRef] [PubMed]
111. Ma, T.; Zhang, J.; Sun, S.; Meng, W.; Zhang, Y.; Wu, J. Current Treatment Methods to Improve the Bioactivity and Bonding Strength of PEEK for Dental Application: A Systematic Review. *Eur. Polym. J.* **2022**, *183*, 111757. [CrossRef]
112. Shi, Y.; Liu, J.; Du, M.; Zhang, S.; Liu, Y.; Yang, H.; Shi, R.; Guo, Y.; Song, F.; Zhao, Y.; et al. Customized Barrier Membrane (Titanium Alloy, Poly Ether-Ether Ketone and Unsintered Hydroxyapatite/Poly-l-Lactide) for Guided Bone Regeneration. *Front. Bioeng. Biotechnol.* **2022**, *10*, 916967. [CrossRef]
113. Papia, E.; Brodde, S.A.C.; Becktor, J.P. Deformation of Polyetheretherketone, PEEK, with Different Thicknesses. *J. Mech. Behav. Biomed. Mater.* **2022**, *125*, 104928. [CrossRef]
114. Alauddin, M.S.; Abdul Hayei, N.A.; Sabarudin, M.A.; Mat Baharin, N.H. Barrier Membrane in Regenerative Therapy: A Narrative Review. *Membranes* **2022**, *12*, 444. [CrossRef]
115. Brum, R.S.; Labes, L.G.; Volpato Câ, M.; Benfatti, C.A.M.; Pimenta, A.L. Strategies to Reduce Biofilm Formation in PEEK Materials Applied to Implant Dentistry-A Comprehensive Review. *Antibiotics* **2020**, *9*, 609. [CrossRef] [PubMed]
116. Wang, M.; Bhardwaj, G.; Webster, T.J. Antibacterial Properties of PEKK for Orthopedic Applications. *Int. J. Nanomed.* **2017**, *12*, 6471–6476. [CrossRef]
117. Wang, Y.; Zhang, S.; Nie, B.E.; Qu, X.; Yue, B. Approaches to Biofunctionalize Polyetheretherketone for Antibacterial: A Review. *Front. Bioeng. Biotechnol.* **2022**, *10*, 895288. [CrossRef]
118. Wei, Z.; Zhang, Z.; Zhu, W.; Weng, X. Polyetheretherketone Development in Bone Tissue Engineering and Orthopedic Surgery. *Front. Bioeng. Biotechnol.* **2023**, *11*, 1207277. [CrossRef]
119. Liao, C.; Li, Y.; Tjong, S.C. Polyetheretherketone and Its Composites for Bone Replacement and Regeneration. *Polymers* **2020**, *12*, 2858. [CrossRef] [PubMed]
120. Hao, X.; Wang, W.; Wang, C.; Han, J.; Zhang, Y. Polyetherketoneketone Mesh for Alveolar Bone Augmentation: Geometric Parameter Design and Finite Element Analysis. *J. Healthc. Eng.* **2023**, *2023*, 8487380. [CrossRef]
121. Zheng, J.; Zhao, H.; Ouyang, Z.; Zhou, X.; Kang, J.; Yang, C.; Sun, C.; Xiong, M.; Fu, M.; Jin, D.; et al. Additively-Manufactured PEEK/HA Porous Scaffolds with Excellent Osteogenesis for Bone Tissue Repairing. *Compos. B Eng.* **2022**, *232*, 109508. [CrossRef]
122. Yuan, B.; Cheng, Q.; Zhao, R.; Zhu, X.; Yang, X.; Yang, X.; Zhang, K.; Song, Y.; Zhang, X. Comparison of Osteointegration Property between PEKK and PEEK: Effects of Surface Structure and Chemistry. *Biomaterials* **2018**, *170*, 116–126. [CrossRef]
123. Peng, S.; Feng, P.; Wu, P.; Huang, W.; Yang, Y.; Guo, W.; Gao, C.; Shuai, C. Graphene Oxide as an Interface Phase between Polyetheretherketone and Hydroxyapatite for Tissue Engineering Scaffolds. *Sci. Rep.* **2017**, *7*, 46604. [CrossRef] [PubMed]
124. Tollemar, V.; Collier, Z.J.; Mohammed, M.K.; Lee, M.J.; Ameer, G.A.; Reid, R.R. Stem Cells, Growth Factors and Scaffolds in Craniofacial Regenerative Medicine. *Genes. Dis.* **2016**, *3*, 56–71. [CrossRef] [PubMed]

125. Baptista, L.S.; Kronemberger, G.S.; Côrtes, I.; Charelli, L.E.; Matsui, R.A.M.; Palhares, T.N.; Sohier, J.; Rossi, A.M.; Granjeiro, J.M. Adult Stem Cells Spheroids to Optimize Cell Colonization in Scaffolds for Cartilage and Bone Tissue Engineering. *Int. J. Mol. Sci.* **2018**, *19*, 1285. [CrossRef]
126. Roskies, M.; Jordan, J.O.; Fang, D.; Abdallah, M.-N.; Hier, M.P.; Mlynarek, A.; Tamimi, F.; Tran, S.D. Improving PEEK Bioactivity for Craniofacial Reconstruction Using a 3D Printed Scaffold Embedded with Mesenchymal Stem Cells. *J. Biomater. Appl.* **2016**, *31*, 132–139. [CrossRef]
127. Alluri, R.; Jakus, A.; Bougioukli, S.; Pannell, W.; Sugiyama, O.; Tang, A.; Shah, R.; Lieberman, J.R. 3D Printed Hyperelastic "bone" Scaffolds and Regional Gene Therapy: A Novel Approach to Bone Healing. *J. Biomed. Mater. Res. A* **2018**, *106*, 1104–1110. [CrossRef] [PubMed]
128. Ragni, E.; Perucca Orfei, C.; Bidossi, A.; De Vecchi, E.; Francaviglia, N.; Romano, A.; Maestretti, G.; Tartara, F.; de Girolamo, L. Superior Osteo-Inductive and Osteo-Conductive Properties of Trabecular Titanium vs. PEEK Scaffolds on Human Mesenchymal Stem Cells: A Proof of Concept for the Use of Fusion Cages. *Int. J. Mol. Sci.* **2021**, *22*, 2379. [CrossRef]
129. Siewert, B.; Plaza-Castro, M.; Sereno, N.; Jarman-Smith, M. Applications of PEEK in the Dental Field. In *Handbook*, 2nd ed.; Kurtz, S., Peek, B.M., Eds.; William Andrew Publishing: Amsterdam, The Netherlands, 2019; pp. 333–342.
130. Montero, J.; Guadilla, Y.; Flores, J.; Pardal-Peláez, B.; Quispe-López, N.; Gómez-Polo, C.; Dib, A. Patient-Centered Treatment Outcomes with Full-Arch PEEK Rehabilitation Supported on Four Immediate or Conventionally Loaded Implants. A Randomized Clinical Trial. *J. Clin. Med.* **2021**, *10*, 4589. [CrossRef]
131. Kwan, J.C.; Kwan, N. Clinical Application of PEEK as a Provisional Fixed Dental Prosthesis Retained by Reciprocated Guide Surfaces of Healing Abutments During Dental Implant Treatment. *Int. J. Oral Maxillofac. Implant.* **2021**, *36*, 581–586. [CrossRef] [PubMed]
132. Tasopoulos, T.; Chatziemmanouil, D.; Kouveliotis, G.; Karaiskou, G.; Wang, J.; Zoidis, P. PEEK Maxillary Obturator Prosthesis Fabrication Using Intraoral Scanning, 3D Printing, and CAD/CAM. *Int. J. Prosthodont.* **2020**, *33*, 333–340. [CrossRef]
133. Boukhris, H.; Hajjami, H.; Ben youssef, S. Clinical Outcomes of Polyetheretherketone (PEEK) Hybrid Prosthesis in All-on-Four Rehabilitation: A Systematic Review Protocol. *F1000Research* **2024**, *13*, 507. [CrossRef]
134. Cárdenas-Serres, C.; Almeida-Parra, F.; Simón-Flors, A.M.; de Leyva-Moreno, P.; Ranz-Colio, Á.; Ley-Urzaiz, L.; Acero-Sanz, J. Custom CAD/CAM Peek Implants for Complex Orbitocranial Reconstruction: Our Experience with 15 Patients. *J. Clin. Med.* **2024**, *13*, 695. [CrossRef]
135. Wang, Z.; Yu, Z.; Wang, Z.; Li, S.; Song, L.; Xu, T.; Shen, G.; Wang, Y.; Huang, T.; Dong, X.; et al. Surface-Activated 3D-Printed PEEK Implant Enhances Anti-Infection and Osteogenesis. *Compos. B Eng.* **2024**, *273*, 111258. [CrossRef]
136. Alyami, M.H. The Applications of 3D-Printing Technology in Prosthodontics: A Review of the Current Literature. *Cureus* **2024**, *16*, e68501. [CrossRef]
137. Sun, M.; Tang, L.; Yang, X.; Lu, J.; He, H.; Lin, J.; He, Y.; Yu, M. Advancements of Biomaterials in Oral Tissue Engineering: Past, Present, and Future. *Beni Suef Univ. J. Basic. Appl. Sci.* **2024**, *13*, 104. [CrossRef]

polymers

MDPI

Article

Generation of Bio-Based, Shape- and Temperature-Stable Three-Dimensional Nonwoven Structures Using Different Polyhydroxyalkanoates

Tim Hiller [1,*,†], Frederik Gutbrod [2,*,†], Louisa Bonten [1], Marc Philip Vocht [1], Mehdi Azimian [1], Julia Resch [2], Christian Bonten [2] and Malte Winnacker [1]

1 German Institutes of Textile and Fiber Research (DITF), Koerschtalstr. 26, D-73770 Denkendorf, Germany
2 Institut für Kunststofftechnik, University of Stuttgart, Pfaffenwaldring 32, D-70569 Stuttgart, Germany
* Correspondence: tim.hiller@ditf.de (T.H.); frederik.gutbrod@ikt.uni-stuttgart.de (F.G.)
† These authors contributed equally to this work.

Abstract: Recent research has shown the potential of polyhydroxyalkanoates (PHAs), particularly poly(3–hydroxybutyrate) (P3HB), to form nonwoven structures with fine fiber diameter distributions ranging from 2.5 µm to 20 µm during the meltblow process. The shortcomings of existing fabrics of this type include high brittleness, low elongation at break (max. 3%), and a lack of flexibility. Furthermore, the high melt adhesion and the special crystallization kinetics of PHAs have commonly been regarded as constraints in filament and nonwoven processing so far. However, these two properties have now been used to elaborate a three-dimensional fiber arrangement on a matrix, resulting in the creation of dimensionally and temperature-stable "nonwoven-parts". Moreover, this study investigated the PHA copolymer poly(3–hydroxybutyrate–co–3–hydroxyhexanoate) (PHBH), revealing a similar processability to P3HB and PHBV in the meltblow process. A significant increase in the (peak load) elongation in the machine direction was observed, reaching values between 5% and 10%, while the tensile strength retained unaltered. The addition of the bio-based plasticizer acetyltributylcitrate (ATBC) to PHBH resulted on an increase in elongation up to 15%. The three-dimensional fabric structure of PHBH exhibited complete resilience to compression, a property that differentiates it from both P3HB and PHBV. However, the addition of the plasticizer to P3HB did not lead to any improvements. This interesting array of properties results in moderate air permeability and hydrophobicity, leading to impermeability to water.

Keywords: sustainability; biopolymers; polyhydroxyalkanoates; poly(3–hydroxybutyrate) (P3HB); poly(3–hydroxybutyrate-co–3–hydroxyvalerate) (PHBV); poly(3–hydroxybutyrat–co–3–hydroxyhexanoat) (PHBH); melt processing; rheology; meltblow(n); nonwovens

Academic Editors: Vishal Gavande, Vasi Shaikh and Won-Ki Lee

Received: 6 December 2024
Revised: 20 December 2024
Accepted: 23 December 2024
Published: 28 December 2024

Citation: Hiller, T.; Gutbrod, F.; Bonten, L.; Vocht, M.P.; Azimian, M.; Resch, J.; Bonten, C.; Winnacker, M. Generation of Bio-Based, Shape- and Temperature-Stable Three-Dimensional Nonwoven Structures Using Different Polyhydroxyalkanoates. *Polymers* **2025**, *17*, 51. https://doi.org/10.3390/polym17010051

1. Introduction

The demand for bio-based and/or biodegradable alternatives to oil-based polymers is one of the most urgent topics of current polymer research. It should be noted that not all bio-based polymers (polymers created on the basis of biomass or monomers derived from biomass) are necessarily biodegradable (decomposing only into CO_2 and water under certain conditions) and the other way around [1]. A promising class of polymers, combining both specifics, are polyhydroxyalkanoates (PHAs). Although over 100 different PHA variations are known today [2,3], poly(3-hydroxybutyrate) (P3HB) and some selected copolymers (mainly PHBV and recently PHBH) retain most of the industrial relevance due to their price and available scale of production. Meanwhile, a variety of biotechnological as well as synthetic methods are available for accessing these bio-polyesters [4].

PHA copolymers are frequently reported to exhibit superior properties to homopolymeric PHAs, primarily due to their ability to achieve lower crystallinities [5], which in turn results in reduced brittleness and stiffness [5–8]. Additionally, the use of copolymers enables more flexible processing due to their lower melt temperatures [5,9–11]. However, the use of copolymers does not entirely circumvent the issues associated with pure PHAs. These include the high costs, lower yields in synthesis [12,13], and the dependence of the property "improvement" on the co-monomer content in the structure [14,15].

The material properties of PHAs are still challenging, which is reasoned in their crystallite melt temperature being located close or slightly below the (thermal) degradation start temperature [16]. The description of the aging of PHAs in the form of chain scission and depolymerization due to thermal and/or mechanical influences represents a significant field of research, to which numerous publications have been dedicated [17–19] and which will only be referred to in this paper. Additional factors, such as high adhesion of the melt and low melt strength and stretchability, limit the usability of PHAs even further [20]. Moreover, the initial crystallization rate is coupled with a strong post-(secondary) crystallization process [21]. The latter results in high degrees of crystallinity (up to 80% [16]) with large spherulites [16,22], which consequently leads to a high brittleness [2,20], which is disadvantageous for industrial applications. In plastics technology, processing methods are therefore used to optimize the material properties. Typical processing tasks include the dispersion of additives and the filling or reinforcement of plastic melts. For this purpose, twin-screw extruders are mainly used. The mixing processes that occur here can be classified as dispersive and distributive mixing processes. Therefore, the screws have a modular design and are assembled in accordance with the specific requirements of the process [23].

Polylactic acid (PLA) is the most frequently reported and most industrialized member of the family of bio-based polymers. Although desirable meltblown fiber characteristics and nonwoven performance have already been achieved [24–31], its biodegradability is limited to industrial conditions (specific environmental conditions, e.g., high-temperature, etc.) [32–35] and is accompanied by the release of toxic residues [36,37]. The use of PHA-based nonwovens offers a number of advantages over PLA-based nonwovens. These include intrinsic hydrophobicity, a more comprehensive biocompatibility, faster biodegradability in flexible environmental conditions, greater flexibility regarding the applied co-monomers and copolymers, and better CO_2 neutrality. In contrast, PLA-based nonwovens may have advantages over PHA-based nonwovens in terms of costs, market availability, and application-specific properties like higher flexibility and temperature resistance [9,38,39].

In previous research, we successfully demonstrated the meltblow processing of different P3HB grades and PHBV on a technical scale [40]. Previously, the meltblow processing of PHAs/P3HB was only reported on lab-scale equipment [41], with a limitation to very coarse fiber diameters (>10 μm) [42] or by use of huge amounts of plasticizers (5–15%), which resulted in maximum fabric elongations of 4% [41]. In our previous research, we achieved fiber diameters below 10 μm [40] with a median fiber diameter down to 2.4 μm. Additionally, the variability of the median diameter was demonstrated to range up to 20 μm through the adjustment of process settings, thereby illustrating the flexibility for potential application fields. The shortcomings of post-crystallization were addressed by developing targeted process settings and time-stable nonwoven fabrics (>1 year). While the fabrics' flexibility was sufficient, the elongations at break remained a significant limitation, reaching a maximum of 3% in both the machine direction (MD) and the cross direction (CD). However, no improvement to fiber diameters and fabric properties could be achieved with PHBV, which may be attributed to the low co–valerate (3–hydroxyvalerate (3HV)) content of industrial-grade PHBV [40].

In this study, our objective was to investigate the potential of PHAs to form a stable three-dimensional (3D) fabric. This was based on the hypothesis that the combination of high melt adhesion and special crystallization kinetics could overcome the limitations commonly associated with PHAs in filament and nonwoven processing. These 3D structures were produced directly on the conveyor belt via the meltblow process and were designed to retain their stability over time and when subjected to certain temperatures. A plant-pot, as a simple yet application-relevant object shape, was chosen as a demonstrator. The produced pots were characterized and compared to purchasable reference materials.

Furthermore, this study examined two approaches to enhance the low elongation and flexibility of PHA nonwovens. Accordingly, the PHA copolymer poly(3–hydroxybutyrate–co–3–hydroxyhexanoate) (PHBH)—containing short-chain 3-hydroxybutyrate (3HB) units and medium-long-chain 3-hydroxyhexanoate (3HHx) units—was investigated regarding its usability for the meltblow process using a technical-scale meltblow line. Nonwoven webs were characterized for their base weight, thickness, average fiber diameter, air permeability, and their mechanical performance (tensile test). Furthermore, the second approach to improve the properties involved the addition of a bio-based plasticizer (acetyltributylcitrate (ATBC)) and a bio-based chain-extender (epoxidized linseed oil (ELO)) to P3HB and PHBH. The compounds were produced using a twin-screw extruder and the materials were characterized based on their rheological and thermal behavior.

2. Materials and Methods

2.1. Materials

2.1.1. Commercial Polymers

P3HB "P316" was obtained from Biomer Biopolyesters (Schwalbach, Germany). Its melt flow index (MFI) is specified as 10 g 10 min^{-1} at 170 °C, 2.16 kg (ISO 1133) [40,43]. This material has a broad molar mass distribution and was comprehensively characterized in an earlier publication [40].

The chemical structure of P3HB is displayed in Figure 1.

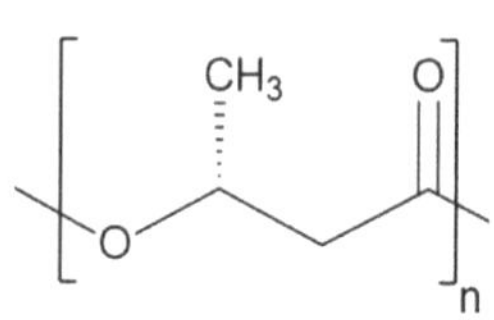

Figure 1. Chemical structure of P3HB.

PHBH (chemical structure, see Figure 2) "BP350–15" from BluePHA®PHA (Beijing, China) was purchased from Helian Polymers BV (DK Belfeld, The Netherlands). According to the manufacturer's specifications, at a 3-hydroxyhexanoate (3HHx) content of 10%, this polymer exhibits a melt flow index of 10–15 g 10 min^{-1} at 165 °C, 5.0 kg (ISO 1133), a melting temperature of 133 °C, a density of 1.19 g cm^{-3}, and a glass transition temperature of −2 °C [44].

2.1.2. Additives

Two different fully bio-based additives were used in order to optimize the material properties. The aim of both was to reduce the viscosity to a range that enabled stable processing in the meltblow process and did not compromise the bio-character. Acetyltributylcitrate (ATBC) (Jungbunzlauer Ladenburg GmbH, Ladenburg, Germany) as well as an epoxidized linseed oil (MERGINAT ELO) (HOBUM Oleochemicals GmbH, Hamburg, Germany) were selected. In addition to its function as a plasticizer, the epoxidized oil provided the potential to react with the hydroxyl and carboxyl groups of the PHAs in order to counteract their degradation process.

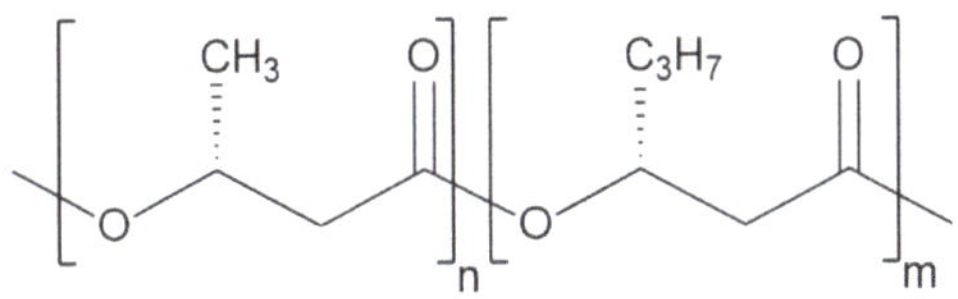

Figure 2. Chemical structure of PHBH.

2.1.3. Reference Nonwoven Fabrics

Meltblown nonwoven fabrics of P3HB (Biomer "P316") and PHBV ("Y1000P", TianAn Biologic Materials Co., Ltd., Ningbo, China) from an earlier study [41] were taken as reference materials. The fabrics were processed using the same machine and setting as for the polymers in this study. The samples are labeled as "P3HB-Ref." and "PHBV-Ref." in the following.

2.1.4. Commercial Plant Seed Pots (Demonstrator References)

Plant seed pots, made from natural fibers, were purchased commercially to provide a direct comparison of the targeted demonstrator application to market-available products which are also of a bio-based and biodegradable nature. Therefore, cellulosic (wooden-based) plant seed pots "Anzuchttöpfe rund 6 cm Ø" and coconut fiber-based pots "Proflora Kokosfaser Anzucht-Pflanztopf 0,30 L" of the same dimensions were obtained from OBI Home and Garden GmbH (Wermelskirchen, Germany).

2.2. Compounding Experiments

A ZSK 26 twin-screw extruder (Coperion GmbH, Stuttgart, Germany) with co-rotating screws was used for the compounding experiments. A proportion of 10 wt.% for the additives was determined on the basis of preliminary tests. Various screw configurations with different proportions of mixing elements and temperature profiles were tested in order to achieve gentle processing with good homogenization. The final screw configuration and temperature profile are shown in Figure 3. In addition to the typical conveying elements, this screw configuration consisted of distributive and disruptive mixing elements as well as a back-pressure element.

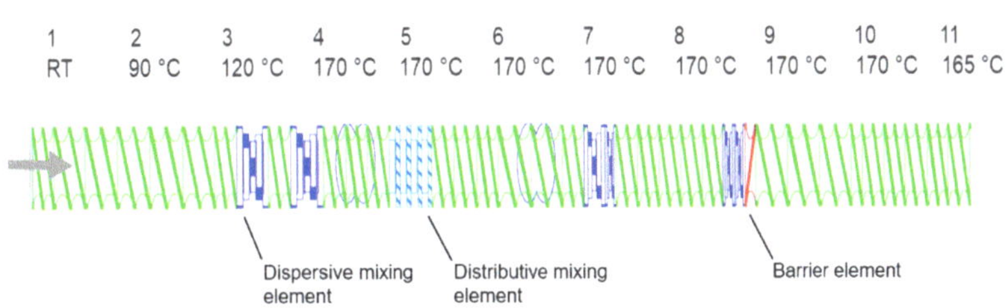

Figure 3. Screw configuration of the ZSK 26 twin-screw extruder and the associated temperature profile for the treatment (the arrow indicates the extrusion direction).

Further processing parameters included a dosing rate of 4 k h^{-1} for the polymer and a rotational speed of the twin-screws of 120 rpm. After compounding, the compounds were dried for at least 24 h at 40 °C in a vacuum chamber.

2.3. Meltblow Process

Nonwoven processing trials were conducted on a technical-scale line with a working width of 500 mm. The line consisted of a single-screw extruder (3 zone screw, ∅ 20 mm × 20 D) from Extrudex GmbH (Mühlacker, Germany) and a gear pump from Mahr Metering Systems GmbH (Göttingen, Germany) with a volume of 0.6 cm^3 rpm^{-1} to melt and transport the polymer to the spinning beam with a maximum throughput of 4 kg h^{-1}. The air system comprised a compressor (Aertronic D12H) from Aerzener Maschinenfabrik GmbH (Aerzen, Germany) with an air volume flow limit of 220 Nm^3 h^{-1} (minimum) and 325 Nm^3 h^{-1} (maximum) in conjunction with a flow heating system produced by Schniewindt GmbH & Co. KG (Neuenrade, Germany). The spinneret was a 561-hole Exxon-type die with a width of 500 mm (28.4 holes per inch (hpi)) and nozzles 0.3 mm in diameter (L/D = 8). The maximal die pressure of the spinneret was set to 50 bar with a safety limit of 45 bar. The set-back between the nozzle tip and the air blades was 1.2 mm and the end gap was set to 2.0 mm for all trials. The conveyor belt manufactured by Siebfabrik Arthur Maurer GmbH & Co. KG (Mühlberg, Germany) was a steel fabric tape in canvas weave with a clip seam, measuring a total width of 0.72 m (no. 16 cm^{-1} linen weave) with a stainless steel (1.4404 AISI 316L) warp and weft wire 0.22 mm in diameter. The maximum take-up velocity was 10 m min^{-1} and the height relative to the die could be adjusted from 200 mm up to 500 mm to vary the die–collector distance (DCD). Below the belt section, where the filaments were laid down, an air-suction box (suction surface of 0.128 m^2, 0.20 m × 0.64 m) with a maximal suction volume of 2900 Nm^3 h^{-1} (maximum flow velocity: 11 m s^{-1}) was placed to remove the process (and secondary) air and to support the web formation on the belt.

Meltblown nonwovens were produced with the polymers listed in Sections 2.1 and 2.2, varying the process temperatures and polymer throughput as the main parameters to reveal a stable process window. The melt temperature was adjusted over the temperature of the die and the spinning head based on the results of the rheological characterization (see Section 2.6) of the respective material with the objective of achieving a zero-shear viscosity at a process temperature of less than 100 Pa s. Furthermore, adjustments were performed during the experiments in order to obtain constant fiber formation at the die and a homogeneous shot-free laydown on the conveyor belt.

The process air throughput was varied between the minimum and maximum output (220–325 Nm^3 h^{-1}) of the compressor in order to define the possible diameter range for each polymer at the respective process setting. Due to the delayed and slow start of the crystallization of PHAs, the process air temperature was maintained 5 K below the melt temperature and the DCD was kept constant at a maximum of 500 mm. This approach was taken to avoid sticking or depletion of the deposit on the conveyor and to reduce fiber-to-fiber bonding [40]. The collector speed was adjusted in accordance with the polymer throughput in order to produce a constant area and base weight of the produced nonwovens, with a target of 100 g m^{-2}. This was intended to ensure the comparability (without influence of the base weight) of web properties under different process settings.

In summary, the following parameters of the entire system were used as variables for this study:

- Polymer throughput: max. 3.5 kg h^{-1};
- Process temperature (polymer, air): potentially up to 420 °C;
- Air throughput: max. 325 Nm^3 h^{-1};
- Collector speed: max. 10 m min^{-1}

2.4. Producution of Three-Dimensional Meltblown Samples

Three-dimensional structures were produced using a "counter-shape", which was placed on the conveyor belt prior to the deposition point of the polymer stream. A plant pot 55 mm (opening) in diameter (50 mm in height; see Figure 4a) was chosen as an application-oriented demonstrator shape, based on commercial biodegradable cellulosic reference plant pots (diameter 6.0 cm, height 65 mm; see Figure 4b).

(**a**) (**b**)

Figure 4. Commercial plant (seed) pots used as the counter-shape and reference samples. (**a**) Plastic plant pot (counter-shape); (**b**) cellulosic seed pot (biodegradable reference sample) [45].

After passing through the stream, the pot was rotated by 90° and placed on the start point again. In total, four passes were carried out with the belt speed adjusted to a base weight of 35 g m^{-2} in order to obtain a total base weight of 150 g m^{-2}. After the last passage, the parts were taken from the conveyor belt and the nonwoven removed from the counterpart (after around 1 min). The process is schematically shown in Figure 5.

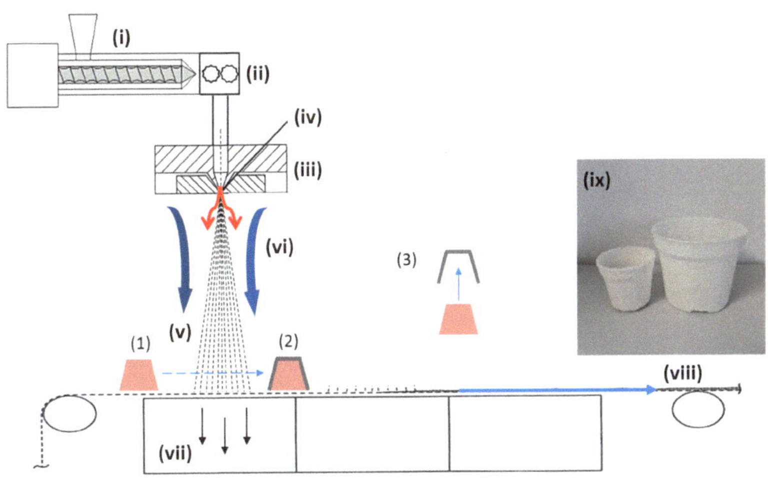

Figure 5. Schematic illustration of the three-dimensional nonwoven deposition on a counter-shape in the meltblow process. Process components: (**i**) extrusion system; (**ii**) gear pump; (**iii**) spin beam; (**iv**) nozzle (red arrows: primary (process) air stream); (**v**) polymer stream; (**vi**) secondary air (blue arrows); (**vii**) air suction; (**viii**) conveyor belt/winding; (**ix**) 3D demonstrator fabrics. Three-dimensional formation process: (**1**) counter-shape; (**1**) → (**2**) deposition of nonwoven layer on the counter-shape in the meltblow stream (**v**); (**3**) removal of 3D-fabric from counter.

2.5. Material Drying and Determination of the Moisture Content

Prior to testing or processing, all materials were subjected to pre-drying in an oven at 80 °C for at least 6 h under fine vacuum condition (<1.8 10^{-1} mbar).

The residual water content for all polymers was determined by means of Karl Fischer titration, which was performed on an "899 Coulometer" and an "885 Compact Oven SC" (both manufactured by Deutsche METROHM GmbH & Co. KG, Filderstadt, Germany) at 140 °C. The resulting water content was required to be <150 ppm, which was achieved for all types of polymers/compounds.

2.6. Polymer Characterization

Shear rheological experiments in time-sweep modes were performed on a Discovery HR-2 rheometer (TA Instruments, New Castle, DE, USA) using a plate–plate geometry at a temperature of 180 °C. The material was placed on the lower plate (25 mm in diameter) and the gap was adjusted to 1.0 mm. Subsequently, the excess material was removed, and the test was carried out under a nitrogen atmosphere (50 mL min^{-1}) with 5% elongation and an angular frequency of 1 rad s^{-1} over a period of 10 min.

Additionally, the thermal properties were determined using different methods. A DSC 2/400 (Mettler-Toledo GmbH, Gießen, Germany) was employed for the DSC analysis, whereby the samples were analyzed in a temperature range from −80 °C to 200 °C at a heating and cooling rate of 10 K min^{-1}. Two heating phases and one cooling phase were carried out, whereby an isothermal phase at 200 °C was omitted in order to reduce the influence of thermal damage on the material. A TGA 3+ (Mettler-Toledo GmbH, Gießen, Germany) was used for the thermogravimetric analysis. The samples were analyzed in a temperature range from 30 °C to 900 °C at 10 K min^{-1} under normal air atmosphere.

2.7. Nonwoven Testing

2.7.1. Fabric Area Base Weight

The area base weight was determined referring to DIN EN ISO29073-1, adjusted by cutting out and weighing 100 cm^2 square sections (10 cm × 10 cm). To consider homogeneity scattering in the cross direction (CD) of the nonwovens, three samples with the dimensions 10 cm × 10 cm were taken in the CD and averaged.

2.7.2. Nonwoven Thickness

The fabric thickness was measured in the samples used for the base weight measurements using a test head (Frank-PTI GmbH, Birkenau, Germany) of 25 cm^2 and a test force of 5 cN cm^{-2}. Eight measurements were conducted diagonally along the sample, determining a median value for thickness (δ).

2.7.3. Air Permeability

In accordance with the base weight sampling, the air permeability was measured on the 10 cm × 10 cm sections in accordance with EN ISO 9237:1995-12 with a sample size of 20 cm^2 and a differential pressure of 200 Pa.

2.7.4. Scanning Electron Microscopy/Fiber Diameter (Distribution)

The fiber diameter distribution was determined by means of scanning electron microscopy (SEM). Therefore, a circular sample was punched out of the nonwoven and placed on the SEM carrier, sputtered in argon plasma (40 s under a vacuum of 0.1 mbar, with a distance of 35 mm, a current of 33 mA, and a voltage of 280 V) with a gold–palladium layer of 10–15 nm. Three SEM micrographs per sample were taken with a magnification of ×500, using a "TM-1000 tabletop electron microscope" of Hitachi High-Tech Corporation (Tokyo, Japan). The accelerating voltage was 15 kV in the "charge-up reduction mode". The magnification was selected to enable the capture of around 40 individual fibers per image. Contrast and brightness were adjusted to obtain an image of straight monochromic fibers in front of a dark monochrome background. To analyze the images with regard to automated fiber diameter distribution, the beta-software "MAVIfiber2d, v.1.1" of Fraunhofer ITWM (Kaiserslautern, Germany) was used [46]. Initially, the images were smoothed by an algorithm and binarized by the software, after which a statistical analysis was conducted on each fiber pixel without segmentation into individual fibers [47,48]. After merging the output of the three images, the mean and median fiber diameter as well as the standard deviation and interquartile range were determined.

2.7.5. Mechanical Properties

Tensile tests of the nonwovens were carried out on an "Instron UPM 4301" of Instron GmbH (Darmstadt, Germany) to determine the tensile strength (σ_m) and the elongation at peak force (ε_m) and at break (ε_B) of the nonwoven fabrics in the MD (machine direction) and CD (cross direction), as well as the Young's modulus (E) as the secant modulus. For each sample, five specimens with a width of 15 mm were cut out in the MD and CD and tested. The 3D demonstrators (pots) were sliced open along one edge to form a flat "sheet" and five specimens were cut out randomly.

The sample thickness was determined individually in accordance with the specifications set forth in DIN EN ISO 9073-2 and the median of five measurements was used for the calculation of the stress from the recorded force. The tests were conducted with 100 mm min^{-1} using a 5 kN measuring head with pneumatic clamps (clamping length of 100 mm). The tenacity was calculated for the sample dimensions, the fabric thickness, and the measured peak force. The median and the standard deviation of all measured properties were employed to facilitate a comparative analysis of the nonwoven characteristics.

2.7.6. X-Ray Diffraction

Wide-angle X-ray diffraction (WAXD) measurements were recorded on a "D/Max Rapid II diffractometer" (Rigaku Corp, Akishima, Japan), equipped with a 0.8 collimator and an image plate detector using monochromatic Cu $K\alpha$ radiation (λ = 0.15406 nm; U_{acc} = 40 V; I_{acc} = 30 mA). A scanning rate of 0.2° min^{-1} and a step size of 0.1° were applied. The measurement time was one hour for all samples under investigation. Background correction was performed using a blank measurement and the resulting scatter images were converted into the corresponding diffractograms using 2θ intensity conversion. The diffraction patterns were analyzed using the PDXL 2 software (version 2.3), and pseudo-Voigt profile fitting was chosen for the evaluation of reflex positions and crystalline fraction determination. The degree of crystallinity χ_c was calculated according to Equation (1),

$$\chi_c = \frac{\sum I_c}{\sum(I_c + I_a)} \tag{1}$$

where I_c and I_a are the integrated intensities of crystalline reflexes and amorphous reflexes, respectively.

The samples were prepared by arranging nonwoven sheets parallel to the carrier.

2.7.7. Thermal Stability

To determine the time and temperature stability, nonwoven fabrics (flat sheets) were tested in relation to their heat shrinkage according to the "drying oven method" (ISO 11501:1995 and GB/T 12027-2004) [49]. Rectangular samples (300 mm × 50 mm in MD) were punched out on the left, in the middle, and on the right side of a nonwoven sample. The specimens were placed free-hanging in an oven at 120 °C. After reaching the specified time of 15 min, the samples were taken out and the size of the samples was measured. The ratio of the dimensional change value to the size before shrinkage was calculated as the percentual shrinkage rate of the sample.

Furthermore, three-dimensional nonwoven demonstrators were placed into an oven with a defined temperature, starting with 100 °C. The parts were kept in the oven for 15 min and the dimensions were measured before and after "tempering". When no changes were observed, the procedure was repeated at a 10 K higher temperature.

2.7.8. Qualitative Evaluation of the Flexibility and Shape Retention of 3D Nonwoven Structures

To evaluate the flexibility of the three-dimensional nonwovens qualitatively, the pots were loaded with a weight of 400 g. After a loading time of one minute, the weight was removed and the shape retention was quantified based on the ratio of the initial to the final pot height (see illustration in Figure 6).

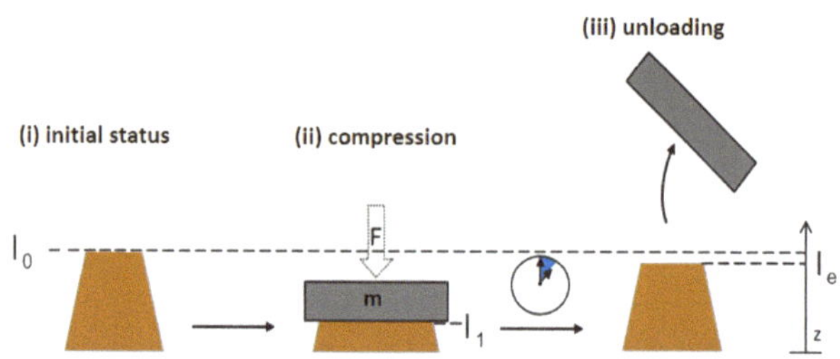

Figure 6. Schematic illustration of the qualitative determination of the flexibility and shape retention of the three-dimensional meltblown structures: (**i**) initial status (height l_0) of the sample; (**ii**) sample compressed to height l_1 by the weight of mass m; (**iii**) sample with height l_e after relaxation after unloading.

Thereby, the original form was labeled as l_0. During loading, the shape was compressed (l_1) due to the weight force (F). When the weight was removed from the shape (l_e), the relieved shape was compared with the previous initial shape. The differences in height were measured in the z-coordinate.

3. Results and Discussion

3.1. Characterization of the Compounds

Figure 7 illustrates the DSC measurements through the use of additives. Multiple peaks, as they appear in the thermograms of P3HB and PHBH, have already been identified in other studies. These may be attributed to a combination of melting, recrystallization during heating, different crystal modifications (polymorphism), different molecular weights, different morphologies, and the physical aging or relaxation of the rigid amorphous part. [50]. A broad and blurred melting range with two peaks was identified for P3HB. According to earlier findings [40], this phenomenon can be attributed to a broad molecular weight distribution, resulting in an inhomogeneous melting behavior.

It can be seen that the melting temperature and crystallization temperature decrease by using both additives. While endothermic peaks at 165.7 °C and 158.2 °C were identified for "pure" P3HB (**a**), these were reduced by around 1–2 K when using 10 wt.% ELO (**b**) and by around 3–5 K when using 10 wt.% ATBC (**c**). This phenomenon can be explained by the improved mobility of the polymer chains due to the additives, whereby the secondary valence forces are reduced allowing for melting at lower temperatures. In comparison, the exothermic peak of 107.2 °C of "pure" P3HB was reduced by 3 K (ATBC) and by 1 K (ELO). The incorporation of additives impedes the formation of tight molecular aggregates, thereby postponing the onset of crystallization. Acetyltributylcitrate exhibits a markedly superior plasticizing effect compared to the epoxidized oil extending the processing window to lower temperatures. For this reason, only acetyltributylcitrate (ATBC) was considered for PHBH.

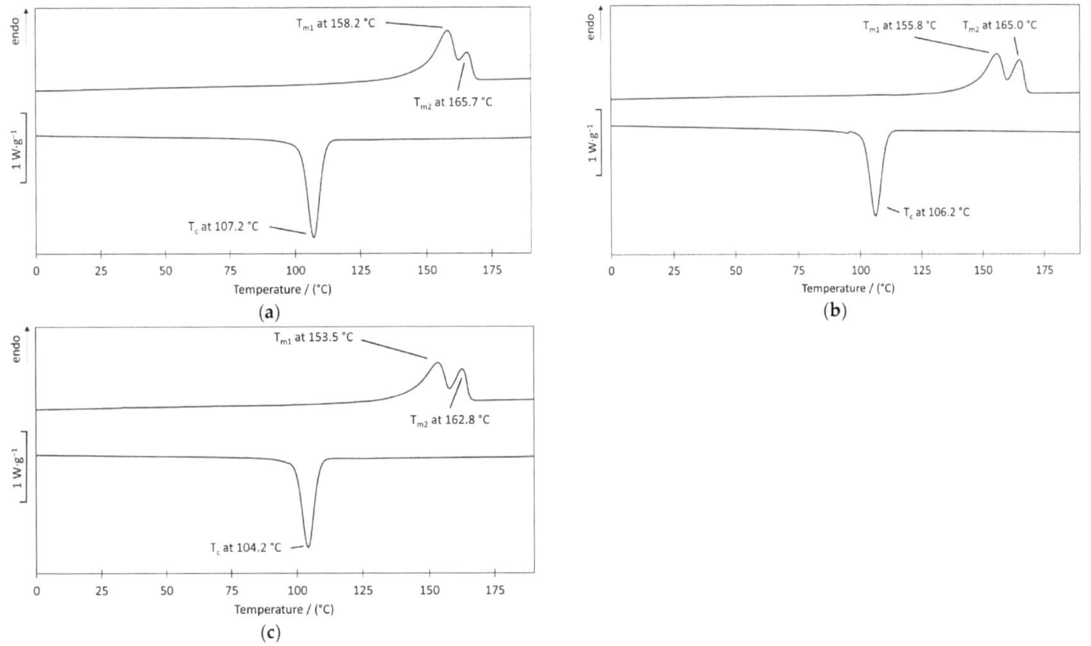

Figure 7. Illustration of the melting temperature Tm (second heating cycle) and crystallization temperature Tc (**a**) of commercially pure P3HB, (**b**) of P3HB with 10 wt.% ELO, and (**c**) of P3HB with 10 wt.% ATBC.

For PHBH, enthalpy relaxation after the T_g and several exothermic effects were observed (Figure 8). This leads to the assumption of slow and incomplete crystallization after the first heating process, with post-crystallization processes and crystal perfection taking place in the subsequent melting process.

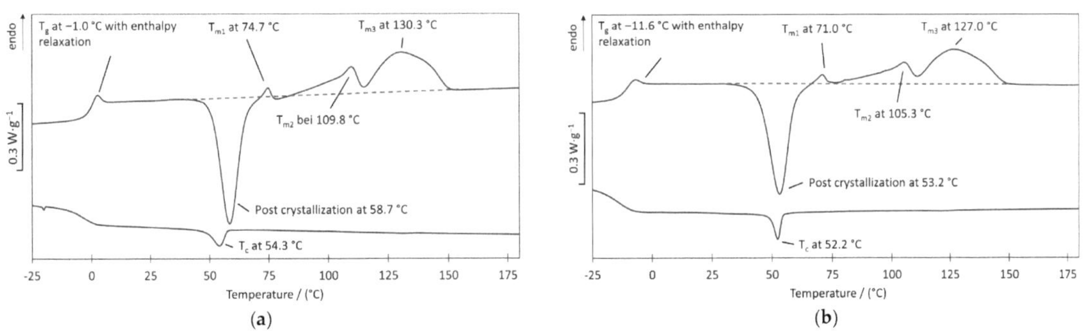

Figure 8. Illustration of the melting temperature T_m (second heating cycle) and crystallization temperatures T_c (**a**) of commercially pure PHBH and (**b**) PHBH with 10 wt.% ATBC (the dashed lines indicate the base line extension of the software for the evaluation of the melting peaks).

Despite the complex behavior, significantly earlier melting was observed compared to P3HB. As PHBH is a copolymer comprising 3HB and 3HHx units, this can be attributed to the steric hindrance caused by the propyl group. Depending on the proportion in the copolymer, this group hinders the dense packing of the polymer chains, which reduces the secondary valence forces and promotes early melting. Since, according to Eraslan et al. [51]

and Volova et al. [52], a PHH homopolymer is an amorphous material, the melting range can be attributed to the crystallites of the 3HB component. With regard to the chain degradation of PHAs due to increased temperatures, the reduced melting temperature allows for processing at lower temperatures. This can counteract chain degradation at elevated temperatures.

Table 1 presents the results of the TGA measurements (onset temperature T_{on}, the temperature at 10% mass loss $T_{10\%}$, and the endset temperature T_{end}) of the degradation stage of the various materials. The decomposition of P3HB takes place over a broader temperature range, which is why a higher temperature with a mass loss of 10% and a higher endset temperature could be identified despite a lower onset temperature. In addition to a broad molecular weight, which could already be observed in [40], this result may also be attributed to the presence of additives that only decompose at elevated temperatures. The use of the citrate ATBC generally resulted in increased degradation at lower temperatures. The degradation stage is characterized by a "smearing" effect, which can be attributed to the mixture of the polymers with the citrate, resulting in a multimodal and broad molecular weight distribution.

Table 1. Results of thermogravimetric analysis.

Material	T_{on}/°C	$T_{10\%}$/°C	T_{end}/°C
P3HB	277.9	284.0	306.4
P3HB + 10 wt.% ELO	275.3	252.0	311.7
P3HB + 10 wt.% ATBC	276.2	244.7	306.8
PHBH	277.1	275.0	295.9
PHBH + 10 wt.% ATBC	276.6	269.0	297.3

The use of an epoxidized oil did not result in less degradation overall compared to the untreated material. Given the assumption that the epoxy group is capable of reacting with the respective chain ends, the usage of an epoxidized oil should delay the progressive depolymerization. Only T_{end} showed a higher temperature, which could be explained by the higher decomposition temperature of the epoxidized oil itself. It was previously found by Park et al. [53,54] that the carboxylic acid groups from the decomposition of PHAs can react with epoxide groups to form networks between the chains and retard degradation.

According to state of the literature, reviewed by Eraslan et al. in 2022 [51], PHBH is said to have increased thermal stability due to its 3HHx content. For PHBH with a 3HHx content of 10 mol.%, the TGA consistently showed the lowest temperatures for the course of decomposition. This is particularly evident in the onset and endset temperatures. Thus, contrary to assumptions, the lowest thermal stability was identified for PHBH.

Figure 9 shows the complex viscosities of the different materials over time. The observations from the DSC measurements are also represented in the rheological analysis. As can be seen, the use of external plasticizers in commercially available PHAs enables melting at lower temperatures and thus reduces viscosity. Therefore, the impact of plasticizers is evident in the vertical shift in the curve towards lower viscosities by allowing the chains to slide away from each other more easily. This development is significant for PHAs, as these react sensitively to elevated temperatures, especially when accompanied by mechanical stress. A reduction in the initial viscosity of approx. 59–64% was achieved with the various materials using ATBC.

In contrast, the epoxidized linseed oil only has a minor influence here too. In addition to the molecular structure, the molar mass can also be cited as the reason for the different behavior of the additives. As acetyltributylcitrate has a significantly lower viscosity than the epoxidized oil, the viscosity is reduced to a greater extent by the addition of the same proportion, irrespective of the softening effect. A lower level of degradation due to reactions with the epoxy groups could not be observed.

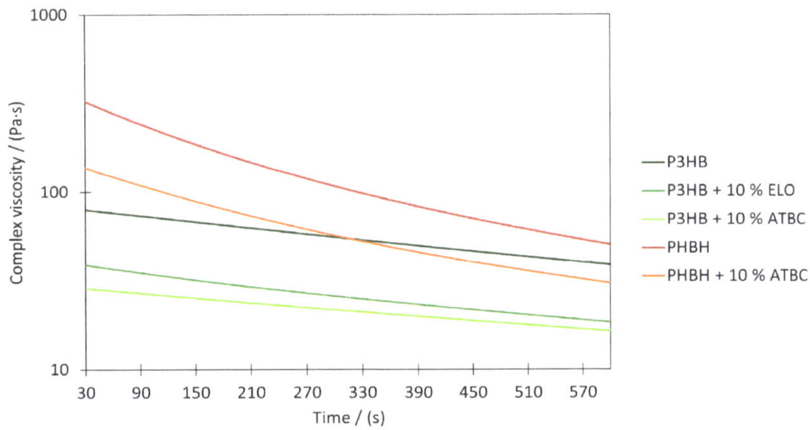

Figure 9. Rheological characterization using time sweep at a temperature of 180 °C, 5% strain, and a frequency of 1 rad s^{-1}.

In addition, the degradation process of the polymers in the rheological analysis also leads to a further drop in viscosity over time. This is particularly pronounced in the case of PHBH, while the viscosity of P3HB decreases to a lesser extent. It is well known that the rheological behavior not only depends on test parameters such as temperature and the constitution of the polymer. With an increasingly broad molecular weight distribution, the low-molecular-weight components also act as a lubricant and promote the flow of the plastic melt [55]. Since, according to [40], P3HB already has a broad molecular weight distribution, this could explain the already low initial viscosity. However, since PHBH has a melting range at lower temperatures than P3HB, a lower viscosity was also anticipated at the set testing conditions. However, factors such as the molar mass itself also play an important role, which is why a higher molar mass can be assumed for PHBH.

3.2. *Investigation of the Meltblown Processability*

Based on the observed differences in the material properties of the three PHA types (P3HB, PHBV, and PHBH) and the PHA compounds (compare Section 3.1 and [40]), differences in their processing behavior and/or the requirement of adaption in the processing parameters (required temperature, limitation in throughput, and accessible fiber diameters) can be assumed. This was already highlighted by the melt flow (rheological) curves and the relevant characteristics of the materials at 180 °C compared in Table 2.

Table 2. Rheological characteristics from the time sweeps (ω = 10 rad s^{-1}, ε = 10%) and MFI data of the different PHAs.

Material	η_{to} (180 °C) /(Pa s)	G'_{t0} (180 °C) /Pa	G''_{t0} (180 °C) /Pa	MFI /(g 10 min^{-1}) [1]
P3HB	137	250	1354	10 [43]
PHBV	434	684	4287	10–25 [56]
PHBH	634	1150	6242	3 [44]
P3HB+ELO	37	30	373	-
P3HB+ATBC	38	33	380	-
PHBH+ATBC	245	252	2405	-

[1] T = 180 °C, 2.16 kg.

While the melt flow index (MFI) is a weak parameter [40,57], the viscosity in the Newtonian (shear-independent) regime serves as an established tool for the prediction of the processability. However, as PHAs are subject to a strong thermal degradation at temperatures close to their melting temperature, the reliability of frequency and temperature sweeps is negated. For this purpose, the start values (η_{to}, G'_{to}, G''_{to}) as unaffected measured values were taken from tests in the time-sweep mode under low-shears condition. These were supplemented by the viscosity value corresponding to a characteristic residence time in the extrusion process (~300 s) η_{300s}. The final resulting processing temperatures of each material and the respective rheological properties are presented and compared in Table 3.

Table 3. Estimation of the process temperature and characteristic shear rheological properties of the different PHAs at the process temperature.

Material	T_{proc} /°C	η_{to} (T_{proc}) /(Pa·s)	η_{300s} (T_{proc}) /(Pa·s)	G'_{t0} (T_{proc}) /Pa	G''_{t0} (T_{proc}) /Pa
P3HB	180	137	67	250	1354
PHBV	200	227	49	199	2262
PHBH	190	325	52	151	3621
P3HB+ELO	176	48	36	47	476
P3HB+ATBC	176	45	37	41	445
PHBH+ATBC	190	62	18	17	623

While the virgin PHA types show higher start viscosities above the processing window for meltblow processing (~<150 Pa s), which one can achieve within a typical extrusion time, the processed compounds "P3HB+ELO", "P3HB+ATBC", and "PHBH+ATBC" already lie below the upper processing limit and thus are in the processable window. Further, it can be noted for PHB that the plasticizing additives cause a lower process temperature due to the reduction in viscosity. However, PHBH, supplemented with ATBC, also shows a significant reduction in viscosity, but no reduction in the processing temperature was possible due to a high process pressure level, although the viscosity dropped significantly over the 300 s residence time.

All processable settings (at the respective lowest possible process temperatures enabling a homogeneous fiber deposition) are given in Table 4.

Table 4. Process settings of the meltblow trials.

Trial-Nr.	T_{melt} /°C	T_{air} /°C	Throughput /(g ho^{-1} min^{-1})	Die-Pressure /bar	Air Volume Flow /(Nm^3 h^{-1})	Limitation(s)
P3HB-Ref–01	180	175	0.077	31.0	220	–
P3HB-Ref–02	180	170	0.077	21.7	220	
P3HB-Ref–03	180	175	0.077	21.8	325	
PHBV-Ref	200	195	0.051	26.4	325	No stable process at lower throughput (degradation dominates) Limitation of max. throughput (die-pressure)
P3HB+ELO–01	176	170	0.039	7.1	220	Edges stick strongly to conveyor belt
P3HB+ELO–02	176	170	0.039	7.1	325	
P3HB+ELO–03	176	170	0.077	12.3	220	
P3HB+ELO–04	176	170	0.077	12.3	325	
P3HB+ATBC–01	176	170	0.039	23.5	220	–
P3HB+ATBC–02	176	170	0.077	25.0	220	The higher the throughput, the more inhomogeneous the deposition
P3HB+ATBC–03	176	170	0.077	25.0	325	
PHBH	190	185	0.032	35.0	325	Limitation of max. throughput (die pressure)
PHBH+ATBC–01	190	185	0.032	47.2	220	High pressure level → throughput limited
PHBH+ATBC–02	190	185	0.032	52.0	325	
PHBH+ATBC–03	190	178	0.039	46.4	240	

As for the virgin polymers, the P3HB compounds allowed a higher flexibility of the throughput, showing the possibility to achieve higher per-hole throughput levels >0.1 g min^{-1}. Unmodified PHBH was limited to low or moderate throughputs at a quite high process temperature of 190 °C, although its melting temperature was intrinsically lower. For this, a too-high mean polymer chain length can be assumed in correlation with the significant higher melt viscosity at 180 °C compared to P3HB and even PHBV, as characterized in a previous work [40]. However, as the processing temperature of PHBH was almost as high as for PHBV, the use of PHBH and the compound PHBH + ATBC resulted in a more stable meltblow process and a more homogeneous fiber deposition due to the higher difference between its processing temperature and its melting temperature.

3.3. Nonwoven Characteristics

Figure 10 shows the SEM micrographs at 100× magnification for the reference nonwovens from [40].

Figure 10. SEM images (100×) of the reference fabrics of P3HB and PHBV: (**a**) P3HB-Ref–01; (**b**) P3HB-Ref–02; (**c**) P3HB-Ref–03; (**d**) PHBV-Ref.

A too-high difference between the air temperature and the temperature of the melt (10 K vs. 5 K) for P3HB–01 (Figure 10b) led to a too-coarse fiber deposition, while a too-high air amount led to a highly nonuniform fiber diameter distribution of fine and coarse fibers (PHB–03; Figure 10c) due to higher air flow turbulences in the process. This was also the case for PHBV (Figure 10d), but at a higher level due to the harsher processing temperatures.

Figure 11 shows the respective SEM micrographs for all produced nonwoven samples from virgin PHBH and the P3HB and PHBH compounds.

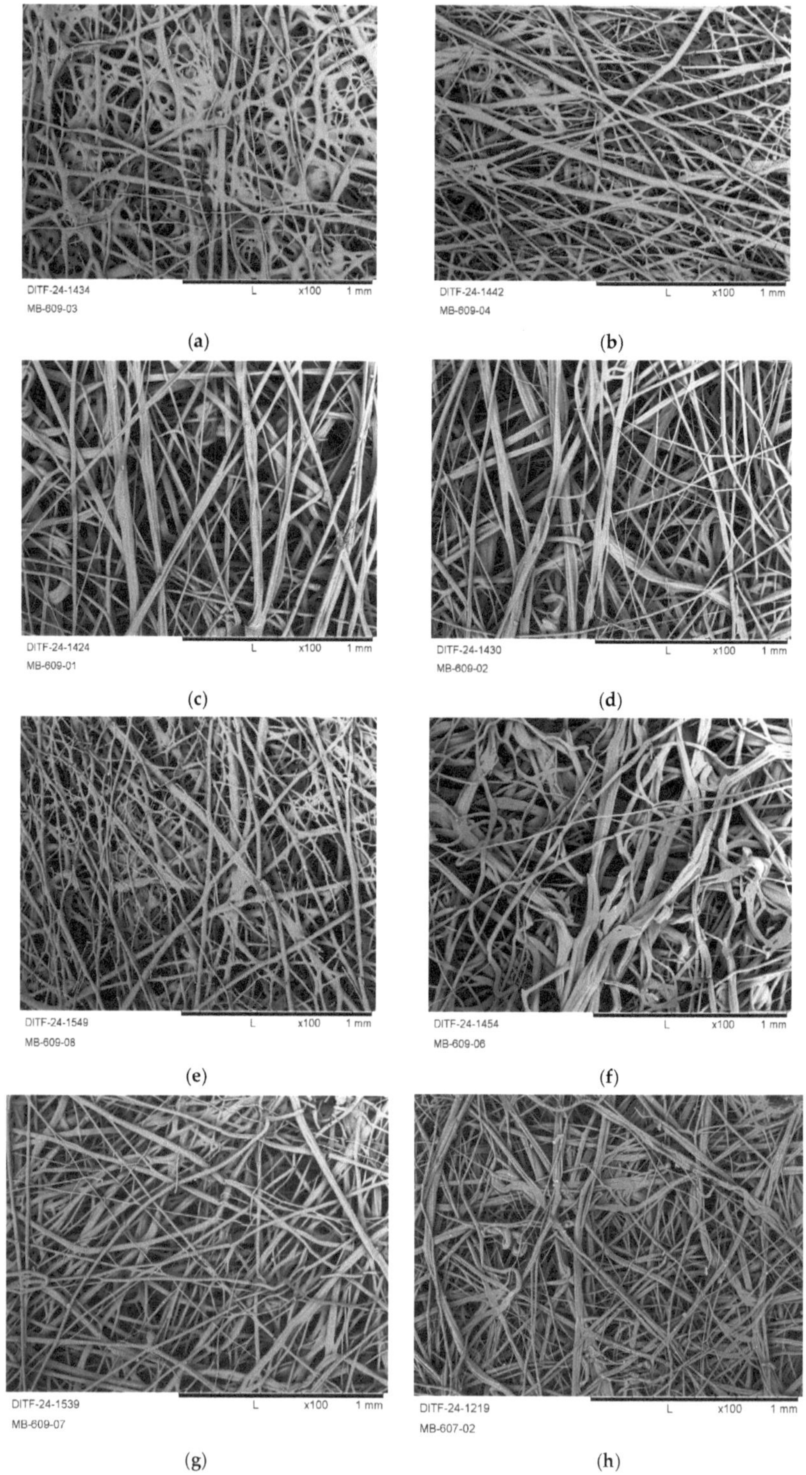

Figure 11. *Cont.*

Figure 11. SEM pictures (100×) of the produced fabrics from P3HB+ELO, P3HB+ATBBC, PHBH and PHBH+ATBC: (**a**) P3HB+ELO–01; (**b**) P3HB+ELO–02; (**c**) P3HB+ELO–03; (**d**) P3HB+ELO–04; (**e**) P3HB+ATBC–01; (**f**) P3HB+ATBC–02; (**g**) P3HB+ATBC–03; (**h**) PHBH; (**i**) PHBH+ATBC–01; (**j**) PHBH+ATBC–02; (**k**) PHBH+ATBC–03.

PHB, compounded with epoxidized linseed oil (ELO), resulted in a deliquescing of the deposited fibers and "thermal branching" (merging of fibers due to flow reasoned by insufficient cooling [27,41]) at a lower throughput of 0.039 g ho^{-1} min^{-1} (P3HB+ELO–01 and P3HB+ELO–02; Figure 11a,b). At a higher throughput of 0.077 g ho^{-1} min^{-1}, the fiber deposition was more homogeneous and the average fiber diameters were similar to those of P3HB from previous works [40,58]. Therefore, a higher difference between process air temperature and melt temperature (12 K vs. 5 K, compare Table 4) successfully achieved a finer and denser fiber deposition (P3HB+ATBC–03, Figure 11k).

The test results of the characterization of the nonwovens are given in Table 5.

Comparing the P3HB compounds to the virgin P3HB showed a similar mechanical performance. A slightly higher tenacity could be achieved with the setting P3HB+ELO–02. However, the elongation of fabrics could not be increased by compounding P3HB with the bio-based plasticizers. Thus the main intended purpose of their use was not successful at this point. Nevertheless, for the compounds of P3HB with ELO, an increase in the Young's modulus (~10%) could be observed when less polymer throughput was applied. This is due to the drag force of the process air acting with higher efficiency on a lower amount of melt (due to higher polymer chain orientation) until solidification. The air permeability was lower for all P3HB compounds, which also correlates with a higher resulting base weight, which derivates by up to two times (>200 g m^{-2}) from the set target value due to a narrower fiber deposition on the conveyor belt (<500 mm vs. 550 mm). A potential reason for this is the formation of a denser air curtain around the nominal lower viscous melt.

Table 5. Characteristics of the produced and reference nonwoven fabrics; bold: samples showing the best/superior properties.

Trial-Nr.	Base Weight /(g m^{-2})	CV [1] /%	Thickness /µm	Fiber Diameter Median /µm	Fiber Diameter Mean /µm	Air Permeability /(L m^{-2} s^{-1})	Tenacity MD/CD /(N mm^{-2})	Elongation [2] MD/CD /%	Modulus [3] MD/CD /(N mm^{-2})
P3HB–Ref–01	93	11	235 ± 19	4.6	7.0	680 ± 100	1.6 ± 0.1/0.9 ± 0.1	4 ± 1/5 ± 1	92 ± 4/47 ± 2
P3HB-Ref–02	94	14	346 ± 43	13.7	16.2	4640 ± 880	0.7 ± 0.1/0.5 ± 0.1	3 ± 1/3 ± 1	39 ± 7/28 ± 4
P3HB-Ref–03	99	16	299 ± 40	7.3	12.7	1550 ± 430	1.4 ± 0/0.7 ± 0.2	3 ± 0/3 ± 1	76 ± 5/38 ± 7
PHBV-Ref	120	8	566 ± 79	3.5	4.8	2030 ± 340	0.8 ± 0.1/0.3 ± 0	1 ± 0/1 ± 1	61 ± 5/31 ± 5
P3HB+ELO–01	220	5	456 ± 55	10.6	16.7	285 ± 14	1.5 ± 0.2/1.1 ± 0.1	2 ± 1/2 ± 1	**116 ± 4/78 ± 5**
P3HB-ELO–02	228	7	434 ± 34	10.5	12.8	230 ± 61	1.8 ± 0.1/1.0 ± 0.1	2 ± 0/2 ± 1	107 ± 5/69 ± 2
P3HB+ELO–03	223	12	562 ± 118	9.3	11.3	863 ± 156	1.1 ± 0.1/0.6 ± 0.1	3 ± 0/2 ± 1	53 ± 6/36 ± 4
P3HB+ELO–04	204	18	504 ± 132	7.3	9.8	537 ± 125	1.3 ± 0.1/1.0 ± 0.1	3 ± 0/2 ± 1	71 ± 6/54 ± 9
P3HB+ATBC–01	188	12	373 ± 52	8.8	12.0	390 ± 19	1.4 ± 0.1/0.8 ± 0.1	2 ± 0/2 ± 0	92 ± 7/56 ± 6
P3HB+ATBC–02	213	12	516 ± 102	14.2	15.9	765 ± 65	1.0 ± 0.1/0.6 ± 0.2	2 ± 0/2 ± 1	69 ± 12/36 ± 7
P3HB+ATBC–03	204	14	454 ± 106	7.4	10.0	543 ± 45	1.3 ± 0.1/0.8 ± 0.2	2 ± 1/2 ± 1	93 ± 17/47 ± 4
PHBH	128	11	390 ± 48	4.5	6.2	700 ± 69	**2.5 ± 0.1/1.2 ± 0.1**	5 ± 0/6 ± 0	100 ± 2/4 ± 1
PHBH+ATBC–01	140	12	335 ± 26	6.5	7.9	**185 ± 32**	1.5 ± 0.1/1.2 ± 0	10 ± 2/**22 ± 3**	60 ± 2/41 ± 2
PHBH+ATBC–02	131	11	343 ± 24	8.8	8.5	411 ± 101	1.6 ± 0/1.0 ± 0.1	**15 ± 2**/14 ± 1	58 ± 5/35 ± 1
PHBH+ATBC–03	120	22	376 ± 69	**3.6**	6.4	270 ± 40	2.2 ± 0.1/1.1 ± 0.1	6 ± 0/10 ± 2	85 ± 5/45 ± 2

[1] Coefficient of variation (standard deviation divided by mean average). [2] Elongation at max. force. [3] Young's modulus.

The fiber diameter range, quantified by the mean and the median of the fiber diameter distribution, is similar to that of the samples presented in previous works [40–42]. The average fiber diameters of the PHBH fabrics, as well as the fabrics of the modified polymers, are slightly higher. However, the sensitivity of the fiber diameter to process parameter changes appears to be reduced. The fiber diameter variation displayed by the samples (ratio of the median to the mean average) is also in a standard range for meltblown nonwovens. In this context, the meltblowing process generally generates broader fiber distributions than other processes, e.g., the melt spinning of yarns. This is due to the complex interactions between the polymer melt and the turbulent air flow [59,60].

Replacing P3HB with PHBH results in a higher tenacity at comparable fiber diameter averages and air permeability values. However, the flexibility of fabrics improved drastically (qualitatively), and the elongation at maximum force could be successfully raised to 5% in the machine direction (MD). The fabrics of the PHBH compounds showed a lower tenacity and became denser, while also showing reduced air permeability, but the elongation further increased to >10%, thus fulfilling this study's hypothesis.

Summarizing the characterization of the nonwovens, the key nonwoven characteristics (tenacity, elongation, modulus, and base-weight-standardized air permeability), selected for the best values for each material, are compared in Figure 12 for the different materials used.

As pointed out before, nonwovens made from virgin PHBH showed the highest tenacity (in both MD and CD), while PHBV was clearly limited in its mechanical strength (Figure 12a). PHBV also showed the lowest elongation, while this parameter was significantly increased for PHBH and further so for the compound of PHBH with ATBC (Figure 12b shows the elongation at break in contradiction to Table 5 at max. force). The Young's modulus (Figure 12c) was again low for PHBV, but superior for P3HB when combined with ELO. The air permeability, standardized to the base weight to ensure comparability (Figure 12d), was significantly higher for both reference materials, virgin P3HB and PHBV, compared to the materials presented in this study, which may be significant for the choice of application. However, virgin PHBH still has a higher ratio of air permeability than the compounds used.

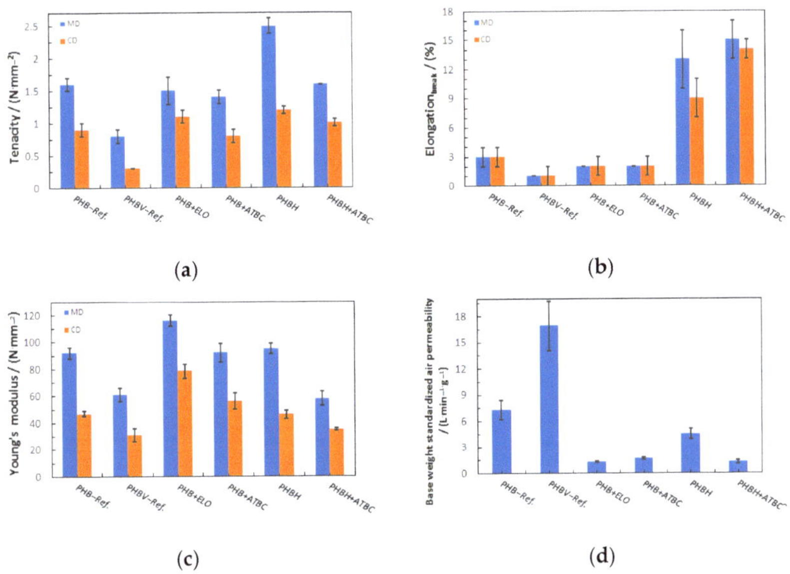

Figure 12. Comparison of the mechanical properties achieved with the different PHAs/compounds in MD (blue) and CD (orange): (**a**) tenacity, (**b**) elongation at break, (**c**) modulus, and (**d**) base-weight-standardized air permeability.

Figure 13 shows the results of the wide-angle X-ray diffraction, performed on a selected sample of each material, in order to obtain information about the crystallization status (degree of crystallinity and crystal structure) of the fabrics.

As already obtained in our previous work [41] (see the samples "PHB–Ref" and "PHBV–Ref" in Figure 8b), all X-ray diffraction images showed a semi-crystalline pattern. Several crystalline reflections occurred at $2\theta \approx 13.5°$ ((020)), 16.9° ((110)), 19.7° ((101)), 21.5° ((111)), 25.3° ((130)), 27.0° ((040)) and 30.0° ((002)) in accordance with the literature [3,40,61].

Differences in the process settings (throughput rates, air amounts, etc.) did not play a role in the crystallization, which is characteristically delayed for PHAs. However, the physics of the meltblow process with high forces from the air stream resulted in high orientation, leading to highly crystallized structures. The crystallinity of all samples was determined by fitting (pseudo-Voigt profiles) the diffractograms in the range of $2\theta = 10$–$40°$. The crystallinity for the P3HB samples was between 71% and 77%, that of OHBV was between 75% and 80%, and that of the PHBH samples was between 60% and 77%. The plasticizers showed limited influence on the peak positions and crystallinity.

In comparison to [3,40], the identified peaks indicate the predominance of orthorhombic α-crystals, with hexagonal β-form crystals (shoulder at $2\theta \approx 20°$, (101)-lattice) also present. This shoulder is indeed more pronounced for P3HB/PHBV and very diffuse for PHBH. Referring to Hufenus et al. [62], β-crystals develop stress introduced in the amorphous P3HB regions between lamellar α-crystals.

The presence of the shoulder at $2\theta \approx 20°$ ((101)-lattice) is typical for introduced chain orientation, such as uniaxial stretching [3,63], stress-induced crystallization [10,64], and the distribution of different crystal sizes. Another reason could be random chain scission during degradation [17,64,65], which is very common in the processing of PHAs, forming crystals with different lamellar thicknesses and thus different melting kinetics. However, the results indicate that the copolymer prefers to form α-crystals rather than β-crystals.

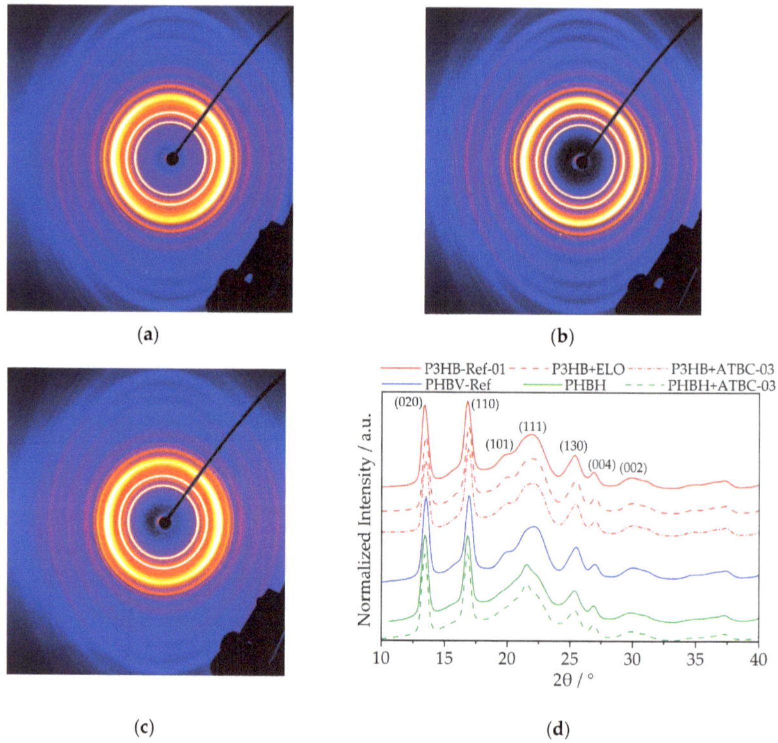

Figure 13. Exemplary X-ray scattering image of (**a**) a P3HB meltblown sample, (**b**) PHBV-ref., (**c**) and a PHBH meltblown sample and (**d**) corresponding X-ray diffraction patterns of different process settings: blue: P3HB; red: PHBV; green: PHBH.

Furthermore, the materials were tested regarding their heat shrinkage. Due to the melting temperature of PHAs, the standard testing temperature (200 °C) was modified to 120 °C. The results are given in Table 6.

Table 6. Results of the heat shrinkage test (120 °C, 15 min) of the meltblown materials; the results comprise all variations of the respective PHA type/compounds.

Material	P3HB	P3HB+ECO	P3HB-ATBC	PHBV	PHBH	PHBH+ATBC
Shrinkage in MD/%	1.7 ± 0.6	1.9 ± 0.6	1.3 ± 0.5	1.6 ± 0.6	1.7 ± 0.8	1.5 ± 0.5
Shrinkage in CD/%			1 ± 1 [1]			

[1] No differences between all samples.

The heat shrinkage of all materials lies in the same range. Furthermore, the dimensional change is at such a low level that heat shrinkage can (almost) be excluded for all PHA types and compounds tested. For PHBH, this behavior of the nonwoven material contradicts the thermal behavior of the granules (see Section 3.1, Figure 9), which already showed two melting peaks of lower enthalpy at 75 °C (T_{m1}) and 110 °C (T_{m2}), before the final melting peak at 130 °C occurred. These peaks could be reproduced by measuring the nonwoven material, but with the absence of a previous recrystallization (see Appendix B). This indicates that the crystal structure corresponding to (T_{m3}) is dominant after meltblow processing due to the high orientation. Furthermore, the enthalpy content of the peak corresponding to T_{m3} (~18 J g^{-1}) is significantly higher towards T_{m2} (2.0 J g^{-1}) and negligible for T_{m1} (0.3 J g^{-1}).

3.4. Processing of Three-Dimensional Nonwoven Structures

The principle of three-dimensional meltblown structures was previously demonstrated by Farer et al. in 2003 [66]. However, they used a small lab-scale die with a width of 78 mm and a six-axis robotic arm and an additional external axis to spray fibers on a 3D-carrier structure/layer. This requires facilitating an independent machine set-up and is not applicable to standard (industrial) meltblow lines.

The high melt adhesion and the delayed crystallization, two parameters often cited as major drawbacks of PHAs, helped to successfully deposit the nonwoven structure on a three-dimensional counter-piece and to remove the resulting 3D-nonwoven structure from it without damage and without loss of the applied shape. Polymers showing a strong initial crystallization, such as polypropylene, do not show this described behavior, as the finest fibers generated in the meltblow process have already cooled too much before they hit the counter piece and lose the applied shape after being withdrawn.

In Table 7, the characteristics of plant seed pots made from P3HB, PHBV, and PHBH are compared to those of two commercial reference pots (of the same dimensions) made from natural fibers (cellulose and coconut fibers).

Table 7. Material characteristics of the three-dimensional PHA nonwoven demonstrators compared to the commercial reference plant pots.

Property	P3HB	PHBH	PHBV	Cellulose [1]	Coconut [1]
Area weight (g m^{-2})	151 ± 12	151 ± 22	148 ± 3	296 ± 18	476 ± 43
Wall thickness (µm)	646 ± 84	409 ± 48	662 ± 49	1597 ± 120	2689 ± 294
Tensile tenacity (N mm^{-2})	1.6 ± 0.1	2.2 ± 0.1	0.8 ± 0.2	1.6 ± 0.4	0.7 ± 0.2
Elongation (%)	4 ± 1	13 ± 3	2 ± 1	3 ± 1	2 ± 0
Youngs modulus (N mm^{-2})	75 ± 19	95 ± 4	49 ± 7	73 ± 13	49 ± 7
Air permeability (L m^{-2} s^{-1})	2010 ± 74	676 ± 75	1060 ± 56	27 ± 3	114 ± 2
Water retention (%)	100 [2]	100 [2]	100 [2]	0 [3]	0 [3]

[1] Commercial references. [2] For at least 50 h after filling. [3] Directly after filling.

Comparing the mechanical properties, the tensile strength of the cellulosic pots could be achieved with P3HB and PHBH and the lower strength of the coconut pots could also be achieved with PHBV. The elongation is low for all materials, including the references, and superior for PHBH. Moreover, the Young's modulus was the highest for PHBH and comparable for P3HB and cellulose fibers, as well as for PHBV and coconut fibers. It is noteworthy that the base weight of the PHA pots is only half that of the cellulosic and around 33% that of the coconut fiber pots, combined with their significantly lower wall thicknesses, while offering at least equal mechanical performance. PHBH has the lowest thickness due to its denser structure. While the air permeability is, as before, high for P3HB, the reference pots are almost impermeable for air. Filling the plant pots with water reveals another interesting feature of the PHAs. Due to their intrinsic hydrophobicity, the water remains in the pots for at least 50 h before the pots become water-permeable.

As with the flat sheet nonwovens, the three-dimensional nonwoven structures were tested with regard to their thermal stability. As all the materials used showed equal results in the heat shrinkage test previously, only P3HB was tested in this test. Plant pot demonstrators were exposed to different temperatures starting at 100 °C for 15 min in an oven. After removing these samples from the oven, they were examined for dimensional changes. Photographs of the 3D-nonwovens after exposure to different temperatures are presented in Figure 14.

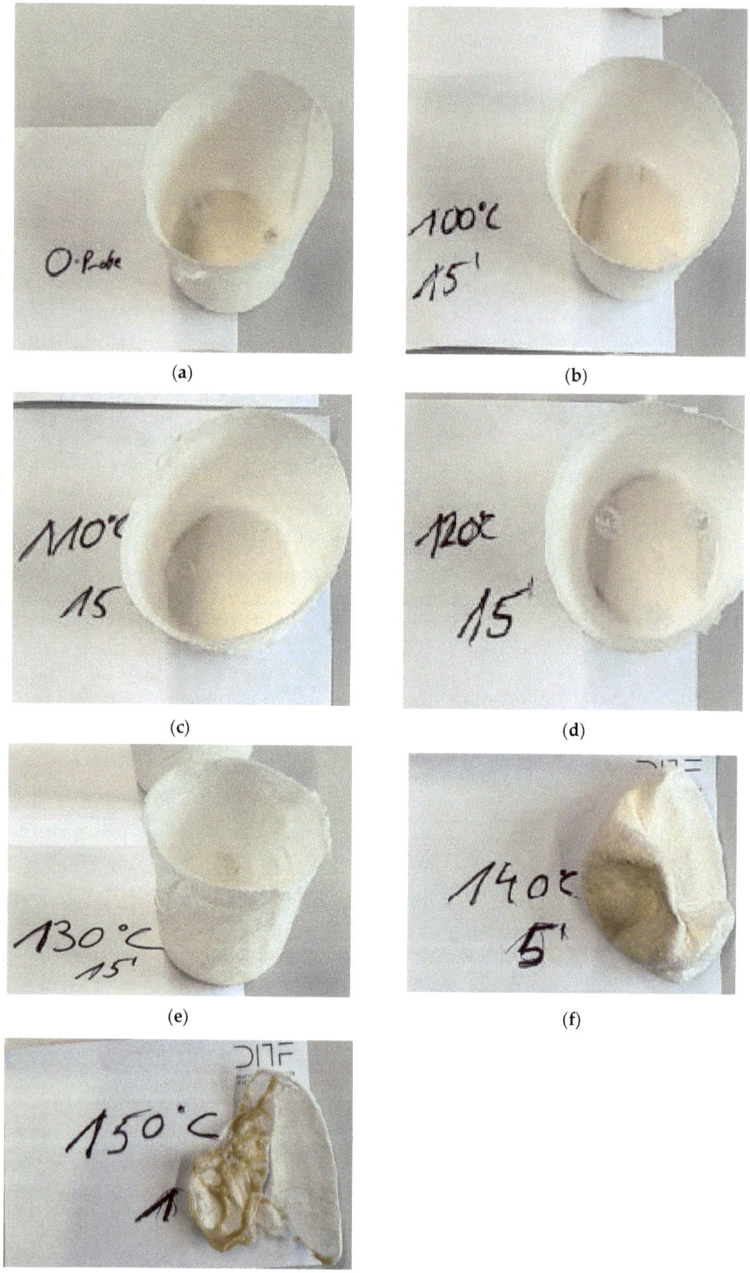

(a) (b) (c) (d) (e) (f) (g)

Figure 14. Results of the tests of 3D P3HB nonwoven samples regarding their thermal stability at different temperatures. (**a**) Reference—untreated; (**b**) 100 °C, 15 min; (**c**) 110 °C, 15 min; (**d**) 120 °C, 15 min; (**e**) 130 °C, 15 min; (**f**) 140 °C, 5 min; (**g**) 150 °C, 1 min.

Starting at 100 °C (Figure 14a) up to 130 °C (Figure 14e), the plant pots showed no change in their shape or their dimensions. The plant pot exposed to 140 °C (Figure 14f) showed loss of its shape after five minutes of exposure and the pot that had been exposed to 150 °C (Figure 14g) showed significant signs of decomposition, while the samples placed in the oven at 160 °C decarbonized entirely without residues (residues sucked away by the suction of the oven or spreading of the melt on the firebrick substrate) after 1 min of exposure to these conditions. As the meltblown fibers were identified to be highly crystallized, this behavior is not based on usual shrinkage mechanisms, but on the onset temperature of the melting region, which lies between 125 °C and 140 °C (see Section 3.1, Figures 7 and 8). The higher surface to volume ratio of fine meltblown fibers can explain their higher affinity for faster melting at lower temperatures compared to the granular material.

In addition to the thermal stability, the structural stability and flexibility of the three-dimensional structures were also tested. For this purpose, plant pots made from P3HB, PHBV, and PHBH were compressed with a load of 400 g and their ability to unfold after compression was quantified as dimensional differences from the original state. The qualitative results are given in Table 8.

Table 8. Results of the qualitative shape stability test (squeezing under load and unfolding); "0": sample remains deformed after compression; "1": full shape resiliency.

Material	PHB [1]	PHBV	PHBH [2]
Result	0.5 [3]	0	1

[1] Including PHB with additives (ELO/ATBC). [2] Including PHBH+ATBC. [3] Pods tear at the opening when loaded, but straighten up after the first time of reloading.

In terms of mechanical characteristics, the PHBH pots show a superior behavior to P3HB (and PHBV). The structure straightens itself out completely after unloading, whereas the P3HB and PHBV pots remain compressed or break easily under compression load. Thus, PHBH reveals a superior flexibility (mechanical resistance) as well as significantly higher ductility (elongation). This flexibility is not only based on elasticity, as the modulus of the PHBH fibers is in the same range as other (brittle) materials, but may be based on the fiber-to-fiber interactions and the fiber network formed in the meltblown deposition. Chemically, this may also be due to the longer alkyl-side chains of the hexanoate building blocks in the polymer chain. These are the reasons for PHBH's lower crystallization [67] and correspondingly the higher proportion of amorphous regions in PHBH, which introduce more flexibility and ductility. Additionally, the higher chain mobility of the copolymer PHBH compared to the homopolymer P3HB, displayed by the lower crystallite melting temperature and especially the lower T_g, contributes to its higher flexibility and ductility at room temperature.

4. Conclusions

In this study, the great potential of polyhydroxyalkanoates (PHAs), especially poly(3–hydroxybutyrate) (P3HB) and poly(3–hydroxybutyrate–co–3–hydroxyhexanoate) (PHBH), to form different nonwoven structures with fine fiber diameter distributions in the meltblow process was demonstrated. The DSC characterizations showed that both P3HB and PHBH exhibit complex thermal and crystalline properties. PHBH generally shows an earlier melting behavior compared to P3HB. The use of additives like epoxidized linseed oil (ELO) and acetyltributylcitrate (ATBC) influenced the thermal behavior by lowering the melting and crystallization temperatures. ATBC proved to be a more effective plasticizer than ELO and extends the processing window to lower processing temperatures. The complex viscosity analysis proved that the results of the thermal characterization are reflected in the rheological properties. Overall, the lower melting temperatures allowed for gentler processing, minimizing thermal chain degradation.

Both polymers were able to produce dimensionally stable three-dimensional nonwoven shapes in a one-step meltblow process, which also showed thermal stability up to 130 °C. Plant seed pots made from P3HB that were produced as a demonstrator application showed equivalent mechanical strength (and elongation) to commercial natural-fiber-based pots, as well as a lower area base weight and wall thickness. Furthermore, another interesting property of PHA nonwoven structures results in moderate air permeability and hydrophobicity, leading to impermeability to water.

Processed P3HB compounds showed greater flexibility in throughput, with values exceeding 0.1 g/min per nozzle. In contrast, unmodified PHBH was limited to lower throughput rates at a high process temperature of 190 °C, despite its lower melting temperature. This is attributed to a higher average polymer chain length and correspondingly a higher melt viscosity compared to P3HB. Nevertheless, PHBH and the PHBH+ATBC compounds led to a more stable meltblown process and more homogeneous fiber deposition due to the larger temperature difference between the process and melting temperatures.

It was also shown that the shape of the three-dimensional fabrics made from PHBH were completely compression-resilient, differing from P3HB and PHBV. However, the addition of plasticizers did not lead to any improvements without introducing inhomogeneities into the fiber deposition. However, the compounding of P3HB and PHBH with ATBC showed the potential to slightly reduce the processing temperatures of P3HB and PHBH, while significantly increasing the elongation and flexibility of PHBH. In contrast, the addition of ATBC resulted in a significant increase in fiber diameter for P3HB, which was generally not observed for PHBH.

As found in previous works, all PHAs showed semi-crystalline WAXS signals, proving that the meltblown process with high forces and strong chain orientation leads to highly crystalline structures without the influence of process parameter variations. The nonwovens made from pure PHBH showed the highest tenacity in both the machine direction (MD) and the cross direction (CD), while PHBV showed significantly lower mechanical strength. PHBV also exhibited the lowest elongation. The Young's modulus was low for PHBV, while it was superior for P3HB with ELO. Furthermore, the use of PHBH led to higher tenacity at comparable fiber diameters and air permeability values. At the same time, fabric flexibility improved significantly, and elongation at maximum force increased in the MD. The PHBH compounds showed lower tenacity and air permeability but further increased elongation, fulfilling the hypothesis of this study. The brittleness of the P3HB nonwovens could not be significantly improved. Further investigations should now be carried out with other biological plasticizers. Furthermore, the mechanical recovery of the plant pots should be investigated over a longer period of stress. Due to the promising mechanical properties of the PHBH nonwovens, the meltblow processing of PHBH should now be further investigated. In the future, further applications for three-dimensional nonwovens made from PHAs can be investigated due to their interesting combination of air permeability and water retention properties, e.g., for use in the medical technology sector as a three-dimensional wound dressing.

Author Contributions: Conceptualization, T.H. and F.G.; methodology, T.H. and F.G.; validation, T.H. and F.G.; formal analysis, T.H., M.P.V. and F.G.; investigation, T.H., L.B., M.P.V. and F.G.; resources, T.H., F.G. and J.R.; data curation, T.H., L.B. and F.G.; writing—original draft preparation, T.H., L.B. and F.G.; writing—review and editing, M.P.V., M.A., M.W., J.R. and C.B.; visualization, T.H., L.B. and F.G.; supervision, M.A., J.R. and C.B.; project administration, T.H. and F.G.; funding acquisition, J.R. and C.B. All authors have read and agreed to the published version of the manuscript.

Funding: This research was funded through state funding for the promotion of innovation projects within the framework of the bioeconomy innovation and investment program for rural areas provided by the federal state of Baden-Württemberg's Ministry of Food, Rural Affairs and Consumer Protection (MLR), represented by the VDI, VDE Innovation + Technik GmbH within the project "Bioabbaubarer Kunststoff für Produktentwicklung PHB2Pro" (Biodegradable plastic for product development), grant number BWIN330173.

Institutional Review Board Statement: Not applicable.

Data Availability Statement: Data available upon request due to privacy restrictions. The data presented in this study are available upon request from the corresponding author. The data are not publicly available due to running project issues.

Acknowledgments: We thank Marina Selent for carrying out the physical nonwoven characterization and Simone Spranz for conducting the air permeability and the heat shrinkage testing of the fabrics. Further, we thank our former DITF colleagues Ingo Windschiegl and Martin Dauner and our former IKT colleague Felix Baumgärtner for their decisive contributions to the project.

Conflicts of Interest: The authors declare no conflicts of interest. The funders had no role in the design of the study; in the collection, analyses, or interpretation of data; in the writing of the manuscript; or in the decision to publish the results.

Appendix A. REM Micrographs of Commercial (Reference) Plant Pots

Figure A1. SEM images of the commercial (reference) plant seed pots: (**a**) coconut fiber-based (×100); (**b**) coconut fiber-based (×500); (**c**) cellulose fiber-based (×100); (**d**) cellulose fiber-based (×500).

Appendix B. DSC Thermogram of PHBH Meltblown Nonwoven Material

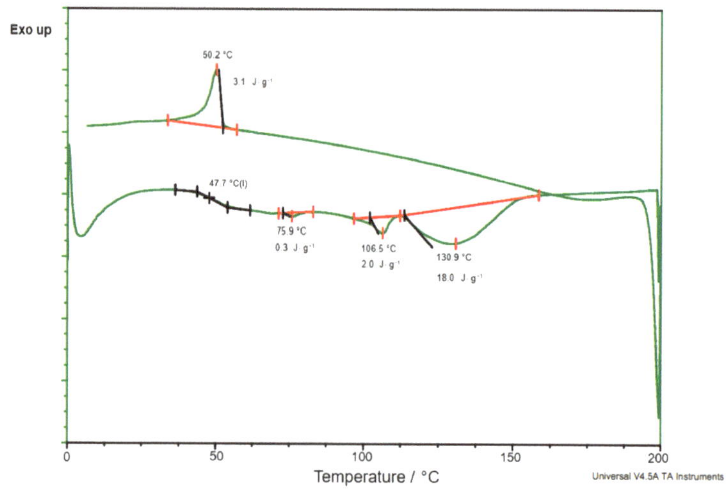

Figure A2. DSC thermogram of MB–PHBH; (red lines: base line extensions of the software for the analysis of the metling peaks, black lines: Tangents on the measurement curve (green line) for the determination of the glass transition temperature.

References

1. Vert, M.; Doi, Y.; Hellwich, K.-H.; Hess, M.; Hodge, P.; Kubisa, P.; Rinaudo, M.; Schué, F. Terminology for biorelated polymers and applications (IUPAC Recommendations 2012). *Pure Appl. Chem.* **2012**, *84*, 377–410. [CrossRef]
2. Varsha, Y.M.; Savitha, R. Overview on Polyhydroxyalkanoates: A Promising Biopol. *J. Microbial. Biochem. Technol.* **2011**, *3*, 99–105.
3. Mottin, A.C.; Ayres, E.; Oréfice, R.L.; Câmara, J.J.D. What changes in poly(3-hydroxybutyrate)(PHB) when processed as electrospun nanofibers or thermo-compression molded film? *Mater. Res.* **2016**, *19*, 57–66. [CrossRef]
4. Winnacker, M. Polyhydroxyalkanoates: Recent advances in their synthesis and appliations. *Eur. J. Lipid Sci. Technol.* **2019**, *121*, 1900101. [CrossRef]
5. Rodriguez-Contreras, A.R. Recent advances in the use of polyhydroyalkanoates in biomedicine. *Bioengineering* **2019**, *6*, 82. [CrossRef]
6. Reddy, C.S.K.; Ghai, R.; Kalia, V.C. Polyhydroxyalkanoates: An overview. *Bioresour. Technol.* **2003**, *87*, 137–146. [CrossRef]
7. Marchessault, R.H. Tender morsels for bacteria: Recent development in microbial polyester. *Trends. Polym. Sci.* **1996**, *5*, 163–168.
8. Köse, G.T.; Kenar, H.; Hasırcı, N.; Hasırcı, V. Macroporous poly(3-hydroxybutyrate-co–3-hydroxyvalerate) matrices for bone tissue engineering. *Biomaterials* **2003**, *24*, 1949–1958. [CrossRef]
9. Hufenus, R.; Reifler, F.A.; Maniura-Weber, K.; Spierings, A.; Zinn, M. Biodegradable bicomponent fibers from renewable sources: Melt-spinning of poly(lactic acid) and poly[(3-hydroxybutyrate)-co-(3-hydroxyvalerate)]. *Macromol. Mater. Eng.* **2012**, *297*, 75–84. [CrossRef]
10. Tanaka, T.; Fujita, M.; Takeuchi, A.; Suzuki, Y.; Uesugi, K.; Ito, K.; Fujisawa, T.; Doi, Y.; Iwate, T. Formation of highly ordered structure in poly[(r)–3-hydroxybutyrate-co-(r)–3-hydroxyvalerate] high-strength fibers. *Macromolecules* **2006**, *39*, 2940–2946. [CrossRef]
11. Chiellini, E.; Solaro, R. Biodegradable polymeric materials. *Adv. Mater.* **1996**, *8*, 305–313. [CrossRef]
12. Qin, Q.; Takarada, W.; Kikutani, T. Fiber structure development of PHBH through stress-induced crystallization in high-speed melt spinning process. *J. Fiber Sci. Technol.* **2017**, *73*, 49–60. [CrossRef]
13. Miyao, Y.; Takarada, W.; Kikutani, T. Improvement of mechanical properties of biodegradable PHBH fibers through high-speed melt spinning process equipped with a liquid isothermal bath. *AIP Conf. Proc.* **2020**, *2289*, 020038.
14. Grigore, M.E.; Grigorescu, R.M.; Iancu, L.; Ion, R.-M.; Zaharia, C.; Andrei, E.R. Methods of synthesis, properties and biomedical applications of polyhydroxyalkanoates: A review. *J. Biomater. Sci. Polym. Ed.* **2019**, *30*, 695–712. [CrossRef]

15. Singh, A.K.; Srivastava, J.K.; Chandel, A.K.; Sharma, L.; Mallick, N.; Singh, S.P. Biomedical applications of microbially engineered polyhydroxyalkanoates: An insight into recent advances, bottlenecks, and solutions. *Appl. Microbiol. Biotechnol.* **2019**, *103*, 2007–2032. [CrossRef]
16. Puppi, D.; Pecorini, G.; Chiellini, F. Biomedical processing of polyhydroxyalkanoates. *Bioengineering* **2019**, *6*, 108. [CrossRef]
17. Erceg, M.; Kovačić, T.; Klarić, I. Thermal degradation of poly(3-hydroxybutyrate) plasticized with acetyl tributyl citrate. *Polym. Degrad. Stab.* **2005**, *90*, 313–318. [CrossRef]
18. Ariffin, H.; Nishida, H.; Shirai, Y.; Hassan, M.A. Determination of multiple thermal degradation mechanisms of poly(3-hydroxybutyrate). *Polym. Deg. Stab.* **2008**, *93*, 1433–1439. [CrossRef]
19. Nguyen, S.; Guyen, S.; Yu, G.; Marchessault, R.H. Thermal degradation of poly(3-hydroxyalkanoates): Preparation of well-defined oligomers. *Biomacromolecules* **2002**, *3*, 219–224. [CrossRef]
20. Van der Walle, G.A.; De Koning, G.J.; Weusthuis, R.A.; Eggink, G. Properties, modifications and applications of biopolyesters. In *Biopolyesters*; Babel, W., Steinbüchel, A., Eds.; Springer: Berlin/Heidelberg, Germany, 2001; Volume 71, pp. 263–293.
21. Kurusu, R.S.; Siliki, C.A.; David, E.; Demarquette, N.R.; Gauthier, C.; Chenal, J.M. Incorporation of plasticizers in sugarcane-based poly(3-hydroxybutyrate)(PHB): Changes in microstructure and properties through ageing and annealing. *Ind. Crops Prod.* **2015**, *72*, 166–174. [CrossRef]
22. Xiang, H.; Chen, Z.; Zheng, N.; Zhang, X.; Zhu, L.; Zhou, Z.; Zhu, M. Melt-spun microbial poly(3-hydroxybutyrate-co-3-hydroxyvalerate) fibers with enhanced toughness: Synergistic effect of heterogeneous nucleation, long-chain branching and drawing process. *Int. J. Biol. Macromol.* **2019**, *122*, 1136–1143. [CrossRef] [PubMed]
23. Bonten, C. *Kunststofftechnik Einführung und Grundlagen*, 3rd ed.; Carl Hanser: Munich, Germany, 2020.
24. Liu, G.; Wang, X.; Yu, J.; Ding, B. Polylactic Acid (PLA) Melt-Blown Nonwovens with Superior Mechanical Properties. *ACS Sustain. Chem. Eng.* **2023**, *11*, 4279–4288. [CrossRef]
25. Zhang, J.; Chen, G.; Bhat, G.S.; Azari, H.; Pen, H. Electret characteristics of melt-blown polylactic acid fabrics for air filtration application. *J. Appl. Polym. Sci.* **2020**, *137*, 48309. [CrossRef]
26. Kara, Y.; Molnár, K. A review of processing strategies to generate melt-blown nano/microfiber mats for high-efficiency filtration applications. *J. Ind. Text.* **2022**, *51* (Suppl. S1), 137S–180S. [CrossRef]
27. Müller, D.H.; Krobjilowski, A. Meltblown fabrics from biodegradable polymers. *Int. Nonwovens J.* **2001**, *1*, 1558925001os-1000106. [CrossRef]
28. Jafari, M.; Shim, E.; Joijode, A. Fabrication of Poly(lactic acid) filter media via the meltblowing process and their filtration performances: A comparative study with polypropylene meltblown. *Sep. Purif. Technol.* **2021**, *260*, 118185. [CrossRef]
29. Hammonds, R.L.; Gazzola, W.H.; Benson, R.S. Physical and thermal characterization of polylactic acid meltblown nonwovens. *J. Appl. Polym. Sci.* **2014**, *131*, 40593. [CrossRef]
30. Gazzola, W.H.; Benson, R.S.; Carver, W. Meltblown Polylactic Acid Nanowebs as a Tissue Engineering Scaffold. *Ann. Plast. Surg.* **2019**, *83*, 716–721. [CrossRef]
31. Liu, G.; Guan, J.; Wang, X.; Yu, J.; Ding, B. Large-Scale Preparation of Mechanically High-Performance and Biodegradable PLA/PHBV Melt-Blown Nonwovens with Nanofibers. *Engineering* **2024**, *38*, 244–252. [CrossRef]
32. Jafari, M. Structure and Filtration Properties of PLA Meltblown Air Filters. Ph.D. Thesis, North Carolina State University, Foxhall, Raleigh, NC, USA, 2017.
33. Messiry, M.E.; Bhat, G.; Eloufy, A.; Latif, S.A.; Ayman, Y. Acoustical absorptive properties of meltblown nonwovens for textile machinery. *Text. Res. J.* **2021**, *91*, 1341–1353. [CrossRef]
34. Feng, S.; Jiao, X.N. The application of PLA resin on nonwovens production. *Adv. Mater. Res.* **2011**, *332*, 1239–1242. [CrossRef]
35. Rosli, N.A.; Karamanlioglu, M.; Kargarzadeh, H.; Ahmad, I. Comprehensive exploration of natural degradation of poly (lactic acid) blends in various degradation media: A review. *Int. J. Biol. Macromol.* **2021**, *187*, 732–741. [CrossRef] [PubMed]
36. Itavaara, M.; Vikman, M. An overview of methods for biodegradability testing of biopolymers and packaging materials. *J. Environ. Polym. Degr.* **1996**, *4*, 29–36. [CrossRef]
37. Santana, V.T.; Gonçalves, S.P.C.; Agnelli, J.A.M.; Martins-Franchetti, S.M. Biodegradation of a polylactic acid/polyvinyl chloride blend in soil. *J. Appl. Polym. Sci.* **2012**, *125*, 536–540. [CrossRef]
38. Lv, S.; Zhang, Y.; Gu, J.; Tan, H. Physicochemical evolutions of starch/poly(lactic acid) composite biodegraded in real soil. *J. Environ. Manag.* **2018**, *228*, 223–231. [CrossRef]
39. Freier, T.; Kunze, C.; Nischan, C.; Kramer, S.; Sternberg, K.; Sass, M.; Hopt, U.T.; Schmitz, K.P. In vitro and in vivo degradation studies for development of a biodegradable patch based on poly(3-hydroxybutyrate). *Biomaterials* **2002**, *23*, 2649. [CrossRef]
40. Höhnemann, T.; Windschiegel, I. Influence of Rheological and Morphological Characteristics of Polyhydroxybutyrate on its Meltblown Process Behavior. *Materials* **2023**, *16*, 6525. [CrossRef]

41. Kann, Y.; Whitehouse, R.S. Production of Nonwoven Materials from Polyhydroxyalkanoate. U.S. Patent US2011025639A1, 20 October 2011.
42. Sójka-Ledakowicz, J.; Łatwińska, M.; Kudzin, M.; Klepacz-Smółka, A. A study on obtaining nonwovens using polyhydroxyalkanoates and the melt-blown technique. *e-Polymers* **2014**, *14*, 373–380. [CrossRef]
43. Personal Phone Call 06/12/2024, Biomer Biopolyesters Specifications. Available online: https://www.biomer.de/IndexD.html (accessed on 6 August 2024).
44. Bluepha® PHA BP350-05 Technical Data Sheet. Available online: https://cdn.shopify.com/s/files/1/0810/0695/2776/files/BluephaA_PHA_BP350-05_Technical_Data_Sheet_V3.1.pdf?v=1710492141 (accessed on 10 June 2024).
45. Anzuchttöpfe (OBI). 2024. Available online: https://www.obi.de/p/1416999/anzuchttoepfe-rund-8-cm%E2%80%9316-stueck (accessed on 8 June 2024).
46. MAVIfiber2d, Fraunhofer ITWM. 2022. Available online: https://www.itwm.fraunhofer.de/de/abteilungen/bv/produkte-und-leistungen/mavi/mavifiber2d.html (accessed on 15 June 2024).
47. Altendorf, H.I.; Dida, S.; Batt, T. Automatische Bestimmung von Faserradienverteilungen. In *Forum Bildverarbeitung*; Kitson, R.E., Ed.; Scientific Publishing: Regensburg, Germany, 2010.
48. Easwaran, P.; Lehmann, M.J.; Wirjadi, O.; Prill, T.; Didas, S.; Redenbach, C. Automatic Fiber Thickness Measurement in SEM Images Validated Using Synthetic Data. *Chem. Eng. Technol.* **2016**, *39*, 395–402. [CrossRef]
49. Labthink: Comparative Analysis on Thermal Shrinkage Test Methods of Plastic Film. 2016. Available online: https://en.labthink.com/en-us/literatures/comparative-analysis-thermal-shrinkage-test-methods.html (accessed on 6 August 2024).
50. Liu, T.; Petermann, J. Multiple melting behavior in isothermally cold-crystallized isotactic polystyrene. *Polymer* **2001**, *42*, 6453–6461. [CrossRef]
51. Eraslan, K.; Aversa, C.; Nofar, M.; Barletta, M.; Gisario, A.; Salehiyan, R.; Goksu, Y.A. Poly(3-hydroxybutyrate-co-3--ydroxyhexanoate)(PHBH): Synthesis, properties, and applications—A review. *Eur. Polym. J.* **2022**, *167*, 111044. [CrossRef]
52. Volova, T.G.; Uspenskaya, M.V.; Kiselev, E.G.; Sukovatyi, A.G.; Zhila, N.O.; Vasiliev, A.D.; Shishatskaya, E.I. Effect of monomers of 3-hydroxyhexanoate on properties of copolymers poly (3-hydroxybutyrate-co 3-hydroxyhexanoate). *Polymer* **2023**, *15*, 2890. [CrossRef] [PubMed]
53. Choi, J.Y.; Lee, J.K.; You, Y.; Park, W.H. Epoxidized polybutadiene as a thermal stabilizer for poly (3-hydroxybutyrate). II. Thermal stabilization of poly (3-hydroxybutyrate) by epoxidized polybutadiene. *Fibers Polym.* **2003**, *4*, 195–198. [CrossRef]
54. Park, W.H.; Lenz, R.W.; Goodwin, S. Epoxidation of bacterial polyesters with unsaturated side chains. I. Production and epoxidation of polyesters from 10-undecenoic acid. *Macromole* **1998**, *31*, 1480–1486. [CrossRef]
55. Schröder, T. *Rheologie der Kusntstoffe: Theorie und Praxis*, 2nd ed.; Carl Hanser Verlag: München, Germany, 2020.
56. TianAn-Biopolymer. ENMAT™ Thermoplastics Resin Y1000P—Technical Data Sheet & Processing Guide (17 February 2021). Available online: https://api.atomler.com/file/110457086 (accessed on 16 June 2024).
57. Höhnemann, T.; Schnebele, J.; Arne, W.; Windschiegl, I. Nanoval Technology—An Intermediate Process between Meltblown and Spunbond. *Materials* **2023**, *16*, 2932. [CrossRef]
58. Höhnemann, T.; Helle, T. Biopolymer aus Mikroalgen–Meltblown-Vliesstoffe aus Polyhydroxyalkanaoten. *Tech. Text./Tech. Text.* **2024**, *65*, 39–41. [CrossRef]
59. Tan, D.H.; Zhou, C.; Ellison, C.J.; Kumar, S.; Macosko, C.W.; Bates, F.S. Meltblown fibers: Influence of viscosity and elasticity on diameter distribution. *J. Non-Newton. Fluid Mech.* **2010**, *165*, 892–900. [CrossRef]
60. Ellison, C.J.; Phatak, A.; Giles, D.W.; Macosko, C.W.; Bates, F.S. Melt blown nanofibers: Fiber diameter distributions and onset of fiber breakup. *Polymer* **2010**, *48*, 3306–3316. [CrossRef]
61. Farrag, Y.; Barral, L.; Gualillo, O.; Moncada, D.; Montero, B.; Rico, M.; Bouza, R. Effect of different plasticizers on thermal, crystalline, and permeability properties of poly (3-hydroxybutyrate-co-3-hydroxyhexanoate) films. *Polymers* **2022**, *14*, 3503. [CrossRef]
62. Hufenus, R.; Reifler, F.A.; Fernández-Ronco, M.P.; Heuberger, M. Molecular orientation in melt-spun poly (3-hydroxybutyrate) fibers: Effect of additives, drawing and stress-annealing. *Europ. Polym. J.* **2015**, *71*, 12–26. [CrossRef]
63. Laycock, B.; Halley, P.; Pratt, S.; Werker, A.; Lant, P. The chemomechanical properties of microbial polyhydroxyalkanoates. *Prog. Polym. Sci.* **2014**, *39*, 536–583. [CrossRef]
64. Lee, S.Y. Plastic bacteria? Progress and prospects for polyhydroxyalkanoate production in bacteria. *Trends Biotechnol.* **1996**, *14*, 431–438. [CrossRef]
65. Fabri, D.; Guan, J.; Cesàro, A. Crystallisation and melting behaviour of poly (3-hydroxybutyrate) in dilute solution: Towards an understanding of physical gels. *Thermochim. Acta* **1998**, *32*, 3–16. [CrossRef]

66. Farer, R.; Seyam, A.M.; Ghosh, T.K.; Batra, S.K.; Grant, E.; Lee, G. Forming shaped/molded structures by integrating meltblowing and robotic technologies. *Text. Res. J.* **2003**, *73*, 15–21. [CrossRef]
67. Selli, F.; Hufenus, R.; Gooneie, A.; Erdoğan, U.H.; Perret, E. Structure–property relationship in melt-spun poly (hydroxybutyrate-co-3-hexanoate) monofilaments. *Polymers* **2022**, *14*, 200. [CrossRef]

polymers

MDPI

Article

The Properties of Thin Films Based on Chitosan/Konjac Glucomannan Blends

Karolina Kulka-Kamińska * and Alina Sionkowska *

Department of Biomaterials and Cosmetic Chemistry, Faculty of Chemistry, Nicolaus Copernicus University, Gagarin 7 Street, 87-100 Torun, Poland
* Correspondence: kkulka@doktorant.umk.pl (K.K.-K.); alinas@umk.pl (A.S.)

Abstract: In this work, blend films were prepared by blending 2% chitosan (CS) and 0.5% konjac glucomannan (KGM) solutions. Five ratios of the blend mixture were implemented (95:5, 80:20, 50:50, 20:80, and 5:95), and a pure CS film and a pure KGM film were also obtained. All the polymeric films were evaluated using FTIR spectroscopy, mechanical testing, SEM and AFM imaging, thermogravimetric analyses, swelling and degradation analyses, and contact angle measurements. The CS/KGM blends were assessed for their miscibility. Additionally, the blend films' properties were evaluated after six months of storage. The proposed blends had good miscibility in a full range of composition proportions. The blend samples, compared to the pure CS film, indicated better structural integrity. The surface structure of the blend films was rather uniform and smooth. The sample CS/KGM 20:80 had the highest roughness value (Rq = 12.60 nm). The KGM addition increased the thermal stability of films. The blend sample CS/KGM 5:95 exhibited the greatest swelling ability, reaching a swelling degree of 946% in the first fifteen minutes of the analysis. Furthermore, the addition of KGM to CS improved the wettability of the film samples. As a result of their good mechanical properties, surface characteristics, and miscibility, the proposed CS/KGM blends are promising materials for topical biomedical and cosmetic applications.

Keywords: chitosan; konjac glucomannan; blends; miscibility; biopolymer film

Citation: Kulka-Kamińska, K.; Sionkowska, A. The Properties of Thin Films Based on Chitosan/Konjac Glucomannan Blends. *Polymers* **2024**, *16*, 3072. https://doi.org/10.3390/polym16213072

Academic Editors: Vishal Gavande, Vasi Shaikh and Won-Ki Lee

Received: 15 October 2024
Revised: 25 October 2024
Accepted: 27 October 2024
Published: 31 October 2024

1. Introduction

The modern industry focuses on ensuring sustainable and holistic development, with an emphasis on safe and environmentally friendly solutions. Biopolymers are a group of biomaterials with significant potential in these areas due to their biodegradability, biocompatibility, and non-toxicity [1,2]. Various modifications of biopolymers are used in the search for new, more efficient materials. A contemporary methodology in numerous disciplines is the creation of composite polymeric materials. These materials are characterized by better physicochemical, mechanical, and stability properties compared to their constituent starting polymers [3].

Two principal techniques for obtaining blends are identified by the extant literature: the first involves mixing in a molten state, while the second utilizes polymeric solutions combined in a suitable solvent [3,4]. Nevertheless, the second method appears to be more pertinent for biopolymers, as a means of preventing the degradation of these natural components when subjected to elevated pressure and temperature during the process of melt mixing [3]. Blend composites are frequently used in the form of film, hydrogel, sponge, or fiber and can be applied as a wound dressing, scaffold, membrane, or even a drug-encapsulating agent [5–8]. An essential quality of blends is their miscibility, which can be evaluated through a few straightforward studies. These include the assessment of the optical homogeneity of the mixture, the determination of the glass transition temperature, and the analysis of molecular-level interactions [3].

The biopolymer that has gained the most popularity over the past two decades, primarily due to its advantageous characteristics, is chitosan, a chitin derivative. The biomacromolecule is composed of 2-acetamido-2-deoxy-D-glucopyranose and 2-amino-2-deoxy-D-glucopyranose units [9]. In addition to exhibiting biodegradability, non-toxicity, and biocompatibility, the biopolymer displays adsorption properties and demonstrates antioxidant and antimicrobial activity. These properties are primarily attributable to the biopolymer's polycationic nature [10,11]. Chitosan is soluble in water only in an acidic environment, whereby the amino groups are protonated. This biomolecule can be derived from a variety of sources of chitin, including fungi, marine organisms, certain algae, insects, and microorganisms [12]. The preparation of chitin can be achieved through two principal methods: chemical and biological [13]. In the pursuit of materials with optimal properties for a given application, chitosan modifications are employed. The presence of functional groups at C-2 (-NH_2), C-3 (secondary -OH), and C-6 (primary -OH) renders chitosan susceptible to chemical modifications [10]. Chitosan derivatives are produced through chemical reactions, while another approach is polymer cross-linking. As previously stated, additional methods of chitosan modification entail the formation of polymer blends [14]. Chitosan-based materials have a wide range of applications in medicine, pharmaceuticals, food, packaging, cosmetics industries, and water purification. This demonstrates the unique properties of chitosan and its high potential as a material.

Konjac glucomannan (KGM) is a neutral polysaccharide that is isolated from Amorphophallus konjac plant tubers. This biopolymer is composed of D-mannose and D-glucose units with a molar ratio of 1.6:1. The units are linked by β-(1,4) glycosidic bonds [15,16]. Additionally, a minor proportion of β-1,3-linkages are present at the mannose C-3 position—the konjac glucomannan chain contains a randomly distributed number of acetyl groups, which represent 5 to 10% [17]. The high affinity of the KGM for water is precisely the merit of the acetyl and hydroxyl groups' presence [18]. KGM displays high viscosity, a robust water-holding capacity, and excellent gel-forming properties in aqueous solutions [19]. Furthermore, it functions as an effective emulsifier and demonstrates remarkable film-forming capabilities [20]. This polysaccharide is a valuable macromolecular raw material used in the food industry, as well as increasingly in the biomedical and cosmetic fields. Nevertheless, the utilization of KGM-based materials in numerous fields is frequently constrained by their inadequate mechanical properties, which can be attributed to the high hydroxyl group content in the polymeric backbone [21].

There are reports in the literature on chitosan/glucomannan blends, which are mainly used for drug-coating applications [22], wound dressing [18], peptide and protein delivery vehicles [23], and membranes [24]. These reports are supplemented by a general survey describing the proposed blend as a potential matrix biomaterial [25], as well as the influence of gamma irradiation on such blend materials as a part of the sterilization process [26].

In this study, we were focused on obtaining a new material based on biopolymers. This paper aimed to evaluate the properties of chitosan/konjac glucomannan blend films. For this purpose, infrared spectroscopy, a mechanical parameters evaluation, and a thermogravimetric analysis were carried out. Additionally, the surface properties and morphology of the materials were evaluated through contact angle measurements, scanning electron microscopy, and atomic force microscopy. The films were prepared from different blend proportions. The characteristics of the blends were compared to the pure chitosan and the pure glucomannan films. The same characterization tools were used for the films' properties evaluation after a six-month period of storage.

2. Materials and Methods

2.1. Materials

Low-molecular-weight chitosan (CS) and konjac glucomannan (KGM) ($M_{vCS} = 7.31 \times 10^5$ g/mol; $M_{vKGM} = 7.71 \times 10^5$ g/mol) were purchased from the POL-AURA company (Dywity, Poland). Acetic acid was acquired from POCH (Gliwice, Poland). Phosphate Buffer Saline (PBS), in the form of tablets, was acquired from Life Technologies Limited (Renfrew, UK).

2.2. Chitosan/Konjac Glucomannan Films Preparation

All of the films prepared during this research were prepared by the solvent casting method. Solutions of 2% (*w*/*v*) chitosan (CS) and 0.5% (*w*/*v*) konjac glucomannan (KGM) were prepared by dissolving the polymers in 0.1 M acetic acid. To prepare the control samples, 25 g of CS pure solution and 25 g of KGM pure solution were poured out separately on polystyrene plates (10 cm × 10 cm). Subsequently, the samples were placed in an incubator, which was set at 37 °C, and allowed to dry. To prepare the blend samples, the CS and the KGM solutions were mixed in weight ratios of 95:5, 80:20, 50:50, 20:80, and 5:95. The blended solutions were stirred for 24 h and then poured onto polystyrene plates. They were dried in the same manner as the control samples. The drying process took 3 to 5 days, depending on the sample. The pictures of prepared solutions and films are presented in Figure 1 and Table 1, respectively.

The films, which were examined after 6 months, were stored during this time period without access to light, at room temperature, at a humidity of approximately 50%.

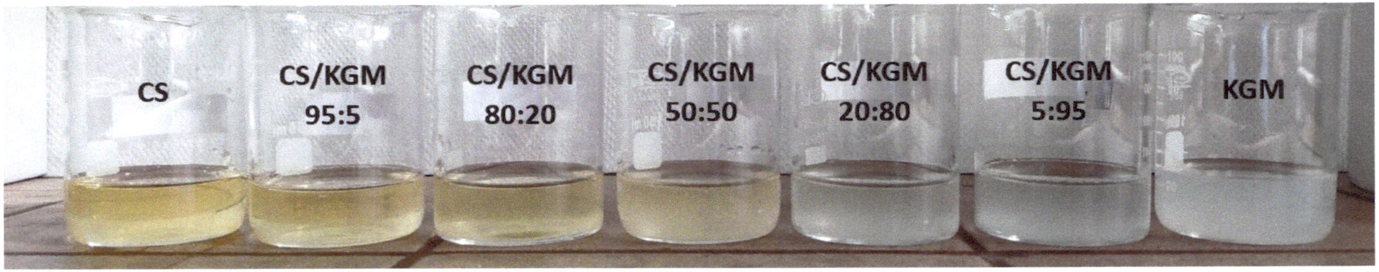

Figure 1. Overview of chitosan (CS), konjac glucomannan (KGM), and their blends' solution appearance.

Table 1. The overview of chitosan (CS), konjac glucomannan (KGM), and their blends' film appearance.

CS	CS/KGM, 95:5	CS/KGM, 80:20	CS/KGM 50:50	CS/KGM 20:80	CS/KGM 5:95	KGM

		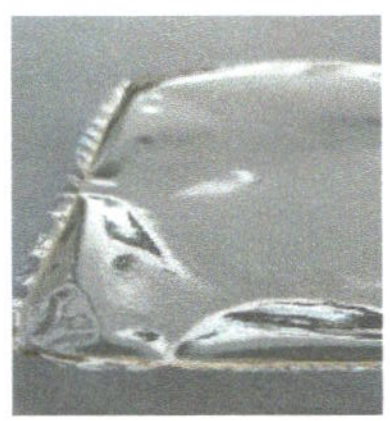			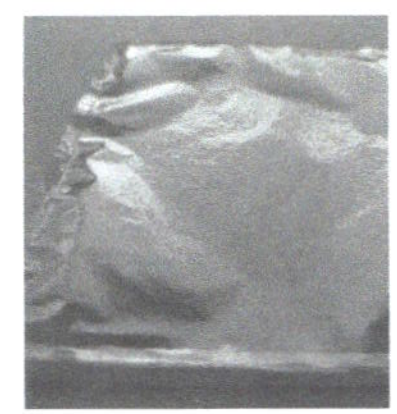	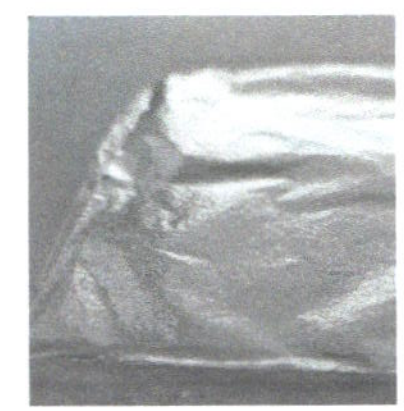

2.3. Fourier Transform Infrared Spectroscopy (FTIR)

The FTIR spectra of all the prepared films were assessed using a Nicolet iS10 spectrophotometer equipped with a diamond attenuated total reflectance (ATR) unit (Thermo Fisher Scientific, Waltham, MA, USA). The samples were scanned in the wavenumber range of 4000–400 cm^{-1} with a 4 cm^{-1} resolution using 64 scans, in accordance with the methodology employed in previous research [27]. The measurements were repeated on the same samples after a period of 6 months. OMNIC (version 9.2.86) and Excel (version 2409) software were used to analyze and process data.

2.4. Mechanical Testing

The mechanical parameters were assessed under room conditions, using a mechanical testing machine (Z.05, Zwick and Roell, Ulm, Germany). The specimens were cut in the same way, maintaining a similar shape, with a constriction of approximately 4 mm [28]. The measurement parameters were as follows: the speed starting position was 50 mm/min, the speed of the initial force was 5 mm/min, and the initial force was 0.1 MPa. The evaluation included the assessment of the Young's modulus, tensile strength, and elongation at the breaking point. The data were collected using the TestXpert II 2017 program and were subsequently presented as average values with standard deviations. A Q-Dixon test was performed to identify and reject outliers. A one-way ANOVA test was applied to determine statistically significant differences between samples; the reference samples were the initial polymers films. The mechanical properties of the 6-month-old samples were also evaluated and compared to the freshly obtained samples.

2.5. Scanning Electron Microscopy (SEM)

The morphology of the obtained films was studied using a scanning electron microscope, manufactured by LEO Electron Microscopy Ltd. (Model 1430 VP, Cambridge, UK,). The samples were covered with gold to provide a conductive surface for the electron beam interaction [29]. The magnification of all of the SEM images presented below is 10,000×.

2.6. Atomic Force Microscopy (AFM)

The surface of the polymer samples was analyzed using an atomic force microscope. The images were obtained by a MultiMode Scanning probe microscope, the NanoScope IIIa (Digital Instruments Veeco Metrology Group, Santa Barbara, CA, USA). The apparatus was operating in a tapping mode, at room temperature, in an air atmosphere [29]. The roughness parameters were calculated from 5.0 μm-by-5.0 μm scanned areas using Gwyddion software (version 2.62).

2.7. Thermogravimetric Analysis

The thermogravimetric assessment was conducted using an SDT 2960 Simultaneous TGA-DTA analyzer from TA Instruments (TA Instruments Manufactures, Eschborn, Germany). The analysis was carried out over a temperature range from 25 °C to 600 °C, at a heating rate of 20 °C/min, in a nitrogen atmosphere [29].

2.8. Swelling and Degradation Properties

The CS/KGM films were cut into squares of similar weights. The samples were dried for 24 h at 45 °C. Each type of sample (five squares in one series) was placed in a container with 50 mL Phosphate Buffer Saline (PBS) at 37 °C [29]. The measurements of the samples' weights were conducted after 15 min, 1 h, 2 h, 4 h, 8 h, 24 h, 48 h, 72 h, 1 week, and 2 weeks. Following an appropriate interval, the samples were taken out from the PBS solution, and, subsequently, the excess fluid was removed using paper. Thereafter, the samples were weighed. The measurements were taken fresh from the collection. They were repeated on samples that were 6 months old. The swelling degrees were calculated using the following equation:

$$\text{Swelling} = (m_t - m_0)/m_0 \times 100\% \ [\%] \quad (1)$$

m_t—the weight of the material after immersion in PBS [g];
m_0—the initial weight of the material [g].

2.9. Contact Angle and Surface Free Energy

The contact angle measurements of the obtained films were used to calculate the surface free energy (γ_s) and its polar (γ_{sp}) and dispersive (γ_{sd}) components, by the Owens–Wendt method [29]. The contact angle values of two liquids (glycerine (G) and diiodomethane (D)) were measured via a goniometer that was equipped with a system for drop shape analyses (DSA 10 produced by Krüss, Hamburg, Germany). The result of the contact angle for each sample is an average value with a standard deviation (SD). The Q-Dixon test was implemented to identify and reject outliers. All the measurements were carried out at a constant temperature value.

3. Results

3.1. Fourier Transform Infrared Spectroscopy (FTIR)

The chemical structure of initial polysaccharides was confirmed with the FTIR analysis. The initial infrared spectra of the pure chitosan and the pure konjac glucomannan films are characteristic of them and show several main bands.

In the konjac glucomannan IR spectra, a characteristic broad peak was observed in the region of 3000–3700 cm^{-1}, with the band of highest intensity at 3316 cm^{-1}, which is related to the O-H stretching vibration [30]. A broader and less-sharp peak occurs in the same wavelength range in the CS spectra, with the highest peak at 3204 cm^{-1}, which also corresponds with O-H and additionally N-H stretching vibrations [31]. These bands may also be connected to the presence of water in the tested samples [32]. The bands at 2872 cm^{-1} and 2881 cm^{-1} in the CS and KGM samples, respectively, confirm the occurrence of C-H bonds that come from -CH_3 and -CH_2 groups being present in the carbohydrates [30].

In the KGM IR spectra, the stretching vibration coming from the C=O of the acetyl groups appears at a wavelength of 1738 cm^{-1} [33].

The peak at 1541 cm^{-1} that occurs only in the CS spectra is attributed to N-H bending vibrations [34]. The bands at around 1651 cm^{-1} and 1316 cm^{-1} confirm the presence of N-acetyl groups, represented by the C=O stretching of amide I and the C-N stretching of amide III, respectively [35]. The spectral region of 1200–900 cm^{-1} is associated with the stretching vibrations of C-O-C, C-C and C-O in the polymers' skeletons [36]. The bands in the range from 1150 cm^{-1} to 896 cm^{-1} indicate the presence of glycosidic bonds in both of the polymers' samples [37–39]. The peaks at 873 and 805 cm^{-1} are connected to the characteristic absorption bands of mannose [38,40].

FTIR spectroscopy is a very useful tool to evaluate the interactions at the molecular level between chemical groups. Intermolecular interactions may indicate the compatibility of different polymer blends. The bands' shifts in the infrared spectrum can indicate good miscibility [40].

The blend films indicate shifts in the 3000–3700 cm^{-1} region that differ from the initial polymers. The wavelength values are increased in comparison with chitosan and decreased in comparison with glucomannan, ranging in between them. Similar observations apply to the C-H absorption band area: there are small shifts in the CS/KGM blend samples. The characteristic band for KGM, at about 1738 cm^{-1}, disappears in the spectra of the blends with a higher content of chitosan; the last blend film with this band is CS/KGM 20:80. Even a small amount of chitosan in the blend sample causes the appearance of the band at about 1540 cm^{-1}, which does not occur in konjac glucomannan. Small changes, including shifts to the lower or higher wavenumbers in the blend films' spectra, are also observed in the 1300–1000 cm^{-1} region. The changes in the FTIR spectra also include the intensity of the bands. The highest intensity is observed in the pure chitosan and the CS/KGM 80:20 samples, and the lowest is observed in the CS/KGM 5:95 film. In light of the FTIR results, which encompass alterations in band intensity and shifts in regions, it can be posited that novel intermolecular hydrogen bonds were established. The principal functional groups engaged in this process were the hydroxyl, amine, and acetyl groups. The FTIR analysis results demonstrate that chitosan and konjac glucomannan exhibit good miscibility. The FTIR spectra for the chitosan, konjac glucomannan, and the blend films are presented in Figure 2.

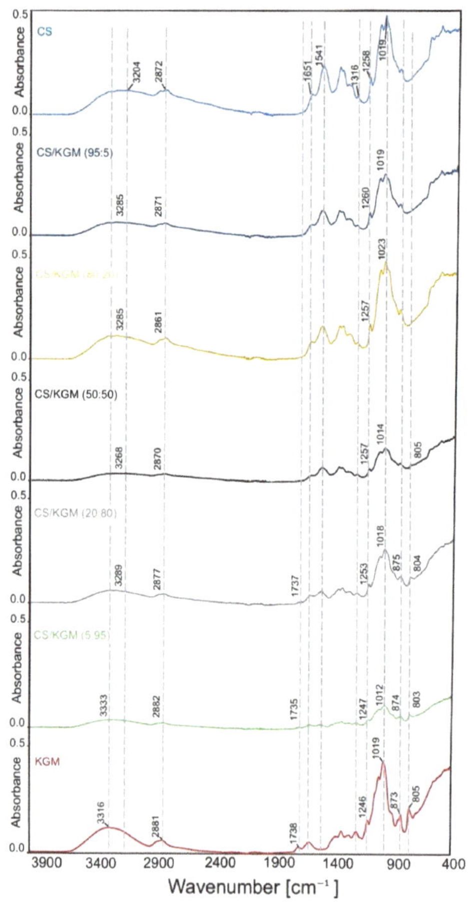

Figure 2. Infrared spectra of chitosan (CS), konjac glucomannan (KGM), and blended films presented in stack.

FTIR After 6 Months of Storage

The properties of the obtained blend films were also examined over time. Significant changes were observed for the pure CS film after the 6-month storage period (Figure 3). The broad band in the 3000–3700 cm^{-1} range shifted to higher wavenumbers in comparison to the initial sample (CS). This may indicate the reorganization of the hydrogen bonds in the material. Further changes include a significant reduction in the intensity of the bands at 1541 cm^{-1} and 1404 cm^{-1}, as well as shifts to the higher values of wavenumber. All these changes may be caused by the water loss and chitosan chain reorganization. The pure KGM sample did not demonstrate any modifications in the appearance and localization of its bands, and only a slight increase of intensity was observed after 6 months of storage.

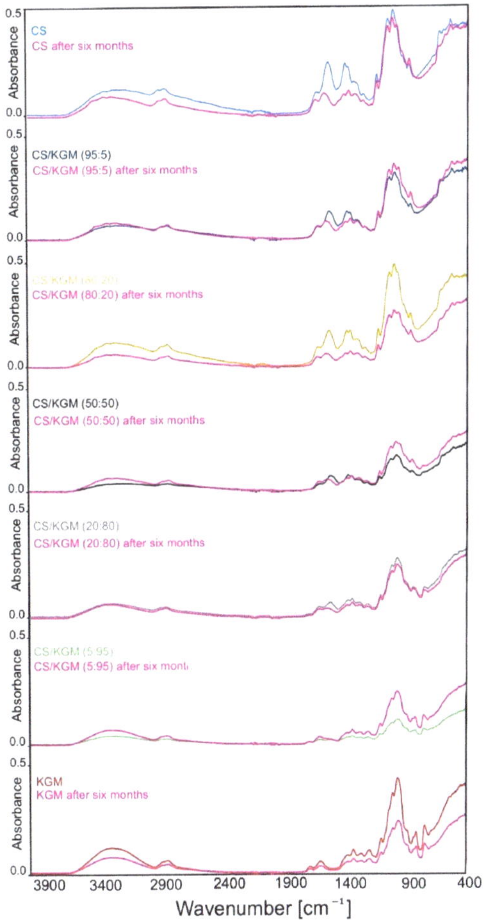

Figure 3. Infrared spectra of chitosan (CS), konjac glucomannan (KGM), and blended films after six months of storage, presented in stack.

The main changes that affected the blend samples included the decrease of individual bands' intensity and shifts in the same region as CS. In the two samples with the highest KGM concentrations (80:20 and 5:95), additionally, an increase of intensity at about 1000 cm^{-1} was observed. In sample 5:95, the band at about 1556 cm^{-1} disappeared in comparison to the freshly obtained sample.

3.2. Mechanical Testing

The proportions of the blend mixtures affected the mechanical properties of the thin films significantly. The tensile test results are shown in Figure 4. The Young's modulus value was the lowest for the pure CS samples. As the KGM content of the blends increased, higher values of this parameter were observed. This indicates the rather low deformability of the KGM films in comparison to CS.

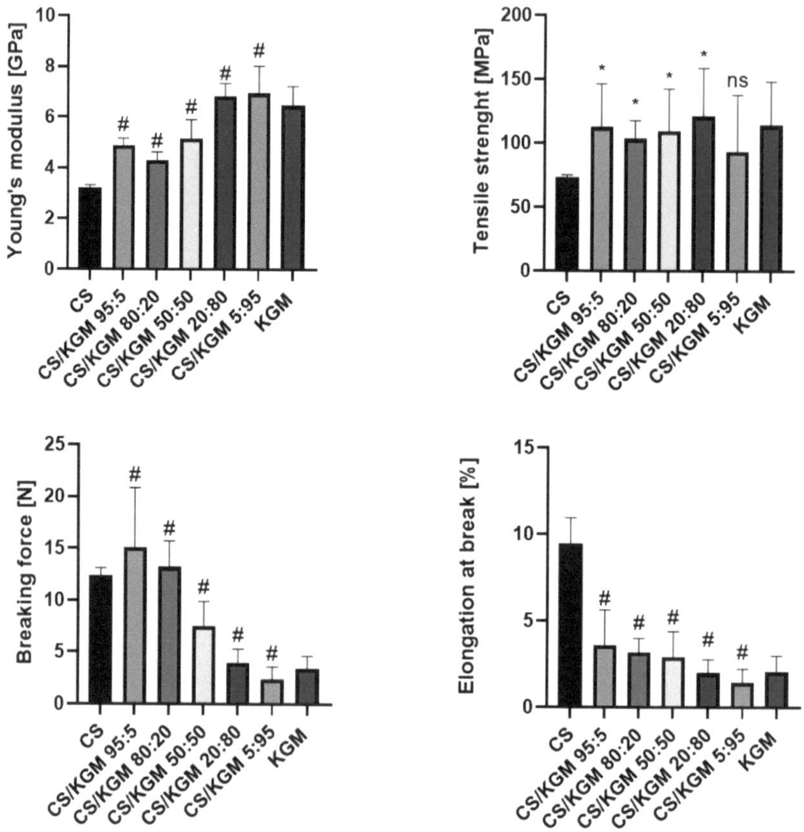

Figure 4. The results of mechanical parameters obtained during tensile tests for pure polymeric and blend films. Data are presented as a mean value with standard deviation. Statistically significant differences are indicated as follows: * $p < 0.05$; # $p < 0.0001$; ns—not significant.

The blend films exhibited variable values of tensile strength, the highest one belonging to CS/KGM 20:80 and the lowest to CS. The highest breaking force values were observed for the CS/KGM 95:5 samples, and the opposite observation was for CS/KGM 5:95. The decreasing trend of the parameter's value with a higher percentage of KGM can be seen.

A similar tendency was observed for the elongation at the breaking point, with the highest value being recorded for CS and lower results being observed as the percentage of this polymer in the blends decreased.

The increased results of the Young's modulus and the lower values of elongation at the breaking point of the blends' films may indicate new hydrogen bond formation between the biopolymers, which may have modified their mechanical properties and increased the structural integrity of the samples.

Mechanical Testing After 6 Months of Storage

The six-month storage period influenced the mechanical properties of the samples (Figure 5). The most significant changes were observed for the pure chitosan film, for which the value of the Young's modulus, tensile strength, and breaking force increased, while elongation at the breaking point decreased compared to the freshly tested sample. These observations indicate a reduction in the elasticity of the film. The passage of time had less impact on the properties of KGM, although the tensile strength decreased significantly. The blend samples also were affected by small changes in their mechanical properties. All those changes reflect the chemical modifications that occurred during storage.

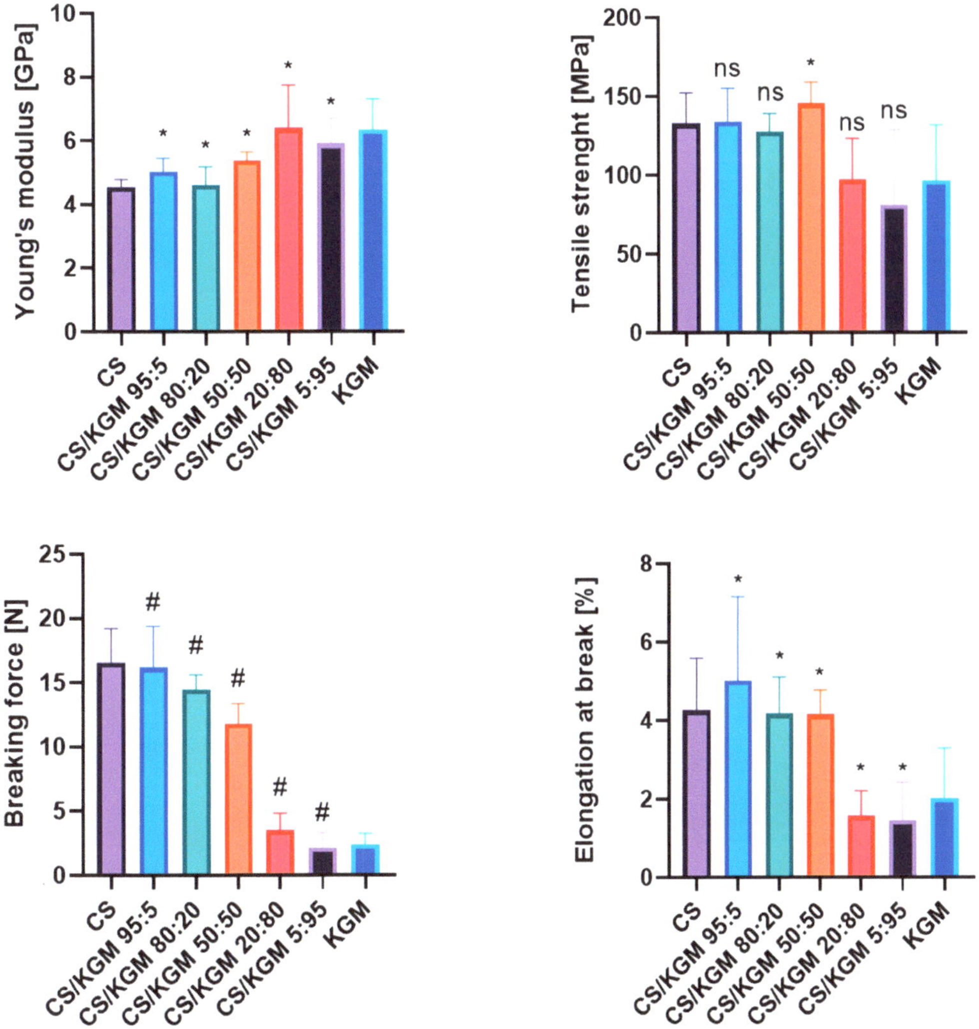

Figure 5. The results of mechanical parameters obtained during tensile tests for 6-month-old pure polymeric and blend films. Data are presented as mean value with standard deviation. Statistically significant differences are indicated as follows: * $p < 0.05$; # $p < 0.0001$; ns—not significant.

3.3. Scanning Electron Microscopy (SEM)

SEM images of CS, KGM, and their blends are presented in Figure 6. The SEM analysis provided information on the surface structure of the films, which were rather uniform and smooth. The chitosan film showed the most surface uniformity. The greater the proportion of KGM in the blending ratio, the less smooth the surface was. The KGM film was characterized by a more granular structure.

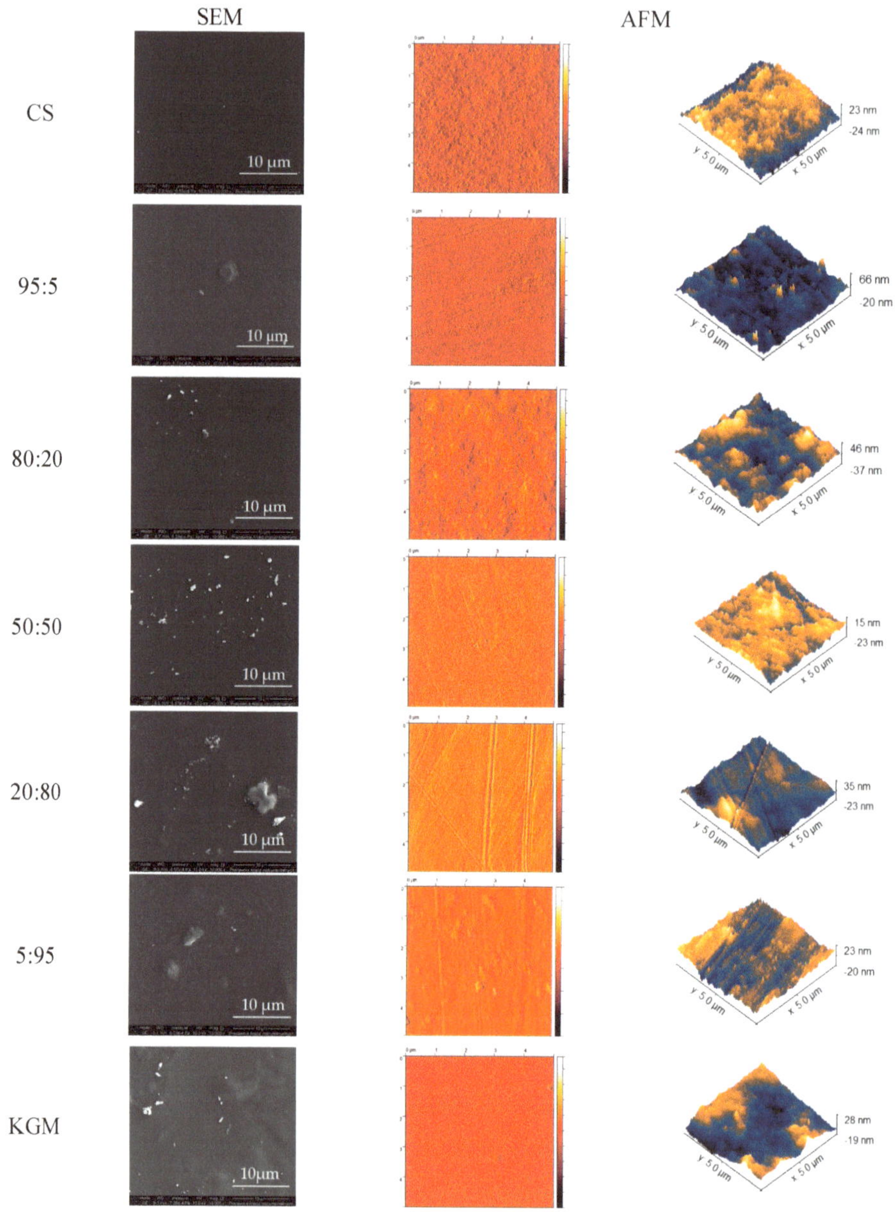

Figure 6. Surface structure analysis – comparison of SEM and AFM images of chitosan (CS), konjac glucomannan (KGM), and blended films.

3.4. *Atomic Force Microscopy (AFM)*

Figure 6 presents the atomic force microscopy (AFM) images of CS, KGM, and their blend films. The topography of the CS sample presents quite a uniform structure: the small, evenly distributed hills can be observed. The microstructure of the KGM sample is more diversified and includes changes in the level of larger areas. It corresponds with the SEM images.

The results of the tested samples' roughness are summarized in Table 2. CS is characterized by the lowest value of roughness; the addition of KGM caused an increase in this parameter. A much higher roughness is visible in the blend proportions CS/KGM 95:5 and 80:20. The blend film with an equal amount of the two polymer solutions exhibited a similar roughness value to the initial CS sample in this parameter. The samples CS/KGM 20:80 and 5:95 had values of roughness between those of the initial polymers, but closer to the KGM. The roughness parameter helps to describe the materials' adhesion properties. The adhesion between two materials is strongly influenced by their surfaces and contact topography. In terms of topical applications, adhesion to the skin is an extremely important parameter, which depends on the mentioned roughness, as well as the other material properties and thickness [41].

Table 2. The roughness values of chitosan (CS), konjac glucomannan (KGM), and blended films.

Sample	R_q [nm]	R_a [nm]
CS	5.64 ± 1.25	4.50 ± 0.97
CS/KGM 95:5	9.60 ± 2.85	6.35 ± 1.04
CS/KGM 80:20	12.60 ± 1.14	9.88 ± 0.77
CS/KGM 50:50	5.95 ± 2.48	4.69 ± 1.92
CS/KGM 20:80	6.57 ± 0.84	4.96 ± 0.46
CS/KGM 5:95	6.17 ± 0.75	4.86 ± 0.50
KGM	6.87 ± 0.94	5.68 ± 0.98

3.5. Thermogravimetric Analysis

The thermal stability and degradation of the prepared blend films can be evaluated by the analysis of the thermal gravimetry (TG) and the differential thermal gravimetry (DTG) curves, which are shown in Figure 7. During the thermal decomposition of each film, two distinct stages were observed. The initial stage of the process entails the release of water molecules and residual solvents; in this instance, acetic acid. The initial stage takes place in a temperature range from 35 °C up to approximately 150–170 °C in all the samples. The smallest weight loss in the first stage was observed for the KGM sample (8.5%), and the highest was observed for the CS/KGM 20:80 film (12.0%); for the rest of the films, the results were close to each other, in the range of 9.5–10.6%. The secondary stage, which was identified as the most significant thermal degradation of polymers at high temperatures, exhibited a more pronounced differentiation. The initial polymers have different decomposition stabilities: KGM is more thermally stable than CS. The T_{max} value for the pure KGM was 326 °C, and additionally, the exothermic peak was highest for this biopolymer, which was connected to the highest weight loss in this step (64.7%). The pure CS T_{max} value was 297 °C. Only the sample CS/KGM 5:95 had a Tmax value higher (305 °C) than that of CS. The rest of the samples had T_{max} values closer to the CS film (Table 3). Some changes, which appeared in the DTG curves of the blend films, may suggest that a hydrogen bonding interaction was established between the chitosan and the konjac glucomannan. Importantly, in the degradation phase, only one peak can be observed in all of the curves, which may indicate good miscibility.

Table 3. Summary of temperature maximum values for decomposition stages for all tested samples.

Sample	T_{max1} [°C]	T_{max2} [°C]
CS	62.87	296.72
CS/KGM 95:5	74.31	296.01
CS/KGM 80:20	65.02	293.86
CS/KGM 50:50	67.16	288.86
CS/KGM 20:80	75.74	285.99
CS/KGM 5:95	65.73	305.30
KGM	50.00	326.04

T_{max}—represents the maximum temperature value for a given stage.

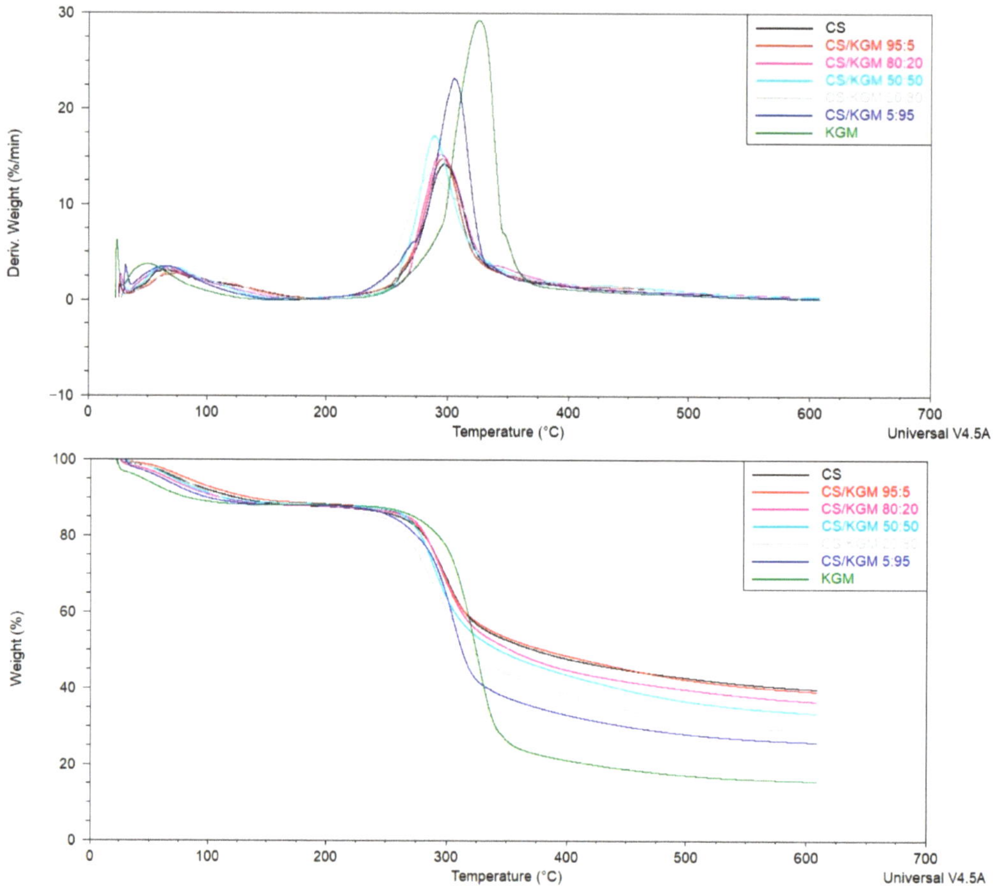

Figure 7. DTG and TG curves of chitosan (CS), konjac glucomannan (KGM), and the blended samples.

3.6. Swelling and Degradation Properties

The samples lost their weight (6.04–9.12%) in a drying process before the swelling measurements, which is connected with water and solvent residue evaporation. The results of the swelling and degradation analysis are collected in Table 4. Each result is presented as an average value with a standard deviation. All of the polymeric samples, besides the pure KGM, could swell and revealed a weight increase after being placed in the PBS solution. The KGM films were dissolved in the first 15 min of measurement. Even a small addition of CS to the KGM film (CS/KGM 5:95) caused a stability increase. The pure CS sample reached a maximum degree of swelling in one hour, after which, the values decreased, which may indicate the onset of the degradation process. Similar observations were noted for all of the samples; their weight reached a maximum at the very beginning of the measurements and then steadily decreased until the end of the experiment. KGM has a higher ability to swell than chitosan; the sample with the highest swelling degree was CS/KGM 5:95. The sample containing equal proportions of the two polymer solutions exhibited the lowest ability to swell.

Table 4. The results of the swelling and degradation analysis for the chitosan (CS), konjac glucomannan (KGM), and blended films, obtained over two weeks (presented as a mean value of weight change [%] with a standard deviation).

Specimen	0.25 h [%]	1 h [%]	2 h [%]	4 h [%]	8 h [%]	24 h [%]	48 h [%]	72 h [%]	168 h [%]	336 h [%]
CS	270 ± 23	279 ± 36	235 ± 16	207 ± 11	189 ± 10	184 ± 5.0	166 ± 9.0	148 ± 5.0	137 ± 4.0	128 ± 7.0
CS/KGM 95:5	236 ± 82	241 ± 100	198 ± 15	173 ± 10	159 ± 12	141 ± 11	134 ± 9.0	124 ± 7.0	119 ± 20	107 ± 20
CS/KGM 80:20	296 ± 44	269 ± 33	230 ± 17	218 ± 10	189 ± 16	177 ± 8.0	152 ± 13	149 ± 9.0	141 ± 3.0	124 ± 8.0
CS/KGM 50:50	162 ± 21	159 ± 27	149 ± 24	144 ± 8.0	144 ± 10	138 ± 9.0	125 ± 15	123 ± 5.0	120 ± 10	122 ± 12
CS/KGM 20:80	310 ± 184	265 ± 117	256 ± 112	238 ± 97	242 ± 91	255 ± 107	247 ± 108	242 ± 109	233 ± 114	229 ± 101
CS/KGM 5:95	946 ± 178	938 ± 95	886 ± 45	841 ± 54	825 ± 52	765 ± 41	761 ± 35	670 ± 51	636 ± 60	587 ± 24
KGM	-	-	-	-	-	-	-	-	-	-

Swelling and Degradation Properties After 6 Months of Storage

All of the samples that were measured after a 6-month period of storage were characterized by much lower swelling degrees (Table 5). These samples reached the maximum weight at a later time, and the degradation time was delayed in comparison to the films that were examined immediately after being obtained. The weight was also more stable over time. This suggests a cross-linking process during the storage period, which influences the durability and swelling ability of the materials.

Table 5. The results of the swelling and degradation analysis for the chitosan (CS), konjac glucomannan (KGM), and blended films (after six months of storage), obtained over two weeks (presented as a mean value of weight change [%] with a standard deviation).

Specimen	0.25 h [%]	1 h [%]	2 h [%]	4 h [%]	8 h [%]	24 h [%]	48 h [%]	72 h [%]	168 h [%]	336 h [%]
CS	54 ± 12	76 ± 20	78 ± 36	78 ± 17	69 ± 13	69 ± 15	72 ± 26	61 ± 15	61 ± 19	62 ± 25
CS/KGM 95:5	87 ± 16	87 ± 3.0	87 ± 4.0	83 ± 3.0	82 ± 3.0	77 ± 4.0	81 ± 1.0	73 ± 4.0	67 ± 2.0	70 ± 5.0
CS/KGM 80:20	85 ± 4.0	87 ± 5.0	87 ± 1.0	82 ± 3.0	84 ± 5.0	84 ± 5.0	78 ± 4.0	76 ± 4.0	77 ± 1.0	73 ± 3.0
CS/KGM 50:50	99 ± 3.0	89 ± 4.0	99 ± 10	105 ± 8.0	102 ± 7.0	83 ± 2.0	88 ± 5.0	96 ± 7.0	86 ± 5.0	90 ± 4.0
CS/KGM 20:80	187 ± 74	168 ± 30	149 ± 14	135 ± 12	137 ± 16	130 ± 3.0	125 ± 5.0	126 ± 8.0	116 ± 6.0	126 ± 8.0
CS/KGM 5:95	373 ± 22	413 ± 45	431 ± 21	430 ± 22	413 ± 26	398 ± 18	410 ± 18	387 ± 23	379 ± 19	367 ± 17
KGM	-	-	-	-	-	-	-	-	-	-

3.7. Contact Angle and Surface Free Energy

The wettability of the CS/KGM blend films was investigated by the measurement of the materials' contact angle using the sitting drop method. This is a very significant feature, which helps to characterize the surface of the material. It is contingent upon the physical and chemical homogeneity of the surface, in addition to the presence of surface roughness or fouling [42].

The initial films that we investigated, which were based on pure chitosan and pure konjac glucomannan, had different hydrophilic affinities (Table 6). The proportion of these two polymers in the blend films significantly affected the values of the contact angle. The CS sample was more hydrophobic, and the addition of 5% of KGM caused the increase of contact angle for glycerine. Further increasing the proportion of KGM in the sample resulted in a decrease in the wetting angle and an increase in the hydrophilic properties.

Table 6. The values of the contact angle and surface free energy for chitosan (CS), konjac glucomannan (KGM), and blended films.

Specimen	Θ^G	Θ^D	γ_s [mJ/m^2]	$\gamma_s{}^d$ [mJ/m^2]	$\gamma_s{}^p$ [mJ/m^2]
CS	90.08 ± 2.93	47.33 ± 9.46	35.50 ± 5.13	35.23 ± 5.41	0.40 ± 0.1
CS/KGM 95:5	97.82 ± 3.00	64.52 ± 2.48	25.97 ± 1.14	25.62 ± 0.89	0.35 ± 0.25
CS/KGM 80:20	89.33 ± 2.87	52.07 ± 6.40	32.73 ± 3.59	31.94 ± 3.56	0.79 ± 0.04
CS/KGM 50:50	81.83 ± 3.53	29.54 ± 7.94	43.88 ± 2.99	43.12 ± 2.73	0.77 ± 0.26
CS/KGM 20:80	75.19 ± 6.94	39.57 ± 8.63	39.30 ± 4.54	35.86 ± 3.14	3.43 ± 1.40
CS/KGM 5:95	76.11 ± 2.57	43.94 ± 3.34	37.32 ± 1.94	33.76 ± 1.51	3.55 ± 0.45
KGM	56.27 ± 3.74	35.24 ± 1.30	45.35 ± 1.69	35.58 ± 0.20	11.77 ± 1.89

The higher the KGM proportion in the blend, the higher the share of the polar component of the surface free energy was. The highest value of this parameter was observed for the pure KGM film, which suggests that hydrogen bonds, dipole–dipole interactions, and induction forces play a significant role on the film surface [43]. Lowering the proportion of KGM in the blends resulted in a reduced occurrence of these interactions. Wettability is a crucial factor regarding many uses, including dermal applications. The enhanced hydrophilicity of the material is not conducive to optimal dermal applications, given the inherent hydrophobic character of the skin, which is largely attributable to the composition of the stratum corneum [44]. However, in the case of an injury to the skin or an open wound, the appropriate hydrophilicity facilitates a more rapid healing process. Furthermore, increased hydrophilicity is linked to an enhanced capacity to absorb wound exudate and facilitate adequate moisture within the wound environment [45].

4. Discussion

Carbohydrates are the most prevalent biomolecules in the natural environment and were demonstrated to confer a multitude of benefits to human health [46]. Polysaccharide-based materials are of significant value in a multitude of applications pertaining to medicine, food, and cosmetics, largely due to their biodegradability, biocompatibility, and liquid absorption properties [14,47]. In addition to their applications in the aforementioned fields, these macromolecules are also employed in the packaging industry and water purification [48–51]. These biopolymers are also environmentally friendly as they come from sustainable and renewable sources [22,52]. Chitosan and konjac glucomannan are polysaccharides that are no exception in this aspect. Chitosan- and glucomannan-based films possess specific mechanical strength and physicochemical properties that are modifiable. The blends of chitosan and glucomannan used in this study allowed the preparation of biopolymer films with different properties compared to the starting polymers.

Changes in the chemical structure of the blend films observed in this research align with observations made by other researchers. Li et al. observed that the broad absorption band at approximately 3440 cm^{-1} exhibited a shift contingent on the konjac glucomannan content of the blend. Moreover, the IR spectrum demonstrates that the band at approximately 1723 cm^{-1}, which is originally present in konjac glucomannan, disappeared, and there was a shift at the wavenumber 1638 cm^{-1}. The aforementioned alterations indicate augmented intermolecular hydrogen bond formation between the biopolymers [53]. Similar observations were made by Neto et al. [18]. The formation of hydrogen bonds between the -OH and -NH_2 groups from chitosan and the -OH and -$COCH_3$ groups in glucomannan, as evidenced by changes in the absorption bands, is the underlying cause of the miscibility observed between the two polymers [22].

The material's mechanical properties are crucial for its application. For instance, products that are applied to the skin should exhibit the requisite tensile strength values (2.5 to 35 MPa) and degrees of elongation at their breaking point (70–78%), which correspond to the tensile properties of healthy human skin, which can withstand some deformations [18]. In the research conducted by Li et al., the highest value for tensile strength was observed for the sample containing 20% chitosan; a pure KGM film had a significantly higher value for this parameter than a CS film [53]. Ye et al. observed the same tendency during CS/KGM films' mechanical examination [25]. The observations from the two studies are similar to those obtained here. A CS/KGM blend film was also examined by Shang et al.; they observed that the tensile strength of a film made using 50:50 proportions of the polymers was very low [54], which remains contrary to our observations. However, the reagents used in their study had different chemical characteristics compared to our polymers. These differences include their molecular weights and the deacetylation degree of chitosan, which may have affected the final properties of material. Fan et al. obtained the fibers in the spinning process that was based on CS/KGM blends in different proportions. The mechanical tests demonstrated that the incorporation of KGM into CS enhances the tensile strength of the fibers. Furthermore, the mechanical properties of the blend can be optimized by controlling the blend conditions [55]. Chen et al. examined the mechanical properties of the chitosan/konjac glucomannan bilayer, and their observations of the degree of elongation at the breaking point show a similar trend to our research. The pure CS film had the highest value of this parameter. The higher the KGM content in the composite, the lower the percentage of elongation at the breaking point was [56]. Pure KGM film materials tend to have poor mechanical properties, and the degree of elongation at the breaking point is one of the indicators of this condition.

The surface morphology of the CS/KGM blends' films was examined by SEM and AFM techniques. Our observations corroborate the findings reported by Xiao et al., whereby the incorporation of KGM into CS resulted in a reduction in the samples' homogeneity, which may suggest a decline in miscibility. However, the samples with a low content of KGM were characterized by smooth and homogenous surfaces [22]. Similar insights were presented by Zou et al. in their research on composite films made of corn starch and konjac glucomannan. The addition of KGM resulted in a more uneven and granular surface structure [33]. Nair et al. also observed small particles on the film surface and suggested that these constituted less soluble fractions of KGM [57].

The thermal stability of the tested blends and the initial biopolymers was evaluated by a thermogravimetric analysis. We demonstrated that the thermal stability of the films is directly proportional to the KGM content of the blend. Neto et al. obtained similar results. All of the blend films that they tested showed the values of the maximum weight loss temperature as being between those of the initial polymers [18]. The same findings were also presented by Xiao et al. and Xu et al. [22,40]. All of these changes suggest that new hydrogen interactions were established between CS and KGM. What is more, the DTG curves from both this research and the mentioned studies confirmed the good miscibility of CS and KGM over all of the composition ranges.

The swelling properties of materials that are designed for skin applications, such as wound adhesive, are extremely important. Two aspects must be considered: providing ideal moisture conditions and absorbing the exudate from the wound, depending on the wound type [58]. The modification of chitosan film by blending with konjac glucomannan increases the swelling ability of the material, especially when evaluated after a long storage period. The results of a water swelling properties analysis of CS/KGM films performed by Xiao et al. showed higher levels of swelling for all the blended samples than for pure chitosan [22]. What is more, a pure KGM sample was dissolved immediately after immersion in liquid, as took place in our study. Fan et al. evaluated the water-retention properties of a CS/KGM blend fiber; the lowest value was observed for pure CS [55]. In a study conducted by Hua et al., blend films composed of chitosan and konjac glucomannan carboxymethylated form were examined. A content of more than 30% of KGM derivative in a film resulted in a notable enhancement of its swelling ability [59]. Konjac glucomannan has a great water absorption capacity, which can be as much as up to 100 g of water per 1 g of konjac [60]. The high swelling degree of the films with high KGM contents that are presented in our study remains consistent with the existing theory. Sample 5:95 achieved 946%, the highest result of all the samples tested. Wu et al. presented a study of a superabsorbent polymer based on KGM. Their product reached almost the maximum swelling value in the first hour of analysis, which is in line with our research [61]. The hydrophilic character of this polysaccharide is the consequence of the hydroxyl group's occurrence in the backbone, which promotes water uptake by the biomaterial. More recently, Chen et al. prepared bilayers of chitosan and konjac glucomannan in different weight ratios. The sample with an equal ratio of CS and KGM exhibited the lowest swelling properties [56]. Although the system tested was a composite, the observations on it were similar to our blend sample of CS/KGM 50:50. One explanation for this phenomenon may be the high proportion of hydrogen bonds formed between the two polymers, which limits hydrophilic interactions with water molecules [56]. The results of the swelling analysis permitted the assessment of the onset of degradation processes, which commenced, for all of the samples, approximately one hour after the start of the analysis. In a study conducted by Tkongachai et al., a film composed of chitosan and collagen was examined. The greatest degradation level was observed at the beginning of the analysis. Following the third day, the percentage of degradation was observed to be lower and was maintained at a comparable point [62]. Over the course of our study, the degradation process started within the first hours of the test period. Furthermore, after the third day, the reduction in the samples' weight was less pronounced and exhibited greater stability over time.

The contact angle measurements confirmed the hydrophilic character of all of the tested films. The lowest values of the contact angle were observed for pure KGM. The higher the KGM content in the blend samples, the more hydrophilic the character of the film was. Similar insights were observed by Strnad et al.; they examined keratin films enriched with konjac glucomannan. Although the keratin film had hydrophobic properties, the addition of KGM caused the change of this state toward hydrophilic characteristics [63]. Qin et al. performed an investigation on the drying temperature's influence on konjac glucomannan/agar blend films. The measurements of the contact angle for pure KGM were the lowest, which remains consistent with the previous observations. Controlling the drying temperature enabled the modification of the water-spreading properties [64].

The results obtained from the materials tested after a specified period of storage demonstrate that the time since the material was obtained has a discernible impact on its properties. These changes include a reduction in their mechanical properties, as evidenced by a loss of flexibility in the samples. The swelling results also indicate a reduction in fluid absorption, which has implications for the functional characteristics of such materials. Depending on the application, these changes can be both positive and negative. In the case of materials such as wound dressings, where the fit to the skin and the ability to absorb fluids is important, the biopolymer materials from this research would not perform as well after prolonged storage.

Biopolymers constitute a diverse class of macromolecules with potential applications in the cosmetic industry. The group includes polysaccharides, which are of particular interest to the cosmetic industry, primarily due to their safety profile. The application of these compounds does not carry the risk of zoonotic diseases, which is a concern in the case of collagen [65]. Polysaccharides were shown to have a range of beneficial effects on the skin, including moisturizing, antioxidant, and anti-ageing properties [66]. In addition to their safety and efficacy, the naturality and eco-friendliness of these compounds are becoming increasingly important considerations for consumers, producers, and researchers [67,68]. Chitosan, due to its polycationic nature, is capable of interacting with the skin and damaged hair [69]. Additionally, konjac glucomannan has been shown to possess a regenerative potential for the skin [15]. Both biopolymers possess film-forming properties. This characteristic is highly beneficial in the development of cosmetic products. Polymeric films can be simply and effectively applied to the skin, for example as a beauty mask or an eye patch. [15,68,70–72].

The same properties that are advantageous in cosmetic applications are of paramount importance in the development of solutions for medical and tissue engineering purposes, for instance, in wound dressings [19,73–78]. Wound adhesives are the first line in wound management; however, traditional medical materials, such as bandages or gauze, have some disadvantages, like poor springiness. Materials based on biopolymers overcome these limitations [79]. Biopolymers, including those used in this study, have a caring or regenerative effect themselves in the aforementioned areas. However, they are frequently employed as a matrix for the incorporation of active substances or drugs. Zeng et al. developed a hydrogel intended for wound healing, consisting of konjac glucomannan and xanthan gum, containing polydopamine nanoparticles. The results of the in vivo study demonstrated the efficacy of the hydrogel formulations in wound treatment, both with and without nanoparticles [79]. Another example is a study conducted by Jiang et al., in which a chitosan/konjac glucomannan matrix incorporated silver nanoparticles [80]. The studies using chitosan and konjac glucomannan as matrices for active compounds for wound healing are much more prevalent, including investigations of the addition of berberine [81], exosome nanoparticles [82], paeoniflorin and zinc nanoparticles [83], gentamicin [84], stevioside-stabilized honokiol [85], or graphene oxide [86]. In the context of the biomedical and cosmetic applications of biopolymers, the role of hydrogels is a fundamental aspect that cannot be overlooked. The -NH_2 and -OH groups present in chitosan and glucomannan, respectively, allow for modifications to be made, including the formation of spatial networks and the creation of hydrogels [87]. Hydrogels possess the remarkable ability to swell and can also act as a reservoir for an active substance, which is the reason for their extensive utilization in drug delivery systems [88] and wound dressings [21]. The blending of two polymers represents the most straightforward physical method of modifying a polymer matrix in this manner [87]. Konjac glucomannan displays a particular potential for use as a hydrogel, given its high swelling capacity. However, its standalone application is constrained by its poor mechanical properties in solutions [21]. The combination with another polymer, in this case chitosan, markedly enhances this parameter. This offers the possibility of further investigation into the potential of the chitosan/konjac glucomannan combination, for example through the utilization of alternative, more complex cross-linking techniques.

The materials presented in this study are in line with modern requirements for cosmetics, wound dressing, or food packaging. Nevertheless, this is merely a preliminary investigation that requires further expansion in the future. However, it represents an excellent point of departure. Although there are already some studies on chitosan and konjac glucomannan in the areas outlined, their blends in the proportions proposed here have not yet been investigated.

5. Conclusions

We successfully obtained blend films based on two biodegradable and biocompatible polysaccharides—chitosan and konjac glucomannan. The implemented method of film production was simple and low-cost. The test methods used in this study made it possible to assess the miscibility of the biopolymers, which was good in all of the composition ranges. The infrared spectra analysis confirmed the chemical structure of pure polymers and enabled the evaluation of any changes found in blend films. It can be posited that the formation of hydrogen bonds was facilitated by the involvement of the -OH, -COO-, -NH, and $COCH_3$ groups. The mechanical properties of the blend samples were modified in comparison to the initial polymers' films. Generally, the addition of KGM to CS increased the structural integrity of the samples. The highest tensile strength value was exhibited by the CS/KGM 20:80 sample. SEM and AFM images showed that the higher the KGM content in the blend composition, the less the homogeneity of the film's surface. The highest roughness was shown by the CS/KGM 80:20 sample. The KGM content in the biopolymeric films increased their thermal stability. The swelling ability of the tested samples was significantly increased by higher KGM concentrations within the blends. The contact angle result revealed the high impact of the KGM content on the hydrophilicity of the samples. In this research, we also evaluated the effects of the passage of time on some of the films' parameters. Such an assessment is extremely important from the point of view of product storage. The infrared spectra showed changes in band intensity and some shifts compared to the previously examined samples. These changes could be connected to water loss, as well as the cross-linking processes. The results of the mechanical parameters assessment and the result of the swelling degree analysis reflect the changes that occurred in the chemical composition of the films.

We can conclude that films based on a combination of chitosan and konjac glucomannan are promising for use as matrices for many applications. The shape of the film predisposes the material to surface and packaging applications. Some aspects of the first-mentioned usage have been taken into account when investigating the properties of the obtained materials. In particular, the swelling and AFM results suggest that the addition of KGM to the samples improves their application properties. Some of the blending ratios proposed in this study and the testing of the properties of the samples after a certain storage period are new approaches. This study contributes to the current research on biomaterials for surface applications, specifically on materials based on the chitosan/konjac glucomannan combination. These biopolymer blend materials can be further modified; this is a prelude to further work in this area.

Author Contributions: Conceptualization, A.S. and K.K.-K.; methodology, K.K.-K. and A.S.; software, K.K.-K.; validation, A.S. and K.K.-K.; formal analysis, A.S. and K.K.-K.; investigation, K.K.-K.; resources, A.S. and K.K.-K.; data curation, K.K.-K.; writing—original draft preparation, K.K.-K.; writing—review and editing, A.S.; visualization, K.K.-K.; supervision, A.S.; project administration, A.S.; funding acquisition, A.S. and K.K.-K. All authors have read and agreed to the published version of the manuscript.

Funding: This research received no external funding.

Institutional Review Board Statement: Not applicable.

Data Availability Statement: The original contributions presented in the study are included in the article; further inquiries can be directed to the corresponding author.

Conflicts of Interest: The authors declare no conflicts of interest.

References

1. Barnawal, J.; Barse, B.; Fais, A.; Delogu, G.L.; Kumar, A. Biopolymer: A Sustainable Material for Food and Medical Applications. *Polymers* **2022**, *14*, 983. [CrossRef] [PubMed]
2. Das, A.; Ringu, T.; Ghosh, S.; Pramanik, N.A. Comprehensive review on recent advances in preparation, physicochemical applications of biopolymers. *Polym. Bull.* **2023**, *80*, 7249–7312. [CrossRef]
3. Grabska-Zielińska, S.; Sionkowska, A. How to Improve Physico-Chemical Properties of Silk Fibroin Materials for Biomedical Applications?—Blending and Cross-linking of Silk Fibroin—A Review. *Materials* **2021**, *14*, 1510. [CrossRef] [PubMed]
4. Rogovina, S.Z.; Vikhoreva, G.A. Polysaccharide-based polymer blends: Method of their production. *Glycoconj. J.* **2006**, *23*, 611–618. [CrossRef] [PubMed]
5. Daud, H.; Ghani, A.; Iqbal, D.N.; Ahmad, N.; Nazir, S.; Muhammad, M.J.; Hussain, E.A.; Nazir, A.; Iqbal, M. Preparation and characterization of guar gum based biopolymeric hydrogels for controlled release of antihypertensive drug. *Arab. J. Chem.* **2021**, *14*, 103111. [CrossRef]
6. Feijoo, P.; Samaniego-Aguilar, K.; Sanchez-Safont, E.; Torres-Giner, S.; Lagaron, J.M.; Gamez-Perez, J.; Cabedo, L. Development and Characterization of Fully Renewable and Biodegradable Polyhydroxyalkanoate Blends with Improved Thermoformability. *Polymers* **2022**, *14*, 2527. [CrossRef]
7. Ilyas, R.A.; Aisyah, H.A.; Nordin, A.H.; Ngadi, N.; Zuhri, M.Y.M.; Asyraf, M.R.M. Natural-Fiber-Reinforced Chitosan, Chitosan Blends and Their Nanocomposites for Various Advances Applications. *Polymers* **2022**, *14*, 874. [CrossRef]
8. Mitura, S.; Sionkowska, A.; Jaiswal, A. Biopolymers for hydrogel in cosmetics: Review. *J. Mater. Sci. Mater. Med.* **2020**, *31*, 50. [CrossRef]
9. Kedir, W.M.; Abdi, G.F.; Goro, M.M.; Tolesa, L.D. Pharmaceutical and drug delivery applications of chitosan biopolymer and its modified nanocomposite: A review. *Heliyon* **2022**, *8*, e10196. [CrossRef]
10. Li, J.; Zhuang, S. Antibacterial activity of chitosan and its derivatives and their interaction mechanism with bacteria: Current state and perspectives. *Eur. Polym. J.* **2020**, *138*, 109984. [CrossRef]
11. Muthu, M.; Gopal, J.; Chun, S.; Devadoss, A.J.P.; Hasan, N.; Sivanesan, I. Crustean Waste-Derived Chitosan: Antioxidant Properties and Future Perspective. *Antioxidants* **2021**, *10*, 228. [CrossRef] [PubMed]
12. Pellis, A.; Guebits, G.M.; Myanhongo, G.S. Chitosan: Sources, Processing and Modification Techniques. *Gels* **2022**, *8*, 393. [CrossRef] [PubMed]
13. Kulka-Kamińska, K.; Sionkowska, A. Chitosan Based Materials in Cosmetic Applications: A Review. *Molecules* **2023**, *28*, 1817. [CrossRef] [PubMed]
14. Seidi, F.; Yazdi, M.K.; Jouyandeh, M.; Dominic, M.; Naeim, H.; Nezhad, M.N.; Bagheri, B.; Habibzadeh, S.; Zarrintaj, P.; Saeb, M.R.; et al. Chitosan-based blends for biomedical applications. *Int. J. Biol. Macromol.* **2021**, *183*, 1818–1850. [CrossRef]
15. Choi, K.H.; Kim, S.T.; Bin, B.H.; Park, P.J. Effect of Konjac Glucomannan (KGM) on the Reconstitution of the Dermal Environment against UVB-Induced Condition. *Nutrients* **2020**, *12*, 2779. [CrossRef]
16. Song, L.; Xie, W.; Zhao, Y.; Lv, X.; Yang, H.; Zeng, Q.; Zheng, Z.; Yang, X. Synthesis, Antimicrobial, moisture Absorption and Retention Activities of Kojic Acid-Grafted Konjac Glucomannan Oligosaccharides. *Polymers* **2019**, *11*, 1979. [CrossRef]
17. Zhou, Y.; Jiang, R.; Perkins, W.S.; Cheng, Y. Morphology evaluation and gelation mechanism of alkali induced konjac glucomannan hydrogel. *Food Chem.* **2019**, *269*, 80–88. [CrossRef]
18. Neto, R.J.G.; Genevro, G.M.; Paulo, L.A.; Lopes, P.S.; de Moraes, M.A.; Beppu, M.M. Characterization and in vitro evaluation of chitosan/konjac glucomannan bilayer film as a wound dressing. *Carbohydr. Polym.* **2019**, *212*, 59–66. [CrossRef]
19. Sun, Y.; Xu, X.; Zhang, Q.; Zhang, D.; Xie, X.; Zhou, H.; Wu, Z.; Liu, R.; Pang, J. Review of Konjac Glucomannan Structure, Properties, Gelation Mechanism, and Application in Medical Biology. *Polymer* **2023**, *15*, 1852. [CrossRef]

20. Da Silva, D.F.; Ferreira, S.B.; Bruschi, M.L.; Britten, M.; Matumoto-Pintro, P.T. Effect of commercial konjac glucomannan and konjac flours on textural, rheological and microstructural properties of low fat processed cheese. *Food Hydrocoll.* **2016**, *60*, 308–316. [CrossRef]
21. Alves, A.; Miguel, S.P.; Araujo, A.R.T.S.; de Jesús Valle, M.J.; Navarro, A.S.; Correia, I.J.; Ribeiro, M.P.; Coutinho, P. Xanthan Gum-Konjac Glucomannan Blend Hydrogel for Wound Healing. *Polymers* **2020**, *12*, 99. [CrossRef]
22. Xiao, C.; Gao, S.; Wang, H.; Zhang, L. Blend Film from Chitosan and Konjac Glucomannan Solutions. *J. Appl. Polym. Sci.* **2000**, *76*, 506–515. [CrossRef]
23. Alonso-Sande, M.; Cuna, M.; Remunan-Lopez, C.; Teijeiro-Osorio, D.; Alonso-Lebrero, J.L.; Alonso, M.J. Formation of New Glucomannan-Chitosan Nanoparticles and Study of Their Ability to Associate and Deliver Proteins. *Macromolecules* **2006**, *39*, 4152–4158. [CrossRef]
24. Lin, W.; Li, Q.; Zhu, T. New chitosan/Konjac glucomannan blending membrane for application in pervaporation dehydration of caprolactam solution. *J. Ind. Eng. Chem.* **2012**, *18*, 934–940. [CrossRef]
25. Ye, X.; Kennedy, J.F.; Li, B.; Xie, B.J. Condensed state structure and biocompatibility of the konjac glucomannan blend films. *Carbohydr. Polym.* **2006**, *64*, 532–538. [CrossRef]
26. Li, B.; Li, J.; Xia, J.; Kennedy, J.F.; Yie, X.; Liu, T.G. Effect of gamma irradiation on the condensed state structure and mechanical properties of konjac glucomannan/chitosan blend films. *Carbohydr. Polym.* **2011**, *83*, 44–51. [CrossRef]
27. Sionkowska, A.; Michalska-Sionkowska, M.; Walczak, M. Preparation and characterization of collagen/hyaluronic acid/chitosan film crosslinked with dialdehyde starch. *Int. J. Biol. Macromol.* **2020**, *149*, 290–295. [CrossRef] [PubMed]
28. Kulka, K.; Szmejkowska, A.; Sionkowska, A.; Wypij, M.; Golińska, P. Materials based on chitosan enriched with zinc nanoparticles for potential applications on the skin. *Eng. Biomater.* **2022**, *167*, 24–31. [CrossRef]
29. Szulc, M.; Lewandowska, K. Characterization of Chitosan Films Modified Using Caffeic Acid and a Neutralization Process. *Materials* **2023**, *16*, 5038. [CrossRef]
30. Da Silva, D.F.; Ogawa, C.Y.L.; Sato, F.; Neto, A.M.; Larsen, F.H.; Matumoto-Pintro, P.T. Chemical and physical characterization of Konjac glucomannan-based powders by FTIR and 13C MAS NMR. *Powder Technol.* **2020**, *361*, 610–616. [CrossRef]
31. Jena, K.; Ananta, S.; Akthar, J.; Patnaik, A.; Das, S.; Singh, J.; Sathyanarayana, K.; Kar, P.K.; Das, B.K.; Hassan, A.; et al. Physical, biochemical and antimicrobial characterization of chitosan prepared from tasar silkworm pupae waste. *Environ. Technol. Innov.* **2023**, *31*, 103200. [CrossRef]
32. Jakubowska, E.; Gierszewska, M.; Szydłowska-Czerniak, A.; Nowaczyk, J.; Olewnik-Kruszkowska, E. Development and characterization of active packaging films based on chitosan, plasticizer, and quercetin for repassed oil storage. *Food Chem.* **2023**, *399*, 135617. [CrossRef] [PubMed]
33. Zou, Y.; Yuan, C.; Cui, B.; Liu, P.; Wu, Z.; Zhao, H. Formation of high amylose corn starch/konjac glucomannan composite film with improved mechanical and barrier properties. *Carbohydr. Polym.* **2021**, *251*, 117039. [CrossRef] [PubMed]
34. Nguyen, T.T.; Thi Dao, U.T.; Thi Bui, Q.P.; Bach, G.L.; Ha Thuc, C.N.; Ha Thuc, H. Enhanced antimicrobial activities and physicochemical properties of edible film based on chitosan incorporated with *Sonneratia caseolaris* (L.) Engl. Leaf extract. *Prog. Org. Coat.* **2020**, *140*, 105487. [CrossRef]
35. Queiroz, M.F.; Melo, K.R.T.; Sabry, D.A.; Sassaki, G.L.; Rocha, H.A.O. Does the Use of Chitosan Contribute to Oxalate Kidney Stone Formation. *Mar. Drugs* **2015**, *13*, 141–158. [CrossRef]
36. Yilmaz, P.; Demirhan, E.; Ozbek, B. Development of *Ficus carica* Linn leaves extract incorporated chitosan films for active food packaging materials and investigation of their properties. *Food Biosci.* **2022**, *46*, 101542. [CrossRef]
37. Hong, T.; Yin, J.-Y.; Nie, S.-P.; Xie, M.-Y. Applications of infrared spectroscopy in polysaccharide structural analysis: Progress, challenge and perspective. *Food Chem. X* **2021**, *12*, 100168. [CrossRef]
38. Lin, W.; Ni, Y.; Wang, L.; Liu, D.; Pang, J. Physicochemical properties of degraded konjac glucomannan prepared by laser assisted with hydrogen peroxide. *Int. J. Biol. Macromol.* **2019**, *129*, 78–83. [CrossRef]
39. Ma, F.; Wang, Z.; Zhao, H.; Tian, S. Plasma Depolymerization of Chitosan in the Presence of Hydrogen Peroxide. *Int. J. Mol. Sci.* **2012**, *13*, 7788–7797. [CrossRef]
40. Xu, X.; Li, B.; Kennedy, J.F.; Xie, B.J.; Huang, M. Characterization of konjac glucomannan-gellan gum blend films and their suitability for release of nisin incorporated therein. *Carbohydr. Polym.* **2007**, *70*, 192–197. [CrossRef]
41. Fischer, S.C.L.; Boyadzhieva, S.; Hensel, R.; Kruttwig, K.; Arzt, E. Adhesion and relaxation of a soft elastomer on surfaces with skin like roughness. *J. Mech. Behav. Biomed. Mater.* **2018**, *80*, 303–310. [CrossRef] [PubMed]
42. Jagodzińska, S.; Pałys, B.; Wawro, D. Effect of chitosan film Surface structure on the contact angle. *Prog. Chem. Appl. Chitin Deriv.* **2021**, *26*, 89–100. [CrossRef]
43. Kristó, K.; Módra, S.; Hornok, V.; Süvegh, K.; Ludasi, K.; Aigner, Z.; Kelemen, A.; Sovány, T.; Pintye-Hódi, K.; Regdon, G. Investigation of Surface Properties and Free Volumes of Chitosan-Based Buccal Mucoadhesive Drug Delivery Films Containing Ascorbic Acid. *Pharmaceutics* **2022**, *14*, 345. [CrossRef] [PubMed]
44. Adamiak, K.; Sionkowska, A. The influence of UV irradiation on fish skin collagen films in the presence of xanthohumol and propanediol. *Spectrochim. Acta Part A Mol. Biomol. Spectrosc.* **2022**, *282*, 121652. [CrossRef] [PubMed]
45. Choundhary, P.; Ramalingam, B.; Das, S.K. Rational design of antimicrobial peptide conjugated graphene-silver nanoparticle loaded chitosan wound dressing. *Int. J. Biol. Macromol.* **2023**, *246*, 125347. [CrossRef]

46. Ruocco, N.; Costantini, S.; Guariniello, S.; Costantini, M. Polysaccharides from the Marine Environment with Pharmacological, Cosmeceutical and Nutriceutical Potential. *Molecules* **2016**, *21*, 551. [CrossRef]
47. Yang, J.; Wang, S. Polysaccharide-Based Multifunctional Hydrogel Bio-Adhesives for Wound Healing: A Review. *Gels* **2023**, *9*, 138. [CrossRef]
48. Jia, D.; Fang, Y.; Yao, K. Water vapor and mechanical properties of konjac glucomannan-chitosan-soy protein isolate edible films. *Food Bioprod. Process.* **2009**, *87*, 7–10. [CrossRef]
49. Ni, Y.; Liu, Y.; Zhang, W.; Shi, S.; Zhu, W.; Wang, R. Advanced konjac glucomannan-based films in food packaging: Classification, preparation, formation mechanism and function. *LWT—Food Sci. Technol.* **2021**, *152*, 112338. [CrossRef]
50. Stefanowska, K.; Woźniak, M.; Dobrucka, R.; Ratajczak, I. Chitosan with Natural Additives as a Potential Food Packaging. *Materials* **2023**, *16*, 1579. [CrossRef]
51. Qi, X.; Tong, X.; Pan, W.; Zeng, Q.; You, S.; Shen, J. Recent advances in polysaccharide-based adsorbents for wastewater treatment. *J. Clean. Prod.* **2021**, *315*, 128221. [CrossRef]
52. Flórez, M.; Guerra-Rodríguez, E.; Cazón, P.; Vázquez, M. Chitosan for food packaging: Recent advances in active and intelligent films. *Food Hydrocoll.* **2022**, *124*, 107328. [CrossRef]
53. Li, B.; Kennedy, J.F.; Peng, J.L.; Yie, X.; Xie, B.J. Preparation and performance evaluation of glucomannan-chitosan-nisin antimicrobial blend film. *Carbohydr. Polym.* **2006**, *65*, 488–494. [CrossRef]
54. Shang, J.; Liu, H.; Qi, C.; Chen, X.; Guo, K. Physical, mechanical properties, and structural characterization of konjac glucomannan-chitosan-polypeptide adhesive blends. *J. Adhes. Sci. Technol.* **2015**, *29*, 2334–2344. [CrossRef]
55. Fan, L.; Zheng, H.; Xu, J.; Huang, J.; Zhang, C. Preparation and properties of Chitosan/Konjac Glucomannan Blend Fibers. *J. Macromol. Sci.* **2007**, *44*, 439–443. [CrossRef]
56. Chen, Y.; Wang, S.; Yang, C.; Zhang, L.; Li, Z.; Jiang, S.; Bai, R.; Ye, X.; Ding, W. Chitosan/konjac glucomannan bilayer films: Physical, structural, and thermal properties. *Int. J. Biol. Macromol.* **2024**, *257*, 128660. [CrossRef]
57. Nair, S.B.; Alummoottil, N.J.; Moothandasserry, S.S. Chitosan-konjac glucomannan cassava starch-nanosilver composite films with moisture resistant and antimicrobial properties for food packaging applications. *Starch/Starke* **2017**, *69*, 1600210. [CrossRef]
58. Ghomi, E.R.; Khalili, S.; Khorsani, S.N.; Neisiany, R.E.; Ramakrishna, S. Wound dressings: Current advances and future directions. *J. Appl. Polym. Sci.* **2019**, *136*, 47738. [CrossRef]
59. Hua, Z.; Yu-min, D. Preparation and Characterization of Chitosan/Carboxymethylated Konjac Glucomannan Blend Films. *Wuhan Univ. J. Nat. Sci.* **2002**, *7*, 107–112. [CrossRef]
60. Koroskenyi, B.; McCarthy, S.P. Synthesis of Acetylated Konjac Glucomannan and Effect of Degree of Acetylation on Water Absorbency. *Biomacromolecules* **2001**, *2*, 824–826. [CrossRef]
61. Wu, J.; Deng, X.; Lin, X. Swelling characteristics of konjac glucomannan superabsorbent synthesized by radiation-induced graft copolymerization. *Radiat. Phys. Chem.* **2013**, *83*, 90–97. [CrossRef]
62. Tkongchai, K.; Chuysinuan, P.; Thanyacharoen, T.; Techasakul, S.; Ummartyotin, S. Integration of collagen into chitosan blend film composites physicochemical property aspects for pharmaceutical materials. *SN Appl. Sci.* **2020**, *2*, 255. [CrossRef]
63. Strnad, S.; Oberhollenzer, Z.; Šauperl, O.; Kreže, T.; Zemljič, L.F. Modifying properties of father keratin bioplastic films using konjac glucomannan. *Cell. Chem. Technol.* **2019**, *53*, 1017–1027. [CrossRef]
64. Qin, J.; Xiao, M.; Wang, S.; Peng, C.; Wu, X.; Jiang, F. Effect of drying temperature on microstructural, mechanical, and water barrier properties of konjac glucomannan/agar film produced at industrial scale. *LWT—Food Sci. Technol.* **2023**, *173*, 114275. [CrossRef]
65. Silvipriya, K.S.; Kumar, K.K.; Bhat, A.R.; Kumar, D.B.; John, A.; Iakshamanan, P. Collagen: Animal Sources and Biomedical Application. *J. Appl. Pharm. Sci.* **2015**, *5*, 123–127. [CrossRef]
66. Wu, Q.; Cheng, N.; Fang, D.; Wang, H.; Rahman, F.-U.; Hao, H.; Zhang, Y. Recent advances on application of polysaccharides in cosmetics. *J. Dermatol. Sci. Cosm. Technol.* **2024**, *1*, 100004. [CrossRef]
67. Santos, V.P.; Marques, N.S.S.; Maia, P.C.S.V.; Lima, M.A.B.; Franco, L.O.; Campos-Takaki, G.M. Seafood Waste as Attractive Source of Chitin and Chitosan Production and Their Applications. *Int. J. Mol. Sci.* **2020**, *21*, 4290. [CrossRef]
68. Afonso, C.R.; Hirano, R.S.; Gaspar, A.L.; Chagas, E.G.L.; Carvalho, R.A.; Silva, F.V.; Leonardi, G.; Lopes, P.; Silva, C.; Yoshida, C. Biodegradable antioxidant chitosan films useful as an anti-aging skin mask. *Int. J. Biol. Macromol.* **2019**, *132*, 1262–1273. [CrossRef]
69. Guzman, E.; Ortega, F.; Rubio, G.R. Chitosan: A Promising Multifunctional Cosmetic Ingredient for Skin and Hair Care. *Cosmetics* **2022**, *9*, 99. [CrossRef]
70. Gaspar, A.L.; Gaspar, A.B.; Contini, L.R.F.; Silva, M.F.; Chagas, E.G.L.; Bahu, J.O.; Concha, V.O.; Carvalho, R.A.; Severino, P.; Souto, E.B.; et al. Lemongrass (*Cymbopogon citratus*)-incorporated chitosan bioactive films for potential skincare applications. *Int. J. Phram.* **2022**, *628*, 122301. [CrossRef]
71. Sionkowska, A.; Lewandowska, K.; Kurzawa, M. Chitosan-Based Films Containing Rutin for Potential Cosmetic Applications. *Polymers* **2023**, *15*, 3224. [CrossRef] [PubMed]
72. Kodolova-Chukontseva, V.V.; Rozova, E.Y.; Dresvyanina, E.N.; Naschekina, Y.A.; Dobrovol'skaya, I.P.; Vlasova, E.N.; Bystrov, S.G.; Popova, E.N.; Maslennikova, T.P.; Yudin, V.E.; et al. New Composite Materials Based on Chitosan Films Reinforced with Chitin Nanofibrils for Cosmetic Applications. *Cosmetics* **2023**, *10*, 51. [CrossRef]
73. Zhou, N.; Zheng, S.; Xie, W.; Cao, G.; Wang, L.; Pang, J. Konjac glucomannan: A review of structure, physicochemical properties, and wound dressing applications. *J. Appl. Polym. Sci.* **2022**, *139*, e51780. [CrossRef]

74. Li, H.; Liang, X.; Chen, Y.; Liu, K.; Fu, X.; Zhang, C.; Wang, X.; Yang, J. Synergy of antioxidant and M2 polarization in polyphenol-modified konjac glucomannan dressing for remodelling wound healing microenvironement. *Bioeng. Transl. Med.* **2023**, *8*, e10398. [CrossRef]
75. Huang, Y.-C.; Chu, H.-W.; Huang, C.-C.; Wu, W.-C.; Tsai, J.-S. Alkali-treated konjac glucomannan film as a novel wound dressing. *Carbohydr. Polym.* **2015**, *117*, 778–787. [CrossRef]
76. Genesi, B.P.; Barbosa, R.M.; Severino, P.; Rodas, A.C.D.; Yoshida, C.M.P.; Mathor, M.B. Aloe vera copaiba oleoresin-loaded chitosan films for wound dressings: Microbial permeation, cytotoxicity, and in vivo proof of concept. *Int. J. Pharm.* **2023**, *634*, 122648. [CrossRef]
77. Colobiatu, L.; Gavan, A.; Potarniche, A.-V.; Rus, V.; Diaconeasa, Z.; Mocan, A.; Tomuta, I.; Mirel, S.; Mihaiu, M. Evaluation of bioactive compounds-loaded chitosan films as a novel and potential diabetic wound dressing material. *React. Funct. Polym.* **2019**, *145*, 104369. [CrossRef]
78. Eulalio, H.Y.C.; Vieira, M.; Fideles, T.B.; Tomas, H.; Silva, S.M.L.; Peniche, C.A.; Fook, M.V.L. Physicochemical Properties and Cell Viability of Shrimp Chitosan Films as Affected by Film Casting Solvents. I-Potential Use as Wound Dressing. *Materials* **2020**, *13*, 5005. [CrossRef]
79. Zeng, Q.; Qian, Y.; Huang, Y.; Ding, F.; Qi, W.; Shen, J. Polydopamine nanoparticle-dotted food gum hydrogel with excellent antibacterial activity and rapid shape adaptability for accelerated bacteria-infected wound healing. *Bioact. Mater.* **2021**, *6*, 2647–2657. [CrossRef]
80. Jiang, J.; Huang, J.; Wu, X.; Ren, Y.; Li, Z.; Ren, J. Controlled release of silver ions from AgNPs using a hydrogel based on konjac glucomannan and chitosan for infected wounds. *Int. J. Biol. Macromol.* **2020**, *149*, 148–157. [CrossRef]
81. Yang, P.; Ju, Y.; Liu, X.; Li, Z.; Liu, H.; Yang, M.; Chen, X.; Lei, L.; Fang, B. Natural self-healing injectable hydrogels loaded with exosomes and berberine for infected wound healing. *Mater Today Bio* **2023**, *23*, 100875. [CrossRef] [PubMed]
82. Qin, D.; Zhang, A.; Wang, N.; Yao, Y.; Chen, X.; Liu, Y. Hydroxybutyl chitosan/oxidized glucomannan self-healing hydrogels as BMSCs-derived exosomes carriers for advanced stretchable wounds. *Appl. Mater. Today* **2022**, *26*, 101342. [CrossRef]
83. Che, L.; Guo, Y.; Chen, L.; Hu, K.; Ruan, L.; Li, P.; Cai, X.; Li, B.; Shou, Q.; Jiang, G. Injectable Zn^{2+} and Paeoniflorin Release Hydrogel for Promoting Wound Healing. *ACS Appl. Bio Mater.* **2023**, *6*, 2184–2195. [CrossRef]
84. Fan, L.; Cheng, C.; Qiao, Y.; Li, F.; Li, W.; Wu, H.; Ren, B. GNPs-CS/KGM as Hemostatic First Aid Wound Dressing with Antibiotic Effect: In Vitro and In Vivo Study. *PLoS ONE* **2013**, *8*, e66890. [CrossRef]
85. Xu, S.; Yan, S.; You, J.; Wu, X. Antibacterial Micelles-Loaded Carboxymethyl Chitosan/Oxidized Konjac Glucomannan Composite Hydrogels for Enhanced Wound Repairing. *ACS Appl. Mater. Interfaces* **2024**, *16*, 13563–13572. [CrossRef] [PubMed]
86. Fan, L.; Yi, J.; Tong, J.; Zhou, X.; Ge, H.; Zou, S.; Wen, H.; Nie, M. Preparation and characterization of oxidized konjac glucomannan/carboxymethyl chitosan/graphene oxide hydrogel. *Int. J. Biol. Macromol.* **2016**, *91*, 358–367. [CrossRef]
87. Chellathurai, M.S.; Chung, L.Y.; Hilles, A.R.; Sofian, Z.M.; Singha, S.; Ghosal, K.; Mahmood, S. Pharmaceutical chitosan hydrogels: A review on its design and applications. *Int. J. Biol. Macromol.* **2024**, *280*, 135775. [CrossRef]
88. Tian, B.; Liu, J. Smart stimuli-responsive chitosan hydrogel for drug delivery: A review. *Int. J. Biol. Macromol.* **2023**, *235*, 123902. [CrossRef]

polymers

MDPI

Article

The Absence of Phasins PhbP2 and PhbP3 in *Azotobacter vinelandii* Determines the Growth and Poly-3-hydroxybutyrate Synthesis

Claudia Aguirre-Zapata [1], Daniel Segura [2], Jessica Ruiz [2], Enrique Galindo [1], Andrés Pérez [3], Alvaro Díaz-Barrera [3] and Carlos Peña [1,*]

1 Departamento de Ingeniería Celular y Biocatálisis, Instituto de Biotecnología, Universidad Nacional Autónoma de México (UNAM), Av. Universidad 2001, Col. Chamilpa, Cuernavaca 62210, Morelos, Mexico; claudia.aguirre@ibt.unam.mx (C.A.-Z.); enrique.galindo@ibt.unam.mx (E.G.)

2 Departamento de Microbiología Molecular, Instituto de Biotecnología, Universidad Nacional Autónoma de México (UNAM), Av. Universidad 2001, Col. Chamilpa, Cuernavaca 62210, Morelos, Mexico; daniel.segura@ibt.unam.mx (D.S.); jessica.ruiz@ibt.unam.mx (J.R.)

3 Escuela de Ingeniería Bioquímica, Pontificia Universidad Católica de Valparaíso, Av. Brasil 2147 Casilla 4059, Valparaíso 2340025, Chile; aaperezb17@gmail.com (A.P.); alvaro.diaz@pucv.cl (A.D.-B.)

* Correspondence: carlos.pena@ibt.unam.mx

Abstract: Phasins are proteins located on the surface of poly-3-hydroxybutyrate (P3HB) granules that affect the metabolism of the polymer, the size and number of the granules, and some also have stress-protecting and growth-promoting effects. This study evaluated the effect of inactivating two new phasins (PhbP2 or PhbP3) on the cellular growth, production, and molecular mass of P3HB in cultures under low or high oxygen transfer rates (OTR). The results revealed that under high OTR_{max} conditions (between 8.1 and 8.9 mmol L^{-1} h^{-1}), the absence of phasins PhbP2 and PhbP3 resulted in a strong negative effect on the growth rate; in contrast, the rates of specific oxygen consumption increased in both cases. This behavior was not observed under a low oxygen transfer rate (3.9 ± 0.71 mol L^{-1} h^{-1}), where cellular growth and oxygen consumption were the same for the different strains evaluated. It was observed that at high OTR, the absence of PhbP3 affected the production of P3HB, decreasing it by 30% at the end of cultivation. In contrast, the molecular weight remained constant over time. In summary, the absence of phasin PhbP3 significantly impacted the growth rate and polymer synthesis, particularly at high maximum oxygen transfer rates (OTR_{max}).

Keywords: poly-3-hydroxybutyrate; molecular mass; phasins; *Azotobacter vinelandii*

Citation: Aguirre-Zapata, C.; Segura, D.; Ruiz, J.; Galindo, E.; Pérez, A.; Díaz-Barrera, A.; Peña, C. The Absence of Phasins PhbP2 and PhbP3 in *Azotobacter vinelandii* Determines the Growth and Poly-3-hydroxybutyrate Synthesis. *Polymers* **2024**, *16*, 2897. https://doi.org/10.3390/polym16202897

Academic Editors: Vishal Gavande, Vasi Shaikh and Won-Ki Lee

Received: 23 August 2024
Revised: 9 October 2024
Accepted: 12 October 2024
Published: 15 October 2024

1. Introduction

Poly-3-hydroxybutyrate (P3HB) is a material recognized as a bioplastic material that belongs to the polyester family of polyhydroxyalkanoates (PHAs). P3HB shares thermomechanical properties with currently used petrochemical plastics, but it is biodegradable and biocompatible [1]. Therefore, P3HB is a material utilized in diverse areas, like biomedical, pharmaceutical, veterinary, food packaging, and cosmetics [2,3].

Many species of bacteria use different carbon sources to accumulate P3HB as a natural mechanism for carbon and energy storage [4]. *Azotobacter vinelandii* is an interesting P3HB production model, able to synthesize this polymer under conditions of nutritional limitation, either of phosphates or mainly of oxygen and using different substrates, such as cane molasses, glucose, sucrose, and compostable waste [5].

P3HB is stored in granules, also named "carbonosomes" due to their complexity acquired through the binding of diverse GAPs (granule-associated proteins), which are P3HB synthases, P3HB depolymerases, regulatory proteins (like PhaR), and phasins [6,7].

Phasins are non-catalytic proteins (10–24 kDa), and are the major group of 2GAPs on the P3HB granule surface. Phasins are very important in controlling the size, number, and distribution of the granules in the cells, in addition to regulating the P3HB synthase and P3HB depolymerase activities [7–9].

It has been observed that the overexpression of some phasin genes is related to increased accumulation of P3HB [8–11]. Likewise, it is known that phasins can also contribute to the regulation of P3HB depolymerization. Phasin PhaP could function as an activator of the soluble PHB depolymerase of *R. rubrum*; surprisingly, PhaP is also a thermoresistant and chemoresistant protein [10]. On the other hand, it was suggested that in *Cupriavidus necator*, phasins modulate the activity of P3HB depolymerase Z1 and Z2, respectively [12].

One of the hypotheses that explains the relationship between phasins and the modulation of the activity of synthases and depolymerases is that, since the phasins are found on the surface of the granules, they interact with P3HB synthases and P3HB depolymerases, causing allosteric changes that modify the activity or binding to the PHB granule [13,14].

On the other hand, it has been proposed that phasins can act as chaperones that help the cell relieve different types of stresses, such as thermal shocks, halophilic environments, solvents and oxidative stress [7,15,16].

A. vinelandii is a P3HB producer bacterium. Three different phasin proteins have been identified in this microorganism. The major protein in the P3HB granules of this bacterium is PhbP1. This protein is involved in the determination of size and number of granules [17]. The second phasin PhbP2 [18], belongs to the same family of phasins as PhbP1. Both proteins have at the same position, a domain Phasin_2 (pfam09361; TIGR01841) that is characteristic of this group of phasins. PhbP1 and PhbP2 phasins are similar in length (186 aa, Mw 20422.82; 179 aa, Mw 19187.69, respectively), although their amino acid sequences are only 31.4% identical. With respect to PhbP3, this protein is also small (203 aa, Theoretical Mw 21789.22), and is also present on the P3HB granules [17]. PhbP3 is not homologous to PhbP1 and PhbP2 (only 15.7% identical to PhbP1 and 20.8% to PhbP2) and it has no significant homology either with other proteins recognized as phasins in other organisms. No conserved domains could be found in its sequence and clear orthologous proteins were found only in *Azotobacter* spp, where they are annotated as hypothetical. However, it is interesting to note that the three phasins have hydrophobic regions in their sequence, as other phasin proteins have, that could be involved in their binding to the P3HB granules. Although their sequences are not very similar, they share some structural characteristics, like regions predominantly of alpha-helical arrangements, with some regions predicted to be disordered or unstructured, which is another characteristic of several phasins. This structural flexibility suggests that it could interact with different targets, including misfolded proteins [15]. It is known that the inactivation of PhbP2 phasin promotes the synthesis of P3HB, and that inactivation of PhbP3 phasin reduces P3HB production in *A. vinelandii* in cultures in shaken flasks in PYS-rich medium [19,20]. However, the mechanism by which PhbP2 and PhbP3 phasins are involved in the regulation of P3HB synthesis and degradation has not been described in detail.

On the other hand, it has been reported that the increase in the agitation rate in bioreactors, and therefore in OTR, can generate an effect known as "sublethal damage", where it is proposed that the high OTR results in a condition of oxidative stress [21,22]. OTR is an important factor in the operation and optimization of biopolymer production. In microorganisms like *A. vinelandii*, a thorough understanding of OTR and culture under oxygen-limited conditions is crucial for P3HB production [23].

Taking the above into consideration, the present study aimed to evaluate the effect of inactivating phasins PhbP2 or PhbP3 on the growth rate, production, and molecular mass of P3HB in cultures at low and high OTR, to better understand their role in the P3HB metabolism of *A. vinelandii* and to contribute to the design of strains capable of improving specific characteristics, such as molecular size.

2. Materials and Methods

2.1. Maintenance and Preservation of Strain

A. vinelandii strain OP, which is non-mucoid, was used as wild type. Strains OP-PhbP2$^-$ and OP-PhbP3$^-$ are mutant derivatives of strain OP containing gene inactivations of the *phbP2* and *phbP3* phasin genes, respectively, and were constructed as follows. In mutant OP-PhbP2$^-$, the gene coding for the phasin named PhbP2 (gene *avin_03930*) was inactivated by insertion of a tetracycline resistance cassette derived from plasmid pHP45 Ω-Tc [24] into the BstXI restriction site within its coding sequence. This mutant allele was introduced into the chromosome of strain OP by double-crossover recombination and selection of tetracycline resistance (genotype *phbP2::Tc*^{R}). In mutant strain OP-PhbP3$^-$ the gene coding for a different phasin named PhbP3 (gene *phbP3*; *avin34720*) was inactivated. This was accomplished by inserting a kanamycin resistance cassette from plasmid pBSL99-Km [25] into the HincII restriction site found in the *phbP3* gene. The resulting mutant allele was introduced into the chromosome of strain OP by gene replacement through a double-crossover recombination and selecting on kanamycin (genotype *phbP3::Km*^{R}). The introduction of each gene inactivation was confirmed by PCR and produced truncated phasins [20]. The strains were stored in cryovials with 40% glycerol at −70 °C.

The OP, OP-PhbP2$^-$, and OP-PhbP3$^-$ strains were grown in Petri dishes with solid PYS medium, which contains sucrose (20 g L^{-1}), peptone (5 g L^{-1}), yeast extract (3 g L^{-1}), agar (18 g L^{-1}) and the corresponding antibiotics selection, tetracycline (60 µg mL^{-1}) and kanamycin (3 µg mL^{-1}). The inoculum was obtained through cells grown in a 500 mL shaken flask with 100 mL of liquid PYS medium (in g L^{-1}: 20 sucrose, 5 peptone and 3 yeast extract) without antibiotic, under 200 rpm and 29 °C, for 24 h until reaching an optical density of 0.16 ± 0.02 at 540 nm (Genesys 10S UV-VIS, Thermo Scientific, Boston, MA, USA), corresponding to 0.08 ± 0.02 g L^{-1} of cell dry weight.

2.2. Bioreactor Cultures

The culture was carried out in a 3 L Applikon (Delft, The Netherlands) bioreactor, equipped with two Rushton turbines (impeller diameter/tank diameter = 0.35), 6 flat-blades and a 7-hole diffuser to provide bubbling aeration. pH was controlled to 7.2 ± 0.1 by the automatic addition of NaOH 2N and HCl 2N through cultivation. Cultures were conducted in PYS medium, at 29 °C using an agitation rate of 300 and 500 rpm for all strains evaluated, with a working volume of 2 L and aeration of 1 vvm. A gas analyzer (Teledyne Analytical Instruments, City of Industry, CA, USA, model 7500) was used to measure O_2 and CO_2 in the gaseous flow at the bioreactor output. The gas analysis equipment was calibrated using nitrogen (auto zero setting) and a reference gas (1% CO_2 and 5% O_2 for SPAN).

The estimation of the oxygen OTR was made from the online analysis of the level of gaseous oxygen at the outlet of the bioreactor [21]:

$$OTR = \frac{M_{O2} * F_G^{in}}{V_R * V_M}\left(X_{O_2}^{in} - X_{O_2}^{out}\right) \quad (1)$$

where M_{O_2} is the molecular mass of oxygen (g mmol^{-1}), F_G^{in} is the volumetric inlet air flow at standard conditions (L h^{-1}), V_R is the working volume (L), V_M is the mol volume of the ideal gas at fraction of oxygen in the inlet air (mol mol^{-1}), $X_{O_2}^{out}$ is the molar fraction of oxygen in the inlet air (mol mol^{-1}), $X_{O_2}^{out}$ is the molar fraction of the oxygen in the outlet gas of the fermenter (mol mol^{-1}).

2.3. Determination of Fermentation Parameters

The specific growth rate (μ) was calculated as described [26].

The μ was calculated at 0 to 24 h of the culture using the following equation:

$$\frac{dX}{dt} = \mu X \tag{2}$$

where μ is the specific growth rate (h^{-1}) and X is the cellular protein concentration (g L^{-1}).

The P3HB volumetric productivity was determined according to:

$$Q_{P3HB} = \frac{P - P_0}{\Delta t} \tag{3}$$

P_0 is the initial P3HB concentration (g L^{-1}), P is the 3HB concentration (g L^{-1}) to 24, 48 or 72 h, and Δt is the time in the period.

The specific oxygen uptake rate (qO_2) was calculated with the following equation:

$$qO_2 = \frac{OTR\left(\text{mmol L}^{-1}\text{h}^{-1}\right)}{cellular\ protein\left(\text{gL}^{-1}\right)} \tag{4}$$

2.4. Biomass and P3HB Quantification

Bacterial growth was measured as protein and measured using the Lowry method [27]. To quantify P3HB, 3–5 mg of dry biomass was taken and 1 mL of concentrated H_2SO_4 was added. The mixture was heated to 90 °C for 1 h. The samples were allowed to cool, then diluted 1:50 with MilliQ water (MilliporeSigma, Burlington, MA, USA) and a 20 µL sample was injected into high-pressure liquid chromatography (HPLC) (Waters Alliance 2695) using an Aminex HPX-87H column (Bio Rad, Philadelphia, PA, USA), using 5 mM H_2SO_4 mobile phase, and a Waters 2996 diode array detector (Milford, MA, USA) was used [19]. The area under the curve was quantified at 220 nm. The standard was prepared through hydrolysis of commercial P3HB (1–0.1 mg mL^{-1}) [23].

2.5. Molecular Mass Determination of P3HB

The biomass contained in 3 or 6 mL of sample was recovered by centrifugation. Subsequently, the biomass was washed with 1 mL of distilled water, resuspended, and centrifuged again at 8060× g for 10 min (Eppendorf, model 5804 R, Hamburg, Germany). A total of 1 mL of acetone was added, and the cell pack was agitated for 10 min. The sample was centrifuged at 8060× g for 10 min, acetone was discarded and 2 mL of chloroform was added to solubilize P3HB, leaving it in contact for 20–24 h at room temperature for subsequent filtration before molecular mass analysis.

The molecular mass distribution was determined by gel permeation chromatography. A Shodex K-807L column (Resonac, Tokyo, Japan) was used, which permits the analysis of samples with molecular mass from 1000 to 20,000 kDa. The column was coupled to HPLC equipment (Waters Alliance 2695, Milford, MA, USA) with a refractive index detector (Waters, 2414, Milford, MA, USA). The injection volume was 50 µL, at a working temperature of 30 °C and a run time of 30 min at a flow of 1 mL min^{-1} using chloroform as a mobile phase. Polystyrene standards were used to elaborate the calibration curve with molecular mass between 2.9×10^3 and $5.6.0 \times 10^6$ Da. Samples were prepared at a concentration of 1–2 mg mL^{-1} and dissolved 24 h before analysis. Each sample was filtered with glass syringes and chloroform-resistant PTFE membranes with a pore size of 0.45 µm (Merk, Millipore, San Francisco, CA, USA, No. SLCR033NB).

Empower Chromatography Data System (Waters) was used for the processing and quantification of the molecular weight and mean molecular mass (MMM) of the samples. From the calibration curve, an equation was obtained to estimate the molecular weight of P3HB depending on the elution volume [23].

The depolymerization rate (Da h^{-1}) was calculated using the following equation:

$$Depolymerization\ rate = \frac{MMM_{72\ h} - MMM_{24\ h}}{48\ h} \quad (5)$$

where MMM_{72} is the mean molecular mass (Da) at 72 h, and MMM_{24} is the mean molecular mass (Da).

3. Results and Discussion

3.1. Growth Kinetics, OTR Profiles and qO_2 in Bioreactor Cultures at 300 and 500 rpm with OP, OP-PhbP2$^-$ and OP-PhbP3$^-$ Strain

From the three different phasin proteins identified in *A. vinelandii*, a physiological role has been established only for the majority phasin PhbP1 [18]. However, the role of PhbP2 and PhbP3 has not been studied. To evaluate the effect of the absence of phasins PhbP2 and PhbP3 under possible oxidative stress and start understanding their role in P3HB metabolism, the OP wild-type strain and its derivative mutants OP-PhbP2$^-$ (gene *phbP2* inactivated) and OP-PhbP3$^-$ (gene *phbP3* inactivated) were tested under different OTR conditions, represented by two agitation rates (300 and 500 rpm). These correspond to low and high OTR. This was analyzed because some phasins have been shown to participate in stress protection due to their chaperone-like activity [15,16,28].

Figure 1a shows the bacterial growth kinetics, measured as protein, of the three strains grown at 500 rpm in PYS medium. In the cultures using the OP-PhbP2$^-$ and OP-PhbP3$^-$ strains, the cellular growth was lower compared to that of the OP strain. In the case of strain OP, the maximal protein concentration was 1.43 ± 0.04 g L^{-1}, whereas the cultures with the mutant strains OP-PhbP2$^-$ and OP-PhbP3$^-$ reached 0.96 ± 0.02 g L^{-1} and 1.08 ± 0.15 g L^{-1}, respectively. The μ was 0.05 ± 0.01 h^{-1} for OP-PhbP2$^-$ and 0.04 ± 0.01 h^{-1} for OP-PhbP3$^-$. In the case of the culture with the parental strain OP, the μ was higher (0.08 ± 0.02 h^{-1}) (Table 1).

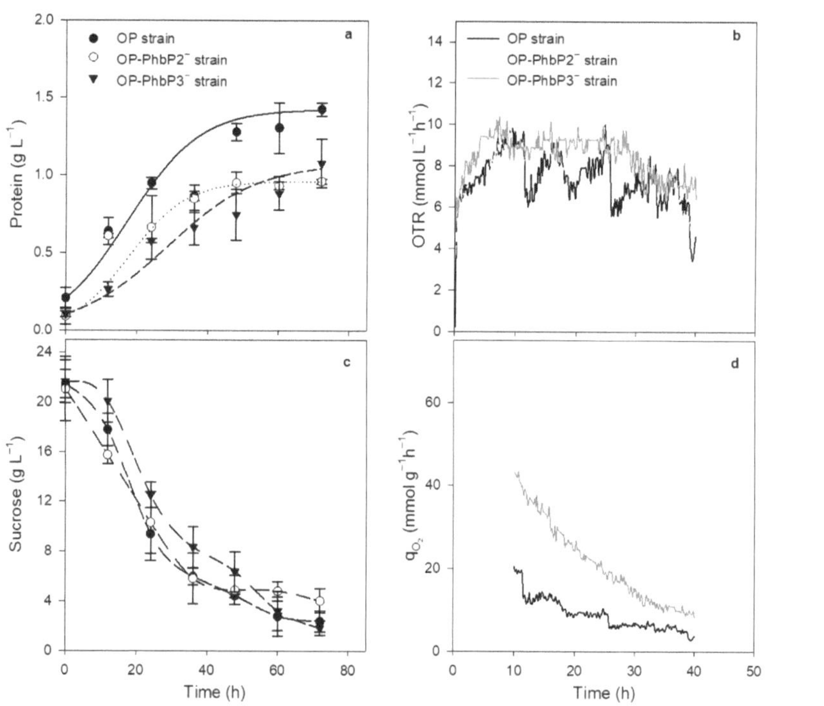

Figure 1. Bacterial growth kinetics (**a**), OTR evolution (**b**), sucrose (**c**) and qO_2 evolution (**d**) for the strains cultured at 500 rpm.

Table 1. Effect of agitation rate on the OTR, qO_2, μ and final protein concentration in cultures of three strains of *A. vinelandii*.

Strain	Agitation Rate (rpm)	OTR_{max} (mmol L^{-1} h^{-1})	qO_2 [a] (mmoL g^{-1} h^{-1})	μ (h^{-1})	Cellular Protein (g L^{-1})
OP	500	8.30 ± 0.56	9.2 ± 1.0	0.080 ± 0.020	1.43 ± 0.04
	300	3.91 ± 0.90	6.9 ± 0.7	0.025 ± 0.001	0.94 ± 0.07
OP-PhbP2⁻	500	8.12 ± 0.24	25.5 ± 1.0	0.050 ± 0.010	0.96 ± 0.02
	300	4.94 ± 0.92	12.1 ± 3.8	0.025 ± 0.001	0.89 ± 0.01
OP-PhbP3⁻	500	8.94 ± 0.79	37.3 ± 1.6	0.040 ± 0.010	1.08 ± 0.15
	300	3.20 ± 1.00	6.9 ± 2.3	0.027 ± 0.001	0.99 ± 0.17

[a] measured at 12 h of cultivation.

Although the μ was lower in the cultures conducted with the mutant strains, the OTR profiles (Figure 1b) were very similar for the three strains evaluated. For example, the OTR increased during the first 10 h of cultivation, reaching maximal values between 8.1 and 8.9 mmol L^{-1} h^{-1} depending on the strain evaluated. Regardless of the strain, the OTR remained constant from 10 to 24 h of cultivation, which was a sign of oxygen limitation in the culture, with a plateau region characteristic of this kind of limitation, as previously described [21,29]. After 24 h, the OTR dropped drastically to a minimal value of 3 mmol L^{-1} h^{-1}.

Figure 1c shows the sucrose during cultivation. For the three strains evaluated, the rate of sucrose consumption at 500 rpm was the same, with values of 0.25 g L^{-1} h. At the end of the culture (72 h), a residual sugar concentration of 2 g L^{-1} was determined for the cultures with the OP strain and OP-PhbP3⁻ strains, and 4.0 g L^{-1} in the cultures with the mutant OP-PhbP2⁻. It is known that in cultures of *A. vinelandii*, the affinity constant (K_s) for sucrose is 0.1 g L^{-1} [30]. Therefore, the cultures were not limited by the carbon source, and thus, it is possible that other nutrients, such as phosphates or trace elements, are responsible for this drop in oxygen transfer rate and oxygen consumption.

The qO_2 was calculated in the range of 12 to 36 h, since in that period, the dissolved oxygen tension remained constant, and therefore, it can be assumed that the oxygen transfer rate is equal to the oxygen uptake rate [30]. As shown in Figure 1d, the highest value of qO_2 was reached with the mutant strains along the culture. In the cultures at 500 rpm, this value was two- to three-fold higher than that obtained with the OP strain during the growth phase (12–24 h). For example, at 12 h of culture, the qO_2 for the culture of the OP-PhbP3⁻ strain was 37.3 ± 1.6 mmoL g^{-1} h^{-1} and 25.5 ± 1.0 mmoL g^{-1} h^{-1} for the culture with mutant OP-PhbP2⁻, whereas for the OP strain, it was 9.2 ± 1.0 mmoL g^{-1} h^{-1} (Table 1).

It has been reported that variations in agitation rate, and therefore in the OTR, could promote an effect known as "semi-lethal damage", where it is proposed that high OTR, similar to those achieved at 500 rpm, results in a condition of oxidative stress [21,22]. On the other hand, it is known that *A. vinelandii* cell walls exhibit a high resistance to mechanical stress; therefore, it seems unlikely that cells of *A. vinelandii* could be damaged by the mechanical stress generated in the bioreactor at 500 rpm [31]. Previous studies have reported that in cultures of *A. vinelandii* at 500 rpm, it is possible to find a situation that causes semi-lethal cell damage [21]. Considering this starting point, and that in the mutant strains the phasins PhbP2 or PhbP3 are not expressed, we propose that these phasins may be playing another function distinct to P3HB metabolism, possibly as chaperone proteins that protect the cells against oxidative stress, as has been shown for other phasins [15,16]. To evaluate this possibility, cultures were carried out at low OTR (300 rpm). As shown in Figure 2a, there were no significant changes in the growth or the qO_2, that being the case for the two mutant strains concerning strain OP. The values of maximal protein concentration were in the range of 0.9 to 1.4 g L^{-1}, with μ 0.025 ± 0.001 h^{-1} for both OP and OP-PhbP2⁻

and 0.027 ± 0.001 h^{-1} for OP-PhbP3^{-}. In the case of qO_2, the values at 12 h of cultivation were in the range of 6.9 to 12.1 mmoL g^{-1} h^{-1} (Table 1). Therefore, it is possible that under conditions of high oxygen transfer, phasins could be involved in participating as chaperone-type proteins that help face a situation of oxidative stress in the cells. However, more studies are necessary to elucidate the mechanisms involved.

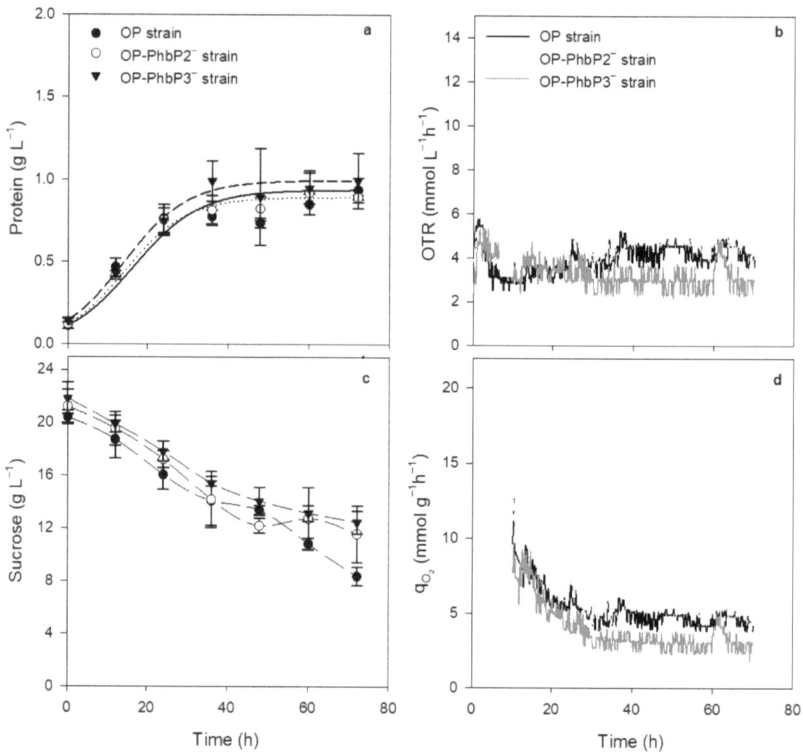

Figure 2. Bacterial growth kinetics (**a**), OTR evolution (**b**), sucrose (**c**) and qO_2 evolution (**d**) for the strains cultured at 300 rpm.

It is known that the phasins exert a stress reduction action, both in P3HB and non-P3HB-synthesizing bacteria, decreasing the induction of heat shock-related genes in *E. coli* [12,13] and promoting protein folding through a chaperone-like mechanism, which suggests an in vivo general protective role [28]. However, it is necessary to carry out more studies that help discern if the suggested effect is carried out genetically or if it is through protein–protein interactions on the granule surface, since in some models, it has been shown that phasins facilitate the anchoring of P3HB synthases to the surface of the granules [28].

3.2. P3HB Production at 300 and 500 rpm Using the OP and Mutant Strains

The accumulation percentage of P3HB and its concentration were also quantified at different culture times. Figure 3a shows the kinetics of P3HB production of the three strains evaluated at 500 rpm. It was observed that the absence of phasin PhbP2 increased the production of P3HB, whereas the absence of phasin PhbP3 resulted in a decrease in the production of the polymer. The maximal concentrations were 4.6 ± 0.5 g L^{-1} for the OP strain, 5.3 ± 0.3 g L^{-1} for the OP-PhbP2^{-} strain, and 3.7 ± 0.6 g L^{-1} for the OP-PhbP3^{-} strain (Table 2).

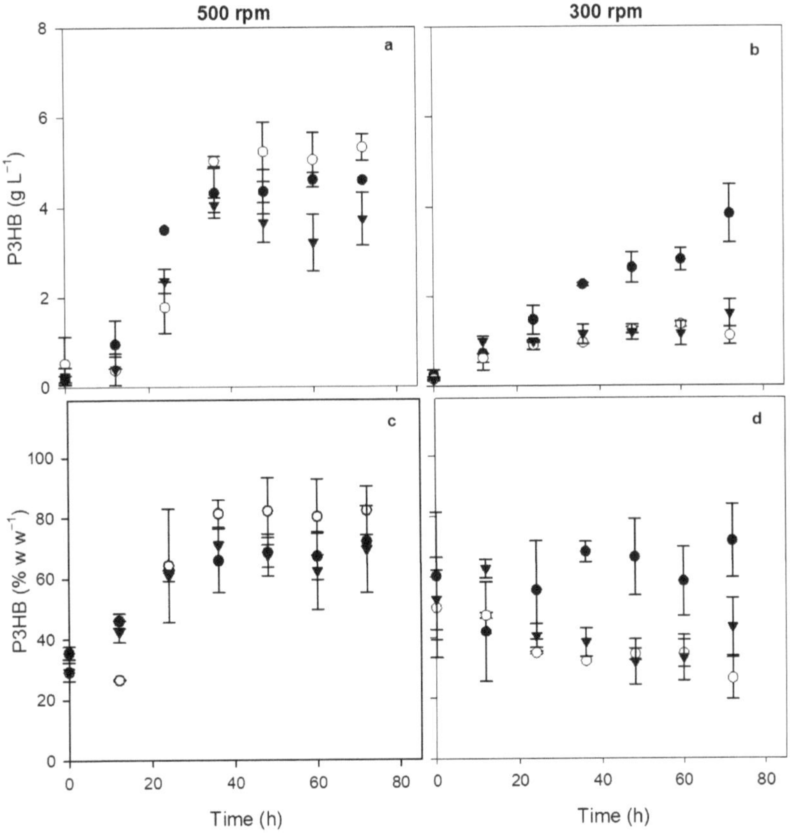

Figure 3. Evolution of the P3HB production (**a**,**b**) and intracellular accumulation (**c**,**d**) in cultures of three strains of *A. vinelandii* at different agitation rates (500 and 300 rpm). OP strain (black circles), OP-PhbP2$^-$ (white circles) and OP-PhbP3$^-$ strain (black triangles).

Table 2. Effect of agitation rate on the P3HB production, accumulation of P3HB and P3HB volumetric productivity for the three strains evaluated of *A. vinelandii*.

Strain	Agitation Rate (rpm)	P3HB$_{max}$ (g L^{-1})	P3HB$_{max}$ (% on Dry Cell Weight)	Q$_{P3HB}$ (g L^{-1} h^{-1})		
				24 h	48 h	72 h
OP	500	4.6 ± 0.5 (72 h)	72.0 ± 0.5	0.138 ± 0.003	0.087 ± 0.009	0.061 ± 0.001
	300	3.8 ± 0.6 (72 h)	69.0 ± 16.4	0.050 ± 0.009	0.049 ± 0.005	0.049 ± 0.008
OP-PhbP2$^-$	500	5.3 ± 0.3 (72 h)	82.3 ± 8.1	0.052 ± 0.001	0.098 ± 0.001	0.067 ± 0.004
	300	1.4 ± 0.1 (60 h)	47.3 ± 1.0	0.031 ± 0.003	0.023 ± 0.001	0.013 ± 0.002
OP-PhbP3$^-$	500	3.7 ± 0.6 (72 h)	71.0 ± 0.5	0.088 ± 0.003	0.071 ± 0.005	0.048 ± 0.006
	300	1.6 ± 0.3 (72 h)	62.7 ± 3.0	0.035 ± 0.001	0.022 ± 0.003	0.020 ± 0.004

Figure 3c shows the accumulation percentage of P3HB at 500 rpm. In general, no significant differences were found in the P3HB accumulation in the different strains evaluated. The OP-PhbP2$^-$ strain showed high levels of P3HB accumulation, reaching values of 82.3 ± 8.1% based on the dry weight of the bacteria. On the other hand, in the cultures with the OP strain, the percentage was 72.0 ± 0.5 w w^{-1}, and 71.0 ± 0.5% w w^{-1} for the OP-PhbP3$^-$ strain (Table 2).

It is important to point out that in the cultures at 300 rpm (OTR about 5 mmol L^{-1} h^{-1}), both the concentration (Figure 3b) and the percentage of polymer accumulation (Figure 3d) were lower with the three strains evaluated than those obtained at 500 rpm. As shown in Figure 3b, the highest production of P3HB was obtained in the cultures with the OP strain, reaching 3.8 ± 0.6 g L^{-1}. In the case the OP-PhbP3$^-$ mutant strain, the P3HB production was 1.6 ± 0.3 g L^{-1}, and for the OP-PhbP2$^-$ strain, it was 1.4 ± 0.1 g L^{-1}. Similarly, the maximal percentage of polymer accumulation was around 70% with the OP strain and 47.3 ± 1.0% w w^{-1} and 62.7 ± 3.0% w w^{-1} with OP-PhbP2$^-$ and OP-PhbP3$^-$, respectively. It is important to point out that at 300 rpm, in the culture with the mutant strains, a decrease in P3HB accumulation according to the culture time was observed. Although we do not have an explanation for this, it is possible that in the mutant strains, there is a slight consumption of P3HB from the beginning of cultivation, which is not necessarily reflected in the concentration of the polymer. It is important to note that the higher Q_{P3HB} was obtained in the cultures conducted at 500 rpm using the OP-PhbP2$^-$ strain (Table 2).

These results contrast with previous reports by other authors, who found that under conditions of low oxygen transfer, such as those obtained at 300 rpm, the synthesis of P3HB is favored, because the carbon source is channeled through the TCA cycle, and much of the carbon source is directed to P3HB production instead of bacterial growth. However, those studies were performed with a different strain (OPNA) that has inactivated the *rsmA* and *ptsN* genes that are involved in the regulation of P3HB synthesis [30].

More recent studies [23] reported a behavior similar to that observed in this study. Those authors found that maximal production (4.2 ± 0.4 g L^{-1}) and accumulation of P3HB (90 ± 5% w w^{-1}) using the strain OP was reached in the cultures grown at 500 rpm (high OTR_{max}). On the contrary, the lowest P3HB production of approximately 1.6 ± 0.3 g L^{-1} (56 ± 2% w w^{-1}) was obtained at 300 rpm (low OTR_{max}).

This increase in the production and accumulation of P3HB at high oxygen transfer may be due to the activation of a cell protection mechanism against oxidative stress, as was previously reported [22]. Those authors found that in cultures at high oxygen transfer (20.2 mmol L^{-1} h^{-1}), *Rhizobium phaseoli* increases both the activity of catalase, an enzyme that acts on hydrogen peroxide, and the production of P3HB, as a strategy to address oxygen stress situations.

On the other hand, it was suggested that in *Cupriavidus necator*, phasins modulate the activity of P3HB depolymerases Z1 and Z2, respectively [32]. In the case of *Azotobacter* sp. FA-8, the phasin PhaPAz is the most abundant P3HB granule-associated protein [16,33]. It was found that this protein displays a growth-promoting effect, also enhancing the polymer production in recombinant P3HB-producing *Escherichia coli* [15].

3.3. Molecular Mass Distributions of Polymers Produced by the OP and Mutant Strains

Figure 4 shows the mean molecular mass data of the three strains evaluated at 500 rpm. Maximal molecular masses of 7080 ± 273 kDa and 6820 ± 237 kDa were reached in the P3HB produced by the OP strain and OP-PhbP2$^-$ at 24 h of cultivation. At the end of the culture (72 h), the mean molecular mass of the polymer decreased in both strains to values of 5480 ± 239 kDa and 4489 ± 281 kDa for the polymers produced, respectively, by the OP strain and OP-PhbP2$^-$. It is interesting to note that in the case of the strain OP-PhbP3$^-$, the mean molecular mass remains constant throughout the cultivation at a value of 7262 ± 1334 kDa at 24 h of cultivation and 6704 ± 981 kDa at 72 h. According to the ANOVA analysis, there are significant differences ($\alpha < 0.05$) in the MMM of the polymer produced by the OP and OP-PhbP2$^-$ in the range of 24 to 72 h, whereas no significant differences were found in the MMM of the P3HB produced by the OP-PhbP3$^-$ in the same period of cultivation (Figure 4).

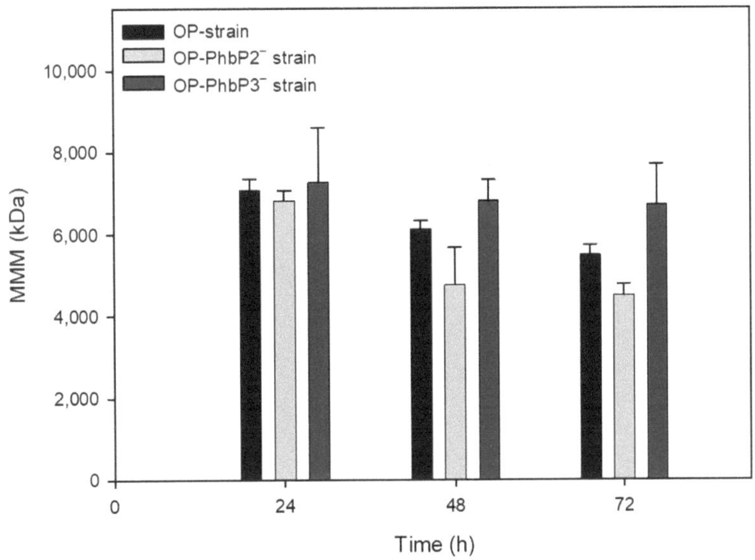

Figure 4. Mean molecular mass (MMM) of P3HB produced for the different strains evaluated in cultures of *A. vinelandii* developed at 500 rpm.

P3HB with different molecular masses exhibits varying mechanical and processing properties, determining their applicability in commercial sectors [2]. Low-molecular-mass P3HB is preferred in applications requiring higher biodegradability and flexibility, such as single-use packaging, controlled drug release systems, and biodegradable sutures. In contrast, higher-molecular-mass P3HB is more suitable for products that demand high strength and durability, such as bioplastics for structural components, textiles, and implantable medical devices [2].

To better understand the changes observed in the molecular mass of P3HB, Figure 5 shows the molecular mass distribution of the P3HB accumulated at 24, 48, and 72 h for the three strains tested.

It is clear that in the case of the OP strain, at 24 h, most of the P3HB fractions are in the range of 1000 to 10,000 kDa, whereas at 48 and 72 h, a significant change in the distribution is found, with a high percentage of molecules in the range of 100 to 1000 kDa. On the other hand, in the polymer synthesized by the OP-PhbP2$^-$ strain, a phenomenon similar to that identified in the OP strain is observed, especially at 72 h, where the percentage of molecules in the range of 100 to 1000 kDa increased significantly. In contrast, in the case of the OP-PhbP3$^-$ strain, no relevant changes were observed in the distribution of molecular mass as a function of culture time and most of the fractions are in the range of 1000 to 10,000 kDa, independently of culture time. This behavior is reflected in the polydispersity index (PI) of the product, finding that in the polymer produced by the OP strain and OP-PhbP2$^-$, the values at the end of the culture are higher than 3 (3.01 $\pm$ 2.04 and 4.93 $\pm$ 0.88, respectively), whereas for the polymer synthesized by the strain OP-PhbP3$^-$, the PI was 1.27 $\pm$ 0.05.

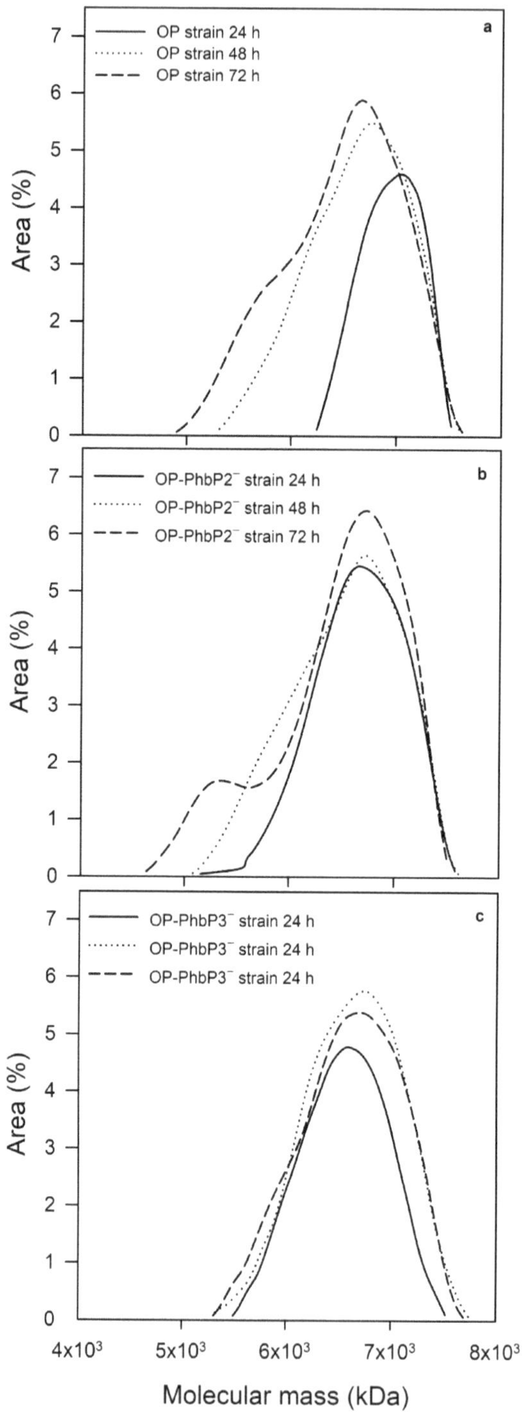

Figure 5. Distribution of molecular mass of P3HB produced by the OP (**a**), OP-PhbP2^{-} (**b**), and OP-PhbP3^{-} (**c**) strains in cultures performed at 500 rpm.

The decrease in the molecular mass throughout the culture observed in the OP and OP-PhbP2$^-$ strains could be related to the activity of P3HB depolymerase during the stationary phase of the cultivation. In the cultures with the OP and OP-PhbP2$^-$ strains, activities of 1.20 ± 0.10 and 1.86 ± 0.20 μg P3HB mg prot^{-1} min^{-1} were detected, respectively, whereas in the OP-PhbP3$^-$ strain, no activity was detected. Previous studies with the OP strain [17] have found an increase between 40 and 50% in the activity of P3HB depolymerase at the end of the exponential growth phase and the stationary phase, under oxygen transfer conditions similar to those used in the present study. In contrast, the P3HB synthase activity decreased by about 50% at the end of the stationary phase [34].

Regarding the results observed in the P3HB synthesized by the OP-PhbP3$^-$ strain, it is possible to hypothesize that the phasin could activate the P3HB depolymerase PhbZ1, which is known to cause a decrease in the molecular mass of the P3HB [17,34]. The absence of phasin PhbP3 would negatively affect the activity of this depolymerase, so less hydrolysis of the polymer would occur, because a similar result to that of mutant OP-PhbP3$^-$ has been observed with the inactivation of PhbZ1, maintaining the MMM along the culture [17]. This kind of effect of phasins on the activity of P3HB depolymerases has been previously reported [13,32].

There are some examples where changes in the expression of different phasins had effects on the molecular mass of the PHAs produced. For example, in *Aeromonas hydrophila,* the over-expression of the phasin PhaPAh reduced the molecular mass of the PHA produced to approximately 50% of that of the wild-type strain. This phenotype was explained by the authors as a possible indirect effect through the PHA synthase, because over-expression of *phaPAh* increased transcription of PhaCAh, and over-expression of *phaC* also negatively affected the molecular mass [35].

On the other hand, the phasin PhaPAh from *Aeromonas hydrophila* seems to regulate the *phaCAh* gene at the transcriptional level [35]. If a similar regulation occurs in *A. vinelandii,* the amount of PhbC would be affected in the phasin mutants, and a correlation between the active synthase concentration and the molecular mass of the PHA produced has been reported, where it was found that the lower the PhaC concentration, the higher the molecular mass of the polymer [36].

Table 3 shows the depolymerization rate of P3HB and Q_{P3HB} in cultures conducted at 500 rpm. It is possible to observe a direct relationship ($r^2 = 0.999$) between the depolymerization rate (determined between 24 and 72 h) and the Q_{P3HB}. For the lowest Q_{P3HB} (0.071 g L^{-1} h^{-1}), obtained in the cultures at 500 rpm using the OP-PhbP3$^-$ mutant strain, a depolymerization rate of 3555 ± 136 Da h^{-1} was reached. On the other hand, in the cultures with the OP strain, the Q_{P3HB} was 0.087 g L^{-1} h^{-1}, with a depolymerization rate 5 times higher (15800 Da h^{-1}) than that with the OP-PhbP3$^-$ strain. The extreme condition occurs with the OP-PhbP2^{-1} strain, where the highest rate of P3HB synthesis was reached (0.098 g L^{-1} h^{-1}), as well as a high rate of depolymerization (24500 Da h^{-1}).

Table 3. Relationship between the P3HB volumetric productivity and depolymerization rate of P3HB in cultures of OP, OP-PhbP2$^-$, and OP-PhbP3$^-$ strains at 500 rpm.

Strain	OTR_{max} (mmol L^{-1} h^{-1})	Q_{P3HB} (g L^{-1} h^{-1})	Depolymerization Rate (Da h^{-1})
OP	8.30 ± 0.56	0.087 ± 0.009	$15{,}800 \pm 1500$
OP-PhbP2$^-$	8.12 ± 0.24	0.098 ± 0.001	$24{,}500 \pm 2500$
OP-PhbP3$^-$	8.94 ± 0.79	0.071 ± 0.005	3555 ± 136

This trend was previously reported by [34], who found a close relationship between the synthesis rate of P3HB and the molecular mass of polymers. These authors reported that when the rate of P3HB synthesis is increased, by manipulating the polymer content in the inoculum, a significant decrease in the molecular mass of the polymer was observed. In addition, this behavior is similar to that observed in recombinant *E. coli* and *Azohydromonas lata* cultures, where an inverse relationship between the P3HB production rate and the molecular mass was reported [17,35,37]. On the other hand, studies in our group [32] have shown that in the case of the cultures developed using an inoculum with 50% of P3HB, the synthase activity was higher when the P3HB productivity was higher (growth exponential phase) and this activity decreases significantly during the stationary phase of growth when the productivity of P3HB decreases [33].

4. Conclusions

Overall, our results demonstrate for the first time that the absence of phasins PhbP2 or PhbP3 leads to a significant reduction in the specific growth rate of mutant strains under high OTR conditions. This behavior could be related to the protective role that some of these proteins have against the effect of oxidative stress generated in cultures under high oxygen transfer. For example, the phasin $PhaP_{Az}$ from *Azotobacter* sp. FA-8, which belongs to the same family of phasins as PhbP2 of *A. vinelandii*, has a stress-reducing action that results in increased growth and higher resistance to superoxide stress. This phasin has chaperone-like activity, so the stress-protective effect could be related with this capacity [15,16]. The PhbP2 phasin could have a similar role, because both PhbP1 and PhbP2 of *A. vinelandii* have a similar predicted structure with several alpha-helical arrangements, alternated with unstructured regions. The regions of predicted structural flexibility could interact with different targets, including misfolded proteins. In the case of PhbP3, it represents a new family of phasins, with no conserved domains shared with PhbP1 and PhbP2 and with poor identity in sequence; however, some structural similarity is observed, which could suggest a similar role. The targets of PhbP2 or PhbP3 are not known, but some of them could be important to sustain normal growth under oxygen stress.

On the other hand, the higher rate of oxygen transfer in the cultures promotes the synthesis of P3HB, doubling of the polymer concentration in the three strains evaluated with respect to that obtained at a low oxygen transfer. Finally, it was observed that the molecular mass of the polymer synthesized by strain OP and OP-PhbP2$^-$ decreased at the end of the culture, whereas in the case of the polymer obtained with the OP-PhbP3$^-$ strain, it remains constant, a situation which could be related to the activity of the depolymerases. This suggests that a P3HB depolymerase could be a target of PhbP3.

Author Contributions: C.A.-Z. carried out the experimental work, the analysis of the results, and the integration of the manuscript. A.P. helped in the elaboration of the figures and the integration of the manuscript. D.S., A.D.-B. and E.G. contributed to the formal analysis of results, as well as the review, editing, and discussion. C.P. contributed to the conceptualization and obtaining the resources to carry out the study. J.R. carried out the experimental work, regarding to the design of mutant strains. All authors have read and agreed to the published version of the manuscript.

Funding: This work was partially financed by grants from the Agencia Nacional de Investigación y Desarrollo (ANID) from Chile through Fondecyt Regular 1231075 and Dirección General de Asuntos del Personal Académico National Autonomous University of Mexico (Grant IG200222).

Institutional Review Board Statement: Not applicable.

Data Availability Statement: The original contributions presented in the study are included in the article, further inquiries can be directed to the corresponding author.

Acknowledgments: The authors thank Celia Flores and Rosa Elia Quiroz for their technical support.

Conflicts of Interest: The authors declare no conflicts of interest.

References

1. Philip, S.; Keshavarz, T.; Roy, I. Polyhydroxyalkanoates: Biodegradable polymers with a range of applications. *J. Chem. Technol. Biotechnol.* **2007**, *82*, 233–247. [CrossRef]
2. Peña, C.; Castillo, T.; García, A.; Millán, M.; Segura, D. Biotechnological strategies to improve the production of microbial poly-(3-hydroxybutyrate): A review of recent research work. *Microbiol. Biotechnol.* **2014**, *7*, 278–293. [CrossRef] [PubMed]
3. Iwata, T. Strong fibers and films of microbial polyesters. *Macromol. Bios.* **2005**, *5*, 689–701. [CrossRef] [PubMed]
4. Magagula, S.I.; Mohapi, M.; Sefadi, J.S.; Mochane, M.J. The production and applications of microbial-derived polyhydroxybutyrates. In *Microbial Polymers: Applications and Ecological Perspectives*, 1st ed.; Vaishnav, A., Choudhary, D.K., Eds.; Springer Nature: Singapore, 2021; pp. 3–43.
5. Andler, R.; Rojas, V.; Pino, V.; Castro, R.I.; Valdés, C.; Kumar, V.; Peña, C.; Díaz-Barrera, A. Efficient production of a polyhydroxyalkanoate by *Azotobacter vinelandii* OP using apple residues as promising feedstock. *Int. J. Biol. Macromol.* **2023**, *242*, 124626. [CrossRef]
6. Jendrossek, D.; Pfeiffer, D. New insights in the formation of polyhydroxyalkanoate granules (carbonosomes) and novel functions of poly(3-hydroxybutyrate). *Environ. Microbiol.* **2014**, *16*, 2357–2373. [CrossRef] [PubMed]
7. Mezzina, M.; Pettinari, J. Phasins, multifaceted polyhydroxyalkanoate granule-associated proteins. *Appl. Environ. Microbiol.* **2016**, *82*, 5060–5067. [CrossRef]
8. Soo-Lee, H.; Lee, H.; Kim, S.; Cho, J.; Suh, M.; Ham, S.; Bhatia, S.; Gurav, R.; Kim, Y.; Lee, E.; et al. Novel phasins from the arctic *Pseudomonas* sp. B14-6 enhances the production of polyhydroxybutyrate and increases inhibitor tolerance. *Int. J. Biol. Macromol.* **2021**, *190*, 722–729. [CrossRef]
9. Tang, R.; Peng, X.; Weng, C.; Han, Y. The overexpression of phasin and regulator genes promoting the synthesis of polyhydroxybutyrate in *Cupriavidus necator* H16 under nonstress conditions. *Appl. Environ. Microbiol.* **2022**, *88*, e01458-21. [CrossRef]
10. Maehara, A.; Ueda, S.; Nakano, H.; Yamane, T. Analyses of a Polyhydroxyalkanoic Acid Granule-Associated 16-Kilodalton Protein and Its Putative Regulator in the *pha* Locus of *Paracoccus denitrificans*. *J. Bacteriol.* **1999**, *181*, 2914–2921. [CrossRef]
11. Lee, H.J.; Jung, H.J.; Kim, B.; Cho, D.H.; Kim, S.H.; Bhatia, S.K.; Gurav, R.; Kim, Y.G.; Jung, S.W.; Park, H.J.; et al. Enhancement of polyhydroxybutyrate production by introduction of heterologous phasin combination in *Escherichia coli*. *Int. J. Biol. Macromol.* **2023**, *225*, 757–766. [CrossRef]
12. Kuchta, K.; Chi, L.; Fuchs, H.; Pötter, M.; Steinbüchel, A. Studies on the influence of phasins on accumulation and degradation of PHB and nanostructure of PHB granules in *Ralstonia eutropha* H16. *Biomacromolecules* **2007**, *8*, 657–662. [CrossRef] [PubMed]
13. Handrick, R.; Technow, U.; Reichart, T.; Reinhardt, S.; Sander, T.; Jendrossek, D. The activator of the *Rhodospirillum rubrum* PHB depolymerase is a polypeptide that is extremely resistant to high temperature (121 °C) and other physical or chemical stresses. *FEMS Microbiol. Lett.* **2004**, *230*, 265–274. [CrossRef]
14. Pötter, M.; Steinbüchel, A. Poly(3-hydroxybutyrate) granule-associated proteins: Impacts on poly(3-hydroxybutyrate) synthesis and degradation. *Biomacromolecules* **2005**, *6*, 552–560. [CrossRef]
15. de Almeida, A.; Catone, M.V.; Rhodius, V.A.; Gross, C.A.; Pettinari, M.J. Unexpected stress-reducing effect of PhaP, a poly(3-hydroxybutyrate) granule-associated protein, in *Escherichia coli*. *Appl. Environ. Microbiol.* **2011**, *77*, 6622–6629. [CrossRef] [PubMed]
16. Mezzina, M.P.; Wetzler, D.E.; de Almeida, A.; Dinjaski, N.; Prieto, M.A.; Pettinari, M.J. A phasin with extra talents: A polyhydroxyalkanoate granule-associated protein has chaperone activity. *Environ. Microbiol.* **2015**, *17*, 1765–1776. [CrossRef] [PubMed]
17. Moreno, S.; Castellanos, M.; Bedoya-Pérez, L.P.; Canales-Herrerías, P.; Espín, G.; Muriel-Millán, L.F. Outer membrane protein I is associated with poly-β-hydroxybutyrate granules and is necessary for optimal polymer accumulation in *Azotobacter vinelandii* on solid medium. *Microbiology* **2019**, *165*, 1107–1116. [CrossRef]
18. Adaya, L.; Millán, M.; Peña, C.; Jendrossek, D.; Espín, G.; Tinoco-Valencia, R.; Guzmán, J.; Pfeiffer, D.; Segura, D. Inactivation of an intracellular poly-3-hydroxybutyrate depolymerase of *Azotobacter vinelandii* allows to obtain a polymer of uniform high molecular mass. *Appl. Microbiol. Biotechnol.* **2018**, *102*, 2693–2707. [CrossRef]
19. Quiroz-Cardoso, R.; Castillo, T.; Galindo, E.; Ruíz Escobedo, J.; Segura, D.; Peña, C. Looking for improved strains of *Azotobacter vineladii* and favorable culture conditions yielding high molecular weight Poly-3-hydoxybutyrate (P3HB). *J. Chem. Technol. Biotechnol.* 2024, *submitted*.
20. Ruíz Escobedo, J. Estudio del Papel de las Proteínas Avin34710 y Avin34720 en el Metabolismo de Polihidroxibutirato (PHB) en la Bacteria *Azotobacter vinelandii*. Master's Thesis, Universidad Nacional Autónoma de México, Mexico City, Mexico, 2020.
21. Díaz-Barrera, A.; Andler, R.; Martínez, I.; Peña, C. Poly-3-hydroxybutyrate production by *Azotobacter vinelandii* strains in batch cultures at different oxygen transfer rates. *J. Chem. Technol. Biotechnol.* **2016**, *91*, 1063–1071. [CrossRef]
22. García-Cabrera, R.I.; Valdez-Cruz, N.A.; Blancas-Cabrera, A.; Trujillo-Roldán, M.A. Oxygen transfer rate affects polyhydroxybutyrate production and oxidative stress response in submerged cultures of *Rhizobium phaseoli*. *Biochem. Eng. J.* **2020**, *162*, 107721. [CrossRef]
23. Gómez-Hernández, E.; Salgado-Lugo, H.; Segura, D.; García, A.; Díaz-Barrera, A.; Peña, C. Production of Poly-3-hydroxybutyrate (P3HB) with ultra-high molecular weight (UHMW) by mutant strains of *Azotobacter vinelandii* under microaerophilic conditions. *Appl. Biochem. Biotechnol.* **2020**, *193*, 79–95. [CrossRef] [PubMed]

24. Fellay, R.; Frey, J.; Krisch, H. 1987. Interposon mutagenesis of soil and water bacteria: A family of DNA fragments designed for in vitro insertional mutagenesis of Gram-negative bacteria. *Gene* **1987**, *52*, 147–154. [CrossRef] [PubMed]
25. Alexeyev, M.F.; Shokolenko, I.N.; Croughan, T.P. Improved antibiotic-resistance gene cassettes and omega elements for *Escherichia coli* vector construction and in vitro deletion/insertion mutagenesis. *Gene* **1995**, *160*, 63–67. [CrossRef] [PubMed]
26. García, A.; Segura, D.; Espín, G.; Galindo, E.; Castillo, T.; Peña, C. High production of poly-β-hydroxybutyrate (PHB) by an *Azotobacter vinelandii* mutant altered in PHB regulation using a fed-batch fermentation process. *Biochem. Eng. J.* **2014**, *82*, 117–123. [CrossRef]
27. Lowry, O.H.; Rosebrough, N.J.; Farr, A.L. Randall Protein measurement with the Folin phenol reagent. *J. Biol. Chem.* **1951**, *193*, 265–275. [CrossRef]
28. Mezzina, M.P.; Wetzler, D.E.; Catone, M.V.; Bucci, H.; Di Paola, M.; Pettinari, M.J. A phasin with many faces: Structural insights on PhaP from *Azotobacter* sp. FA8. *PLoS ONE* **2014**, *9*, e103012. [CrossRef]
29. Castillo, T.; Heinzle, E.; Peifer, S.; Schneider, K.; Peña, C. Oxygen supply strongly influences metabolic fluxes, the production of poly(3-hydroxybutyrate) and alginate, and the degree of acetylation of alginate in *Azotobacter vinelandii. Proc. Biochem.* **2013**, *48*, 995–1003. [CrossRef]
30. García, A.; Ferrer, P.; Albiol, J.; Castillo, T.; Segura, D.; Peña, C. Metabolic flux analysis and the NAD(P)H/NAD(P)(+) ratios in chemostat cultures of *Azotobacter vinelandii. Microb. Cell Fact.* **2018**, *17*, 10. [CrossRef]
31. Pena, C.; Trujillo-Roldán, M.A.; Galindo, E. Influence of dissolved oxygen tension and agitation speed on alginate production and its molecular weight in cultures of *Azotobacter vinelandii. Enzym. Microb. Technol.* **2000**, *27*, 390–398. [CrossRef]
32. Bresan, S.; Sznajder, A.; Hauf, W.; Forchhammer, K.; Pfeiffer, D.; Jendrossek, D. Polyhydroxyalkanoate (PHA) granules have no phospholipids. *Sci. Rep.* **2016**, *6*, 26612. [CrossRef]
33. Pettinari, J.M.; Chaneton, L.; Vazquez, G.; Steinbuchel, A.; Mendez, B.S. Insertion sequence-like elements associated with putative polyhydroxybutyrate regulatory genes in *Azotobacter* sp. FA8. *Plasmid* **2003**, *50*, 36–44. [CrossRef] [PubMed]
34. Millán, M.; Segura, D.; Galindo, E.; Peña, C. Molecular mass of poly-3-hydroxybutyrate (P3HB) produced by *Azotobacter vinelandii* is determined by the ratio of synthesis and degradation under fixed dissolved oxygen tension. *Process. Biochem.* **2016**, *51*, 950–958. [CrossRef]
35. Tian, S.J.; Lai, W.J.; Zheng, Z.; Wang, H.X.; Chen, G.Q. Effect of over-expression of phasin gene from *Aeromonas hydrophila* on biosynthesis of copolyesters of 3- hydroxybutyrate and 3-hydroxyhexanoate. *FEMS Microbiol. Lett.* **2005**, *244*, 19–25. [CrossRef] [PubMed]
36. Hiroe, A.; Ushimaru, K.; Tsuge, T. Characterization of polyhydroxyalkanoate (PHA) synthase derived from *Delftia acidovorans* DS-17 and the influence of PHA production in *Escherichia coli. J. Biosci. Bioeng.* **2013**, *115*, 633–638. [CrossRef]
37. Millán, M.; Salazar, M.; Segura, D.; Castillo, T.; Díaz-Barrera, A.; Peña, C. Molecular mass of Poly-3-hydroxybutyrate (P3HB) produced by *Azotobacter vinelandii* is influenced by the polymer content in the inoculum. *J. Biotechnol.* **2017**, *259*, 50–55. [CrossRef]

polymers

MDPI

Article

Deep Learning for Cell Migration in Nonwoven Materials and Evaluating Gene Transfer Effects following AAV6-ND4 Transduction

Ilya I. Larin [1,*], Rimma O. Shatalova [1], Victor S. Laktyushkin [2], Stanislav A. Rybtsov [2], Evgeniy V. Lapshin [1], Daniil V. Shevyrev [2], Alexander V. Karabelsky [1], Alexander P. Moskalets [3], Dmitry V. Klinov [3,4] and Dimitry A. Ivanov [4,5]

1 Center for Translational Medicine, Sirius University of Science and Technology, Krasnodar Territory Sirius, 1 Olympic Ave., Sirius 354340, Russia
2 Resource Center for Cell Technology and Immunology, Sirius University of Science and Technology, Krasnodar Territory Sirius, 1 Olympic Ave., Sirius 354340, Russia
3 Federal Research and Clinical Center of Physical-Chemical Medicine, Federal Medical Biological Agency, Moscow 119435, Russia
4 Center for Genetics and Life Sciences, Sirius University of Science and Technology, Krasnodar Territory Sirius, 1 Olympic Ave., Sirius 354340, Russia
5 Institut de Sciences des Matériaux de Mulhouse—IS2M, CNRS UMR 7361, F-68057 Mulhouse, France
* Correspondence: larin.ii@talantiuspeh.ru

Abstract: Studying cell settlement in the three-dimensional structure of synthetic biomaterials over time is of great interest in research and clinical translation for the development of artificial tissues and organs. Tracking cells as physical objects improves our understanding of the processes of migration, homing, and cell division during colonisation of the artificial environment. In this study, the 3D environment had a direct effect on the behaviour of biological objects. Recently, deep learning-based algorithms have shown significant benefits for cell segmentation tasks and, furthermore, for biomaterial design optimisation. We analysed the primary LHON fibroblasts in an artificial 3D environment after adeno-associated virus transduction. Application of these tools to model cell homing in biomaterials and to monitor cell morphology, migration and proliferation indirectly demonstrated restoration of the normal cell phenotype after gene manipulation by AAV transduction. Following the 3Rs principles of reducing the use of living organisms in research, modeling the formation of tissues and organs by reconstructing the behaviour of different cell types on artificial materials facilitates drug testing, the study of inherited and inflammatory diseases, and wound healing. These studies on the composition and algorithms for creating biomaterials to model the formation of cell layers were inspired by the principles of biomimicry.

Keywords: electrospun; mats; neural network; fibroblasts; transduction; AAV; cell homing and migration

Citation: Larin, I.I.; Shatalova, R.O.; Laktyushkin, V.S.; Rybtsov, S.A.; Lapshin, E.V.; Shevyrev, D.V.; Karabelsky, A.V.; Moskalets, A.P.; Klinov, D.V.; Ivanov, D.A. Deep Learning for Cell Migration in Nonwoven Materials and Evaluating Gene Transfer Effects following AAV6-ND4 Transduction. *Polymers* **2024**, *16*, 1187. https://doi.org/10.3390/polym16091187

Academic Editors: Vishal Gavande, Vasi Shaikh and Won-Ki Lee

Received: 28 March 2024
Revised: 15 April 2024
Accepted: 16 April 2024
Published: 24 April 2024

1. Introduction

The field of biomedical material research plays a crucial role in the progress of modern high-tech medicine [1]. The demand for these materials is rising due to an ageing population, increased risks of developing senile diseases, and the spread of such civilisational diseases as diabetes mellitus, osteoporosis, and cardiovascular diseases. All of this escalates the prevalence of chronic trophic ulcers, impairs wound healing, and increases the risk of surgical interventions and injuries [2,3].

In preclinical research, the most promising strategy for wound healing is the development of methods to generate an extracellular matrix (ECM) [4]. Fibroblasts, as the largest cell population in the dermis, are critical for maintaining the composition and stability of the ECM. They regulate the production and degradation of collagen, elastin, and other

ECM components. On the other hand, the extracellular matrix has the ability to control the growth and movement of dermal fibroblasts and also affect their transdifferentiation into various cell types, influencing both the function and morphology of fibroblasts (see Figure 1) [5].

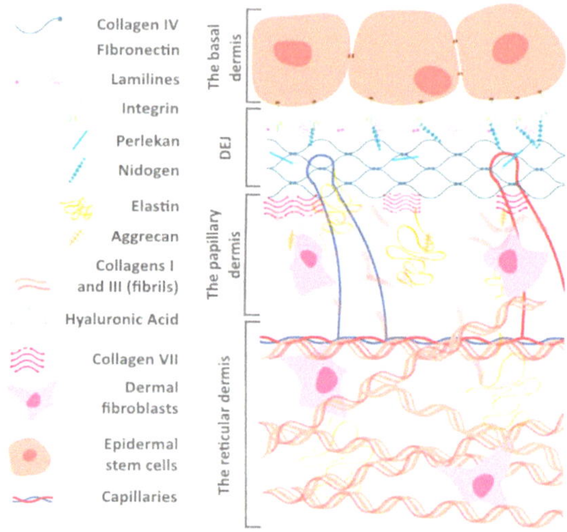

Figure 1. The multilayered arrangement of the main components of the dermoepidermal junction (DEJ) and dermis demonstrates the interaction of cell populations and the extracellular matrix (ECM).

Functional polymer materials with unique properties have the potential to improve medical devices and drugs, leading to more effective treatments [6]. These novel polymers are being produced in biomedical research using a simple and technologically advanced electrospun process [7]. The morphology and proliferation of eukaryotic cells depend on their microenvironment, as the local orientation of the electro spun fibres has a pronounced influence on cells [8].

Artificial scaffolds provide a reliable surface for cell attachment, ECM formation, transdifferentiation, and functional tissue formation. Scaffolds can be created from different materials, which include natural, synthetic, or composite materials. Techniques such as pore-forming leaching, electrospinning, molecular self-assembly, and phase separation can be used to create these scaffolds. A potential solution for overcoming specific limitations of individual materials, such as inadequate mechanical properties, weak cell adhesion, and inappropriate biodegradation rates, is the use of composite biomaterials [9]. Electrospun materials offer significant benefits for tissue engineering due to their ability to closely imitate natural scaffold structures as well as sustain cell migration and proliferation. Dermis scaffold implants accelerate the healing process of partially or completely damaged skin caused by thermal or chemical exposure [10]. Furthermore, the three-dimensional scaffold matrix promotes tissue regeneration while preventing the formation of new wound lesions during the initial healing stages. The low cytotoxicity and adhesive properties of the scaffold's structural features facilitate cell migration along the fibres within the wound. Thus, the key feature of nonwoven polymer materials in medical applications is their ability to support cell migration and homing for tissue formation. For materials scientists in biotechnology, the establishment of reliable methods for assessing cell migration, proliferation, and homing is a challenging task.

Operando research in biomaterials requires advanced imaging techniques, real-time monitoring tools, and specialised experimental set-ups to observe and analyse materials in their functional states in natural environments [11].

The migration of fibroblast cells is pivotal in the wound recovery process, bone formation, and implant healing. The application potential and utility of nanofibre mats in tissue engineering will be maximised if they promote cell monolayer and ECM formation. Model materials such as PLA-gelatine are useful for extending and modifying the fibre surface topology. Convolutional or deep neural network (CNN) analysis, by monitoring nuclei as an indicator of cell migration, differentiation, and homing in different microenvironments, influences gelatine release as an important factor for cellular colonisation of mats. By cell migration, we mean the change in the number of cells in the volume of an artificial, synthetic, or native implant per unit of time and per unit of area. The proposed CNN model effectively assesses cellular stress based on migration rates, homing, and nuclear morphology. We have used primary fibroblasts with a mutation in the ND4 gene, which is used as a cell model for Leber Hereditary Optic Neuropathy (LHON) [12]. These fibroblast cells were previously used to generate iPSCs that differentiated along retinal ganglion cells (RGCs) [13]. In this way, we are creating disease models and drug development for gene therapy that exploit allotopic expression of the ND4 gene through MTS signalling [14,15]. We aimed to investigate whether the migration and homing of ND4 mutated primary fibroblasts would be altered after transduction by a therapeutic AAV vector compared with a GFP-expressing control. This is important because AAV vector transduction has been reported to be toxic to cells [16] and may induce cellular stress by activating the antiviral immune response.

2. Materials and Methods

2.1. Electrospun Technique

Poly-(D,L)-lactide (PLA, Ingeo 4032d, NatureWorks, 650 Industrial Park Drive, PO Box 564 Blair, NE 68008, USA) -gelatine (ServiceBio, Wuhan, China) solutions of 100 mg/mL and 200 mg/mL of 1,1,1,3,3,3-hexafluoro-2-propanol (HFiP, 105228-2KG, Sigma-Aldrich, St. Louis, MO, USA), respectively. The PLA-gelatine solutions were composed through combination at 1:1 and 1:3 vol. % ratios. For instance, mat A is PLA-gelatin (100 mg/mL) (3:1), mat B is PLA-gelatine (100 mg/mL) (1:1), mat C is PLA-gelatine (200 mg/mL) (3:1), and mat D is PLA-gelatine (200 mg/mL) (1:1). Electrospinning was conducted with an acceleration voltage of 30 kV and a flow rate of 1 mL/h using an HSW NORM-JECT syringe (Henke Sass Wolf, Noerten-Hardenberg, Germany) at a distance of 30 cm from the counter electrode on a MECC NF 500 (MECC CO, Kyoto, Japan) at a temperature of 25 °C and 45–55% humidity. A 22 gauge stainless steel needle was employed as the nozzle, and an emitting electrode of positive polarity, alongside a ground electrode, was connected to a flat static collector. A polymethylmethacrylate (PMMA) dielectric collector was utilised to acquire freestanding mats positioned 10 cm from the counter-electrode. The mats that were 150–300 μm thick were deposited into Petri dishes 9 cm in diameter (Perint, St. Petersburg, Russia). Subsequently, the dishes were incubated for 24 h at 30 °C.

2.2. Scanning Electron Microscopy

The surface morphology and topography of the electrospun nanofibres, as well as morphology of the primary fibroblasts, underwent examination via a scanning electron microscope (Carl Zeiss Crossbeam 550, Oberkochen, Germany). The SEM was operated at an accelerating voltage of 8 kV and a current of 150 pA. A Quorum (Q150T S/E/ES+, London, UK) was used to deposit a 10 nm gold-palladium (80:20 at.%) coating. Around 100 random measurements using ImageJ image analysis software (National Institutes of Health, Bethesda, MD, USA) were taken to estimate the diameter and distribution [17]. Two detectors such as SE2 and InLens were employed to measure the cell area. The SE2 detector is apt for visualising the cell surface topology and cell boundaries, while the InLens detector allows the outline of a cell's nucleus to be visible. The cell morphology was analysed by quantifying the cell area and cell proliferation.

The primary human skin fibroblasts were prepared for SEM through three steps: fixation, dehydration, and plating. The cells were fixed to a 4% formalin solution for

30 min at room temperature and then washed twice with DPBS. The fixed cells on the mat holder were subjected to stepwise exposition in 500 μL ethanol-water solution for 5 min at 10%, 20%, 30%, 40%, 50%, 60%, 70%, 70%, 80%, 90%, and 96% (the 70% and 96% steps were prolonged to 10 min) [18]. The samples were incubated in a 1:1 mixture of hexamethyldisilazane (HMDS)-ethanol for 10 min and subsequently subjected to three rounds of incubation in 100% HMDS (300 μL). The first two incubations lasted for 10 min each, while the third incubation was extended overnight to ensure complete evaporation of the HMDS.

2.3. Dynamic Mechanical Analysis

Dynamic mechanical analysis (DMA) was carried out using an RSA-G2 (TA Instruments, New Castle, DE, USA). The test was performed at a constant tensile rate of 5×10^{-3} mm/min in accordance with ASTM D638 without any preload until the sample broke [19]. The measurements were taken at a temperature of 37 °C. The samples had an aspect ratio of 1:6 and a thickness that varied from 150 to 300 μm. Determining the mechanical properties of thin nanofibre composites is challenging due to the limitations of anchorage and the application of excessive strain. These samples were incubated at 37 °C in dPBS for a fixed period of time. They were mounted in a special frame to avoid mechanical deformation. We tested at least three samples from each batch of electrospun mats.

2.4. Laser Scanning Confocal Microscopy

Primary fibroblast proliferation and migration were analysed after 1, 4, and 7 days. An LSM 980 confocal laser scanning microscope (Carl Zeiss, Oberkochen, Germany) with a Plan-Apochromat 20×/0.8 objective was used to assess proliferation and cell migration. The laser power at 405 nm was set to 0.5–1% nominal power. The cells were fixed with a 4% solution of buffered cold formalin for 15 min, followed by washing twice with DPBS. Then, the cell nuclei were stained with Hoechst 33342 (Thermo Fisher Scientific, Waltham, MA, USA) and propidium iodide.

2.5. AAV Production

Briefly, a suspension of HEK293 cells (ECACC 85120602) was transfected with pAAV expression plasmid (ND4 or GFP), pHelper Vector (Part No. 340202, Cell Biolabs Inc., San Diego, CA, USA), and pAAV-RC6 Vector (Part No. VPK-426) with a molar ratio of 2:2:5, respectively. Then, pAAV-GFP plasmid (Cat. number VP-401, CellBiolabs, San Diego, CA, USA) was used to produce AAV6-GFP vector. To produce an AAV vector for allotopic expression of ND4, the pAAV-COX8-ND4 plasmid, expressing the ND4 gene with a COX8 mitochondrial localisation signal at the 5′ end under the control of the CMV promoter, was used (Supplementary Figure S1). DNA (1.5 μg per 1 million cells) was mixed with PEI MAX (Linear polymer, MW 40′000, Polysciences Inc., Warrington, PA, USA) at a mass ratio of 1:5, respectively, to a final volume of 5%. After transfection, the cells were cultivated in BalanCD HEK293 medium (Irvine Scientific, Santa Ana, CA, USA) at 37 °C, 5% CO_2, and 100 rpm for 120 h. The cells were then lysed (0.05% Tween-20), nuclease-treated, and centrifuged at 3000× g for 10 min, and the supernatant was filtered by 0.22 μm membranes. The concentrate was then purified by affinity chromatography (AAVX, Thermo Fisher Scientific). The number of viral genomes was determined by qPCR. The obtained sample was subsequently filtered by a 0.22 μm syringe filter (TPP 99722).

2.6. Fibroblast Transduction by AAV

The fibroblasts were transduced with 30,000 viral genomes per cell (MOI), and 500,000 cells were transferred to 3 cm TC-treated plates. After 17–18 h of incubation (37 °C, 5% CO_2), the medium was collected, the cells were trypsinised according to the protocol described above, and after 36 h, the transduced cells were plated onto the mats. The efficiency of transduction was estimated by GFP expression and reached 50% GFP+ cells after 17–18 h of culturing by fluorescence intensity on Axio Vert.A1 (Carl Zeiss, Germany).

2.7. Convolutional Neural Network (CNN)

The U-Net [20] trained on the primary fibroblasts on the mats (5140 images) and the DSB dataset [21] detected nuclei in the PNG files using predefined metrics like the IoU and dice loss. Augmentation boosts the dataset in such a manner that reproducing raw images similar to the original produces a more robust model with a lower overfitting risk or better generalisation capabilities [22]. The intersection over union (IOU) can help assess a model's performance on training and validation datasets for nuclei vision tasks. It shows whether the model can effectively generalise and be applied to new data. Fibroblast nuclei masks were manually designed for the model training. A training IOU is a metric on a training dataset that indicates the model's segmentation quality of objects in the images used in training. A high training IOU demonstrates that the model is proficiently trained and efficiently identifies the objects in the images. A validation IOU is a metric value assigned to a validation dataset that is not utilised for model training.

All testing and training of the neural network were carried out using an NVIDIA RTX 3060 graphics card (256 bits, 8 GB memory). Our study utilised two separate cell datasets comprising microscopic images from distinct biological fields. The first dataset consisted of 108 CZI files, each containing 14 layers and boasting a resolution of 13,926 × 13,926 ppi (Figure 4a).

3. Results and Discussion

3.1. Mechanical Properties of Microenvironment

The SEM images (Figure 2a) showed a dense layer of composite fibres on the mats. The average fibre diameter of the mats was 566 ± 123 nm for mat A, 690 ± 124 nm for mat B, 820 ± 255 nm for mat C, and 1440 ± 400 nm for mat D. The average diameter and standard deviation of a minimum of 200 fibres per sample are presented. Higher concentrations of gelatine in the fibres showed an irregular morphology, with more ribbons and thinner spider-like or reticular structures. Fibre diameter reduction was a variable parameter and was not investigated further.

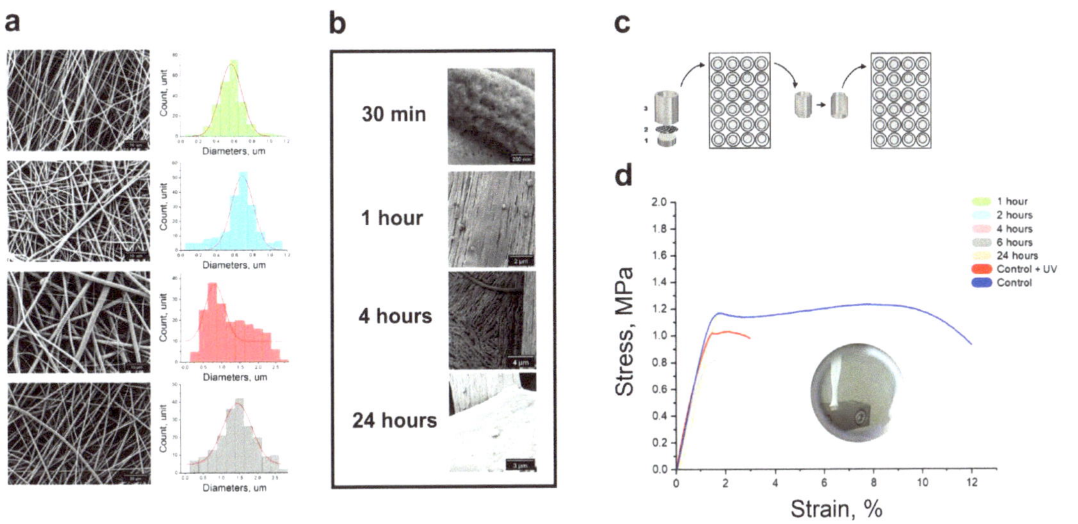

Figure 2. Compositions A, B, C, and D (from top to bottom). (**a**) SEM images of microstructure and diameter distribution of fibres. (**b**) Release of gelatine from fibres. (**c**) Fibrous mat-mounted cell culture with 1 cm^2 inner diameter for PMMA holders. (**d**) Stress–strain curves for mat composition B.

The fibres underwent considerable topographical changes after being incubated in an aqueous medium for 30 min. The initial roughness of the fibre was substituted with deep furrows. Moreover, the surfaces of thin tapes were separated into plates, which combined with the fibres.

As the incubation time increased, grooves resembling cracks emerged on the fibres, which were still visible even after 24 h (Figure 2b). Images (Figure 2b) are provided for visual assessment of the fibre topography at 200 nm, 2 μm, 4 μm, and 3 μm magnifications after the 24 h incubation period. These observations are an indication that PLA-gelatine mats are readily subject to prolonged degradation. This allows for a positive prediction of its potential for long-term incubation in cell culture. The degradation level of the mats should match the rate of new tissue formation, ensuring optimal conditions for successful regeneration. Rapid mat degradation can inhibit cell proliferation, while slow mat degradation can hinder the restoration of biological tissue function [23]. Therefore, maintaining a balance between the rate of mat degradation and new tissue formation is critical.

The material must have adequate strength to resist manipulation during implantation and preservation under both storage and physiological conditions. The skin's tensile strength is approximately 1–27 MPa [22] during the first and second stages of deformation, which promotes wound healing at the microscopic level, with mechanical signalling being critical for cell growth, transdifferentiation, and ultimately tissue engineering. The mechanical properties of the mats decreased from 1.6 MPa to 0.7 MPa as the strain decreased from 3% to 1.5% due to gelatine release during incubation (as presented in Figure 2d). This was accompanied by a reduction in fibre diameter [Table 1]. The tensile strength of the samples that were sterilised with UV radiation before cell seeding decreased to 0.8 MPa. UV irradiation can break down polymer chains into smaller fragments. During the experiment, the samples were subjected to a 14 W/cm^2 ultraviolet lamp. Following 24 h of incubation, the mechanical properties of the samples remained soft and within the range of the native tissue strength.

Table 1. Stress–strain values for each of the incubated mats.

	σ_{break}	σ_σ	ε_{break}	σ_ε
1 h	1.75	0.23	2.81	0.68
2 h	1.39	0.25	2.86	0.22
4 h	1.14	0.24	2.03	0.41
6 h	0.83	0.12	1.67	0.40
24 h	1.04	0.07	2.27	0.20
Control + UV	1.03	0.43	2.54	1.13
Control	1.18	0.40	10.17	1.19

3.2. *Proliferation and Changes in Nuclear Morphology during Cell Migration and Homing*

The viability of primary cells declines after a few passages in vitro, which is a disadvantage compared with immortalised cells [24]. However, primary human skin fibroblasts can retain characteristics identical to cells of intact tissue over several passages. Primary fibroblasts are better suited to closely mimicking real-life processes and conditions, including cell migration and regeneration. We used a fibroblast culture that resembled intact tissue human fibroblasts regarding their morphological features and proliferation rate (Figure 3a). The fibroblasts without transduction (Figure 3p) and with GFP transduction (Figure 3q) are shown on a culture plate (imaging by Incucyte S3, Sartorius, Germany).

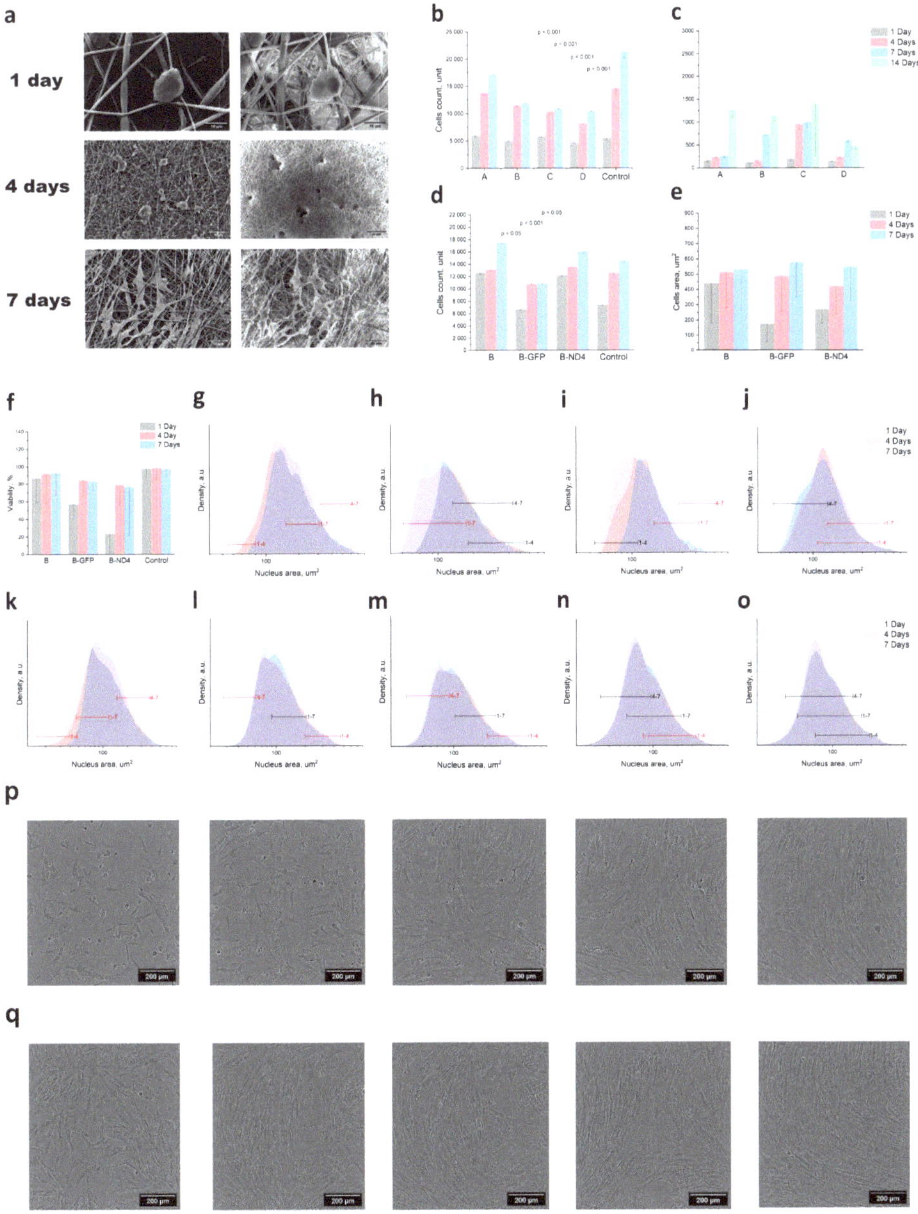

Figure 3. (**a**) SEM images of the morphology of the primary fibroblasts on mat B, cultured for 7 days. (**b**) Cell numbers on different mats through DL (2-SE; *p* values established by Freedman comparison). (**c**) Cell area per mat under SEM (SD). (**d**) Number of AAV-transduced GFP and ND4 cells through DL (2-SE; *p* values established by Freedman comparison). (**e**) Cell area per transduced cell on mat B under SEM (SD). (**f**) Cell viability (Kolmogorov test). (**g–j**) Cell nucleus area on the culture plate, mat B, transduced GFP, and ND4 (from left to right, with mean difference from Tukey comparison (red = significant)). (**k–o**) Cell nucleus area on the culture plate and mats A, B, C, and D (from left to right, with mean difference from Tukey comparison (red = significant)). (**p**) Cells without transduction (1–5 days). (**q**) Transduced cells (1–5 days). All parameters showed statistical significance ($p < 0.05$).

Nonwoven materials facilitate cell attachment and cell division and guide cell migration and differentiation, providing the necessary conditions for tissue engineering [25]. As previously found, the AAV6 serotype effectively transduces primary cell cultures, including fibroblasts and keratinocytes [26]. The experiments were carried out in several stages: assessment of cell proliferation and migration on nonwoven materials, selection of the best nonwoven material according to these parameters, and transduction of cells with AAV-GPP or AAV-ND4 to assess migration and proliferation on the selected nonwoven substrate. Analysis of the cell morphology is of particular interest, since changes in cell shape can be a sign of a change in the fibroblast phenotype (Figure 3a). We observed that on the fourth day of the experiment, on all mats, the cells acquired a spindle-shaped appearance and multidirectional filopodia.

On the electrospun mats, the proliferative activity of the primary fibroblasts was assessed on days 1, 4, and 7 of the experiments by measuring the stained cell nuclei using ML from the LSCM images. According to the preliminary experiments, cell proliferation depended on the size of the mat fibres. Mats B and C had the same fibre diameter, and their cell activity was also quite similar. Mat A showed the most favorable dynamics of proliferative activity and the largest fibre diameter.

Fibroblasts were cultured for seven days to evaluate cell growth on the electrospun mats. After static seeding on the first day, the cells were attached well to both the mat and the culture dish. However, cell growth was higher on all mats except mat D compared with the culture dish (Figure 3b). The primary human skin fibroblasts grew well on mats A, B, and C, while on mats B and C, they demonstrated the highest rates of proliferation (see Figure 3c). During long-term cultivation, fibroblasts were distributed over the surface of the mat and infiltrated them. On the seventh day of cultivation, the surface of the nonwoven material was completely covered with primary human fibroblasts.

The AAV-transduced cells on mat B had similar levels of proliferative activity, as shown in (Figure 3g) and (Figure 3k). The number of transduced cells on mat B increased linearly in all groups (Figure 3d). However, AAV transduction had toxic effects on the cells. On day 7, the cells transduced with AAV-GFP showed a decrease in their proliferation rate, possibly due to the cellular toxicity of AAV or due to the negative effects of GFP, as previously shown [27]. In contrast, the transduction by AVV containing ND4 resulted in a proliferation rate that was comparable to the wild-type cells on the cultural plate. This implies a potential positive effect of ND4 expression on cellular function. According to SEM analysis, on mat B, there were no significant differences in nuclear area variance between the GFP-transduced cells and ND4-transduced cells. On day 7, multiple filopodia running in different directions were observed in all groups.

Cytotoxicity assessment of the electrospun mats (see Figure 3f) showed no significant variation in human skin fibroblast viability when comparing individual mats with the culture plastic. In terms of the number of living cells per nanofibre mat and viability, mat B showed the most satisfactory results. There was no significant difference in the number of live cells between mat B and the control samples. However, the total cell count showed significantly different pairs in the B-GFP and B-ND4 cases according to one-way ANOVA.

The aim of our study was to test whether the area of the nuclei could be used to estimate cell migration through fibrillar surfaces into the nonwoven volume. We also analysed the effects of cell transduction on cell behaviour on the mats. Of significant interest is the study of the topography of substrates for cultivation and its influence on variations in nuclear morphology and the rates of cell proliferation and differentiation [28]. The morphology of the stress fibres of the cytoskeleton plays a crucial role in the deformation and division of the cell nucleus. This phenomenon is directly intertwined with the cellular mechanotransduction signaling and gene regulatory mechanism during cell interaction with the microenvironment [29]. Fibroblast reprogramming can be enhanced using specific cues from the three-dimensional substrate [30]. We found that the area and morphology of the nuclei on the surface and within the nonwoven substrate could fluctuate [31].

Each graph showed a minimum of 15,000 nuclei distributed log-normally. The changes in cell volume and nuclei morphology triggered several intracellular signalling pathways [32]. The nuclei on the culture plate (Figure 3g,k) showed a consistent increase in nuclei area. By day 7, the major-to-minor axis ratio of the nucleus should have been 2–2.5, compared with the 1:1 ratio on day 1. The nuclear area increased, and the diameter ratio increased due to the high proliferative activity of the cells. The different passages may explain the observed difference in nuclear area between experiments g and k on the cultural plate. The cultured primary fibroblasts carrying m.11778G>A on the nonwoven mats exhibited a notable decrease in nuclear area, as demonstrated in the experiment series (Figure 3h,m) compared with the culture plastic (Figure 3h,k). The area of the cell nuclei decreased (Figure 3l,m) on the nonwoven A and B substrates with diameters (Figure 2a) due to the three-dimensional topology of the mats and the hindrance of actin polymerisation. Nevertheless, the total cell area (Figure 3c) increased, which is intriguing. Actin polymerisation promotes morphological transformations in cell proliferation during the attachment procedure [33]. Three factors, namely the highly coiled state of the nuclei in detached cells, pressure differences on the nuclear envelope, and mechanical effects from the cytoplasm, are responsible for the decrease in nuclear volume or area during cell detachment [34]. The fibre topology and morphology influence cell division and drug release. For example, in thin fibres, the release of drugs is accelerated. We used type-B gelatin as a useful model substance to modify the fibre surface topology. The release time of 80% of the gelatin was about 4 days.

The size of the nucleation area remained constant on nonwoven substrates C and D (Figure 3n,o), although their shapes and diameters differed significantly from A and B (Figure 2a). In contrast, the nuclear area of the AAV-GFP-transduced cells fluctuated. The nuclear area decreased by day 4 and increased by day 7 compared with the cells on mat B. The cells transduced with AAV-ND4 were expected to restore ND4 gene function, which would return the cells to a control phenotype with normal functional activity. When comparing the ND4-transduced cells on mat B, we found that their nuclear dynamics were similar to those of the cells on mats C and D (Figure 3n,o). However, the presence of young cells observed on mat B (Figure 3g) indicates that the fibre topology may also contribute to providing a favorable spatial configuration for the cells. Nevertheless, this does not conclusively establish a correlation between the nuclear area, substrate spatial configuration, and cell cycle.

Despite research on various techniques and biological applications of nonwoven polymer composites, only a minority of scientists have investigated lateral cell migration on non-native structures. Therefore, we propose using machine learning tools for counting and segmenting diverse structures, as well as assessing cell migration.

3.3. Model of Cell Migration on Three-Dimensional Systems

The movement of various cell types in response to the microenvironment has been previously addressed [35,36]. The validation IOU value aids in estimating the model's performance on fresh, novel data. The more the validation IOU value approximates to one, the better the model generalises pre-existing knowledge and detects objects in new images (Figure 4c). We generated a summary image of the dataset layer by layer, with a step of 7 μm for each mat in the z direction (Figure 4d). Each layer was divided into blocks of multiples of 1024 × 1024 pixels to count the nuclei of the primary fibroblasts. Objects detected in focus were entered into a CSV file. Morphological transformations were applied to each image layer to segment the nuclei into relevant metrics (Figure 4f). The SSIM metric was used to compare each nucleus. Only the fibroblast nucleus in focus was assigned to a specific layer (Figure 4e), and it was not counted in subsequent layers. The fibroblast nuclei that were found were encircled with different colours for visualisation (Figure 4b).

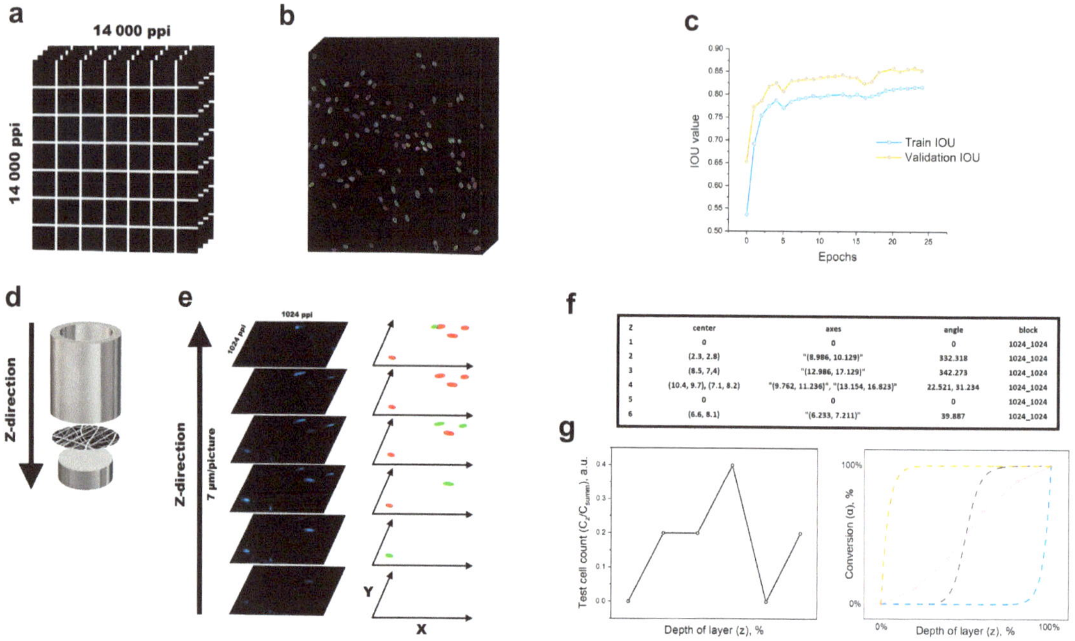

Figure 4. (**a**) Laser-scanning confocal image. (**b**) Part of validation image U-net + aMT. (**c**) IOU score. (**d**) Scanning in z direction on mat holder. (**e**) SSIM per layer. (**f**) Metrics for nuclei. (**g**) Cell count per layer and conversion plot.

As mentioned above, we could procure the coordinates of nuclei and their corresponding measurements for each layer (Figure 4f) by analysing the test images (Figure 4e, from bottom to top). The results indicate a detection of five nuclei in six test images. This ratio of cell numbers in each layer provides reliable insight into the behaviour of the chosen culture on the nonwoven substrate.

The data are displayed on a graph with a y axis (C_z/C_{summ}) and x axis (z). C_z represents the number of cells per layer, and C_{summ} represents the total number of cells in the material (refer to Figure 4g, left). Conversion curves are frequently employed to illustrate the kinetics of n-order model processes in the DSC technique applied at different heating rates [37,38]. The conversion value (α) indicates the fraction of cells in the material layer in this scenario. It corresponds to the red curve in the test data (Figure 4g, right). The cells will settle on the mat holder's upper layers of the nonwoven substrate. Ideally, the conversion curve should correspond to the yellow dotted exponential curve in (Figure 4g, right). As the mass of cells shifts within the material under examination, the curve should take the form of a blue dashed exponential curve. In general, the cells on the surface remain active and do not disappear from observation if the nonwoven material is spatially positioned in a suitable configuration. In the case of electrospun materials, the free volume fraction typically ranges from 23 to 93%, which correlates with Feigenbaum's second constant per layer [39–41]. As a result, the cell mass conversion curve assumes a Hill function form, as indicated by the grey dashed curve on the right-hand side of Figure 4g.

Nonwoven materials have characteristics similar to the skin frame in terms of fibre diameter and the presence of free volume for migration and cell colonisation [42]. Figure 5a (i = A, ii = B, iii = C, and iv = D) provides the cell counts for each layer of nonwoven materials of varying compositions. The primary fibroblasts migrated freely in the mats along and between the fibres of compositions A, B, and C. Each data point was generated from three independent prepared samples. On day 7, there was an increase in the number of cells in the upper layers of material D. This observation of the cell dynamics is significant for further justification of the approach.

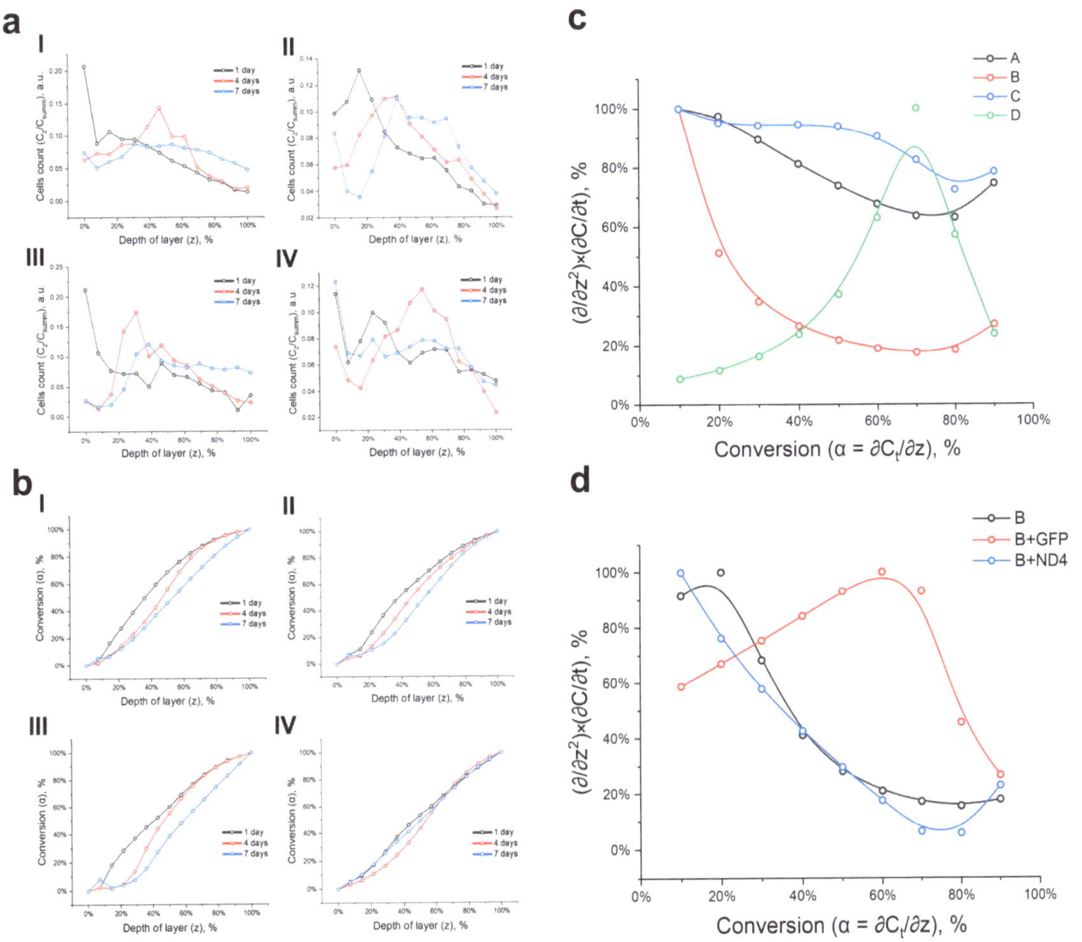

Figure 5. (**a**) Number of cells in each layer per mat. (**b**) Conversion curves of cells per mat. (**c**) Cell migration plot per mat. (**d**) Plot of transduced cell migration on mat B.

The number of cells for each composition and diameter distribution increased exponentially, as shown in the graphs for day 1. These conversion curves (Figure 5b, where i = A, ii = B, iii = C, and iv = D) illustrate this information clearly. As the cell mass increased in the deeper layers, the curve took on the shape of a hill. Based on the proliferative activity plot (Figure 3b,d), this indicates the rate of cell division in a specific microenvironment. The rate was assessed from the repopulation time to day 1 (t_1), from day 1 to day 4 (t_2), and from day 4 to day 7 (t_3). To establish the value on the *x* axis (z) for each day's conversion curve, with 10% α increments, we had to multiply it by the proliferation rate (C_{summ}/t_d) for each location. We plotted the curve on the axes of $(\alpha/z^2) \times (C_{summ}/t_d)$ (α), performed a linear fit, and determined the tangent of the slope $(1/z^2) \times (C_{summ}/t_d)$ for each step. Subsequently, we plotted the obtained curves for the mats of all compositions (Figure 5c).

This unusual format can demonstrate the distinct characteristics of diverse cell cultures on nonwoven substrates, suggest a method for evaluating cell migration potential in various materials, and serve as a valuable tool in assessing migration potential in diverse materials. It can be inferred from the graph that half of the cells in sample composition B accounted for approximately 75% of the migration activity $[(1/z^2) \times (C_{summ}/t_d)]$. The study of cell migration on distinct nonwoven materials can be likened to the study of cells stimulated by ND4, which can improve functional activity, migration, and homing in the presence of nanofibres (Figure 5d). The graph demonstrates comparable migration activity between cells with gene dysfunction and cells transduced with the ND4 transgene. The graphs demonstrate how cells behave on a nonwoven material for gene transfer efficacy. This includes their proliferation activity per day, which is dependent on the cell density and is based on a significant amount of data.

A study using a CNN found differences between the cellular dynamics of the ND4 and GFP transduced cells. Restoring ND4 expression normalises cell mitochondrial function, which restores the cell distribution, becoming similar to that of wild-type cells. In addition, these studies will allow us to study the effect of AAV-mediated transduction and transgene expression on cell division, cell viability, and the study of the cellular immune response, differentiation, and functional activity [43,44].

4. Conclusions

4.1. Fibroblast Mechanobiology

Our findings enhance the comprehension of fibroblast mechanobiology. Furthermore, our deep learning-based migration evaluation method has the potential to forecast and regulate cell–substrate interactions, which are extensively utilised in the regenerative medicine and tissue engineering fields. CNNs have surpassed human accuracy in image classification. It is believed that CNNs will greatly impact research on feature and pattern detection in tissue repair tasks.

Numerous rare skin diseases require approbation of therapies on biomimetic systems. In order to achieve optimal therapeutic outcomes, topical gels with different drug delivery methods and substrates coated with autologous or allogeneic cells should take into account the microenvironmental behaviour under different external factors. The article presents a method of evaluating cell migration and cell homing potential, specifically with primary fibroblasts as an example. One approach to assessing cell behaviour involves analysing the nucleus area with a sufficiently large dataset to achieve statistical significance. Studying the behaviour of biological systems in artificial environments could aid researchers in developing multidimensional systems to replace various tissues and organs.

4.2. Conclusion and Future Perspectives

In the future, this research will be repeated on fibroblasts with diverse mutations, as well as animal fibroblasts. We will scrutinise the consequences of distinct transgene expressions on cell phenotypic traits simultaneously. Thus, further research can provide valuable insights for improving corrective techniques and implementing advanced CNN methodologies to treat orphan diseases.

Supplementary Materials: The following supporting information can be downloaded at: https://www.mdpi.com/article/10.3390/polym16091187/s1, Figure S1: Supplementary Image 1.

Author Contributions: Methodology, I.I.L., R.O.S., A.V.K., V.S.L., D.V.S., A.P.M. and D.V.K.; software, I.I.L. and V.S.L.; validation, R.O.S., I.I.L. and E.V.L.; formal analysis, S.A.R., D.V.S., I.I.L., D.A.I. and A.V.K.; resources, R.O.S., E.V.L., A.V.K., D.A.I. and V.S.L.; data curation, I.I.L. and R.O.S.; writing—original draft, I.I.L. and R.O.S.; writing—review and editing, I.I.L., S.A.R. and A.V.K.; visualisation, I.I.L.; supervision, A.V.K. and S.A.R.; project administration, A.V.K.; funding acquisition, A.V.K. and D.A.I. All authors have read and agreed to the published version of the manuscript.

Funding: This work was supported by the Ministry of Science and Higher Education of the Russian Federation (Agreement No. 075-10-2021-093; Project Nos. GTH-RND-2112 and BMA-RND-2006).

Data Availability Statement: Data are contained within the article and Supplementary Materials; Training dataset: https://drive.google.com/file/d/1lEWe5-SMsUIyQ6YiYEAbwusTY22_XOQb/view?usp=sharing Available until 25 November 2024.

Acknowledgments: The authors thank Roman V. Dementiev (PJSC Sberbank) for assistance with writing the neural network and training and validation of the experimental data.

Conflicts of Interest: The authors declare no conflicts of interest.

References

1. Stegemann, J.P.; Wagner, W.R.; Göbel, U.; Perets, E.; Leong, K.W.; Leach, J.K.; Gilbert, J. The Year 2022 in biomaterials research: A perspective from the editors of six leading journals. *J. Biomed. Mater. Res. Part A* **2023**, *111*, 1298–1308. [CrossRef] [PubMed]
2. Stupin, V.A.; Gabitov, R.B.; Sinelnikova, T.G.; Silina, E.V. Biological mechanisms of chronic wound and diabetic foot healing: The role of collagen. *Serbian J. Exp. Clin. Res.* **2018**, *19*, 373–382. [CrossRef]
3. Pina, S.; Oliveira, J.M.; Reis, R.L. Natural-based nanocomposites for bone tissue engineering and regenerative medicine: A review. *Adv. Mater.* **2015**, *27*, 1143–1169. [CrossRef] [PubMed]
4. Diller, R.B.; Tabor, A.J. The role of the extracellular matrix (ECM) in wound healing: A review. *Biomimetics* **2022**, *7*, 87. [CrossRef] [PubMed]
5. Barrientos, S.; Stojadinovic, O.; Golinko, M.S.; Brem, H.; Tomic-Canic, M. Growth factors and cytokines in wound healing. *Wound Repair Regen.* **2008**, *16*, 585–601. [CrossRef]
6. Przekora, A. Current trends in fabrication of biomaterials for bone and cartilage regeneration: Materials modifications and biophysical stimulations. *Int. J. Mol. Sci.* **2019**, *20*, 435. [CrossRef]
7. Manea, L.R.; Hristian, L.; Leon, A.L.; Popa, A. Recent advances of basic materials to obtain electrospun polymeric nanofibers for medical applications. In *IOP Conference Series: Materials Science and Engineering*; IOP Publishing: Bristol, UK, 2016; Volume 145, p. 032006. [CrossRef]
8. Bogdanova, A.S.; Sokolova, A.I.; Pavlova, E.R.; Klinov, D.V.; Bagrov, D.V. Investigation of cellular morphology and proliferation on patterned electrospun PLA-gelatin mats. *J. Biol. Phys.* **2021**, *47*, 205–214. [CrossRef]
9. Liu, S.; Yu, J.-M.; Gan, Y.-C.; Qiu, X.-Z.; Gao, Z.-C.; Wang, H.; Chen, S.-X.; Xiong, Y.; Liu, G.-H.; Lin, S.-E.; et al. Biomimetic natural biomaterials for tissue engineering and regenerative medicine: New biosynthesis methods, recent advances, and emerging applications. *Mil. Med. Res.* **2023**, *10*, 16. [CrossRef] [PubMed]
10. Wu, Y.Z.; Huang, C.H.; Zhang, X.Y.; Shen, G.L. The clinical application effects of artificial dermis scaffold and autologous split-thickness skin composite grafts combined with vacuum-assisted closure in refractory wounds. *Int. Wound J.* **2023**, *20*, 2113–2120. [CrossRef]
11. Žunar, B.; Ito, T.; Mosrin, C.; Sugahara, Y.; Bénédetti, H.; Guégan, R.; Vallée, B. Confocal imaging of biomarkers at a single-cell resolution: Quantifying 'living' in 3D-printable engineered living material based on Pluronic F-127 and yeast Saccharomyces cerevisiae. *Biomater. Res.* **2022**, *26*, 85. [CrossRef]
12. Man, P.Y.W.; Turnbull, D.M.; Chinnery, P.F. Leber hereditary optic neuropathy. *J. Med. Genet.* **2002**, *39*, 162–169. [CrossRef] [PubMed] [PubMed Central]
13. Wu, Y.-R.; Wang, A.-G.; Chen, Y.-T.; Yarmishyn, A.A.; Buddhakosai, W.; Yang, T.-C.; Hwang, D.-K.; Yang, Y.-P.; Shen, C.-N.; Lee, H.-C.; et al. Bioactivity and gene expression profiles of hiPSC-generated retinal ganglion cells in MT-ND4 mutated Leber's hereditary optic neuropathy. *Exp. Cell Res.* **2018**, *363*, 299–309. [CrossRef] [PubMed]
14. Omura, T. Mitochondria-targeting sequence, a multi-role sorting sequence recognized at all steps of protein import into mitochondria. *J. Biochem.* **1998**, *123*, 1010–1016. [CrossRef] [PubMed]
15. Bonnet, C.; Augustin, S.; Ellouze, S.; Bénit, P.; Bouaita, A.; Rustin, P.; Sahel, J.-A.; Corral-Debrinski, M. The optimized allotopic expression of ND1 or ND4 genes restores respiratory chain complex I activity in fibroblasts harboring mutations in these genes. *Biochim. Biophys. Acta (BBA)-Mol. Cell Res.* **2008**, *1783*, 1707–1717. [CrossRef] [PubMed]
16. Howard, D.B.; Powers, K.; Wang, Y.; Harvey, B.K. Tropism and toxicity of adeno-associated viral vector serotypes 1, 2, 5, 6, 7, 8, and 9 in rat neurons and glia in vitro. *Virology* **2008**, *372*, 24–34. [CrossRef] [PubMed] [PubMed Central]
17. Schindelin, J.; Arganda-Carreras, I.; Frise, E.; Kaynig, V.; Longair, M.; Pietzsch, T.; Preibisch, S.; Rueden, C.; Saalfeld, S.; Schmid, B.; et al. Fiji: An open-source platform for biological-image analysis. *Nat. Methods* **2012**, *9*, 676–682. [CrossRef] [PubMed]
18. Zhu, X.; Cui, W.; Li, X.; Jin, Y. Electrospun fibrous mats with high porosity as potential scaffolds for skin tissue engineering. *Biomacromolecules* **2008**, *9*, 1795–1801. [CrossRef] [PubMed]
19. *ASTM D638*; Standard Test Method for Tensile Properties of Plastics. ASTM International: West Conshohocken, PA, USA, 2022.
20. Ronneberger, O.; Fischer, P.; Brox, T. U-net: Convolutional networks for biomedical image segmentation. In *Medical Image Computing and Computer-Assisted Intervention–MICCAI 2015, Proceedings of the 18th International Conference, Munich, Germany, 5–9 October 2015*; Part III 18; Springer International Publishing: Berlin/Heidelberg, Germany, 2015; pp. 234–241.
21. Data Science Bowl. 2018. Available online: https://www.kaggle.com/c/data-science-bowl-2018 (accessed on 21 December 2021).
22. Mumuni, A.; Mumuni, F. Data augmentation: A comprehensive survey of modern approaches. *Array* **2022**, *16*, 100258. [CrossRef]
23. Joodaki, H.; Panzer, M.B. Skin mechanical properties and modeling: A review. *Proc. Inst. Mech. Eng. Part H J. Eng. Med.* **2018**, *232*, 323–343. [CrossRef]

24. Carter, M.; Shieh, J. Cell culture techniques. In *Guide to Research Techniques in Neuroscience*; Elsevier: Amsterdam, The Netherlands, 2015; pp. 295–310.
25. Bambole, V.; Yakhmi, J.V. Tissue engineering: Use of electrospinning technique for recreating physiological functions. In *Nanobiomaterials in Soft Tissue Engineering*; William Andrew Publishing: Norwich, NY, USA, 2016; pp. 387–455.
26. Ellis, B.L.; Hirsch, M.L.; Barker, J.C.; Connelly, J.P.; Steininger, R.J.; Porteus, M.H. A survey of ex vivo/in vitro transduction efficiency of mammalian primary cells and cell lines with Nine natural adeno-associated virus (AAV1-9) and one engineered adeno-associated virus serotype. *Virol. J.* **2013**, *10*, 74. [CrossRef]
27. Ansari, A.M.; Ahmed, A.K.; Matsangos, A.E.; Lay, F.; Born, L.J.; Marti, G.; Harmon, J.W.; Sun, Z. Cellular GFP toxicity and immunogenicity: Potential confounders in in vivo cell tracking experiments. *Stem Cell Rev. Rep.* **2016**, *12*, 553–559. [CrossRef] [PubMed]
28. Mamaghani, P.; Donald, A.M. The impact of fibronection stripe patterns on the cellular and nuclear morphology of fibroblasts. *bioRxiv* **2018**. [CrossRef]
29. Hanson, L.; Zhao, W.; Lou, H.-Y.; Lin, Z.C.; Lee, S.W.; Chowdary, P.; Cui, Y.; Cui, B. Vertical nanopillars for in situ probing of nuclear mechanics in adherent cells. *Nat. Nanotechnol.* **2015**, *10*, 554–562. [CrossRef] [PubMed]
30. Kulangara, K.; Adler, A.F.; Wang, H.; Chellappan, M.; Hammett, E.; Yasuda, R.; Leong, K.W. The effect of substrate topography on direct reprogramming of fibroblasts to induced neurons. *Biomaterials* **2014**, *35*, 5327–5336. [CrossRef] [PubMed]
31. Qin, S.; Ricotta, V.; Simon, M.; Clark, R.A.F.; Rafailovich, M.H. Continual cell deformation induced via attachment to oriented fibers enhances fibroblast cell migration. *PLoS ONE* **2015**, *10*, e0119094. [CrossRef] [PubMed]
32. Jiang, H.; Sun, S.X. Cellular pressure and volume regulation and implications for cell mechanics. *Biophys. J.* **2013**, *105*, 609–619. [CrossRef] [PubMed]
33. Parsons, J.T.; Horwitz, A.R.; Schwartz, M.A. Cell adhesion: Integrating cytoskeletal dynamics and cellular tension. *Nat. Rev. Mol. Cell Biol.* **2010**, *11*, 633–643. [CrossRef]
34. Kim, D.-H.; Li, B.; Si, F.; Philips, J.; Wirtz, D.; Sun, S.X. Volume regulation and shape bifurcation in the cell nucleus. *J. Cell Sci.* **2015**, *128*, 3375–3385. [CrossRef] [PubMed]
35. Luthria, G.; Li, R.; Wang, S.; Prytyskach, M.; Kohler, R.H.; Lauffenburger, D.A.; Mitchison, T.J.; Weissleder, R.; Miller, M.A. In vivo microscopy reveals macrophage polarization locally promotes coherent microtubule dynamics in migrating cancer cells. *Nat. Commun.* **2020**, *11*, 3521. [CrossRef]
36. Stringer, C.; Wang, T.; Michaelos, M.; Pachitariu, M. Cellpose: A generalist algorithm for cellular segmentation. *Nat. Methods* **2021**, *18*, 100–106. [CrossRef]
37. Borchardt, H.J.; Daniels, F. Differential thermal analysis of inorganic hydrates. *J. Phys. Chem.* **1957**, *61*, 917–921. [CrossRef]
38. Ozawa, T.J. Calorimetric Measurements of Grain growth in Ultrafine-Grained nichel. *Therm. Anal.* **1970**, *2*, 301. [CrossRef]
39. Wang, T.; Chen, Y.; Dong, W.; Liu, Y.; Shi, L.; Chen, R.; Pan, T. Fractal Characteristics of Porosity of Electrospun Nanofiber Membranes. *Math. Probl. Eng.* **2020**, *2020*, 2503154. [CrossRef]
40. Chen, Y.; Qi, L.; Liu, Y.; Zhang, L.; Zhu, R.; Chen, R. Study of fractal dimension of electrospun nanofibres non-woven fabrics and the relationship between it and air resistance. *J. Innov. Technol. Educ.* **2015**, *2*, 15–28. [CrossRef]
41. Wang, T.; Dong, W.; Chen, Y.; Pan, T.; Chen, R. Study on porosity of electrospun nanofiber membrane by neural network. *Appl. Math. Sci.* **2018**, *12*, 1059–1074. [CrossRef]
42. Hong, Z.-X.; Zhu, S.-T.; Li, H.; Luo, J.-Z.; Yang, Y.; An, Y.; Wang, X.; Wang, K. Bioengineered skin organoids: From development to applications. *Mil. Med. Res.* **2023**, *10*, 40. [CrossRef]
43. Veron, P.; Allo, V.; Rivière, C.; Bernard, J.; Douar, A.-M.; Masurier, C. Major subsets of human dendritic cells are efficiently transduced by self-complementary adeno-associated virus vectors 1 and 2. *J. Virol.* **2007**, *81*, 5385–5394. [CrossRef]
44. Zhang, Y.; Chirmule, N.; Gao, G.-P.; Wilson, J. CD40 ligand-dependent activation of cytotoxic T lymphocytes by adeno-associated virus vectors in vivo: Role of immature dendritic cells. *J. Virol.* **2000**, *74*, 8003–8010. [CrossRef]

polymers

Article

Crystallization of Polylactic Acid with Organic Nucleating Agents under Quiescent Conditions

Peng Gao [1,2,*], Saeed Alanazi [1] and Davide Masato [1,*]

1 Department of Plastics Engineering, University of Massachusetts Lowell, Lowell, MA 01854, USA; saeed_alanazi@student.uml.edu
2 Department of Engineering and Design, Western Washington University, Bellingham, WA 98225, USA
* Correspondence: gaop@wwu.edu (P.G.); davide_masato@uml.edu (D.M.)

Abstract: Polylactic acid (PLA) is a versatile and sustainable polymer used in various applications. This research explores the use of orotic acid (OA) and ethylene bis-stearamide (EBS) as nucleating agents to enhance the quiescent crystallization of PLA within the temperature range of 80 °C to 140 °C. Different blends were produced via melt processing before analyzing via DSC, XRD, and SEM. Our results show that both nucleating agents significantly accelerated the crystallization process and reduced the incubation time and the crystallization half-time. The most promising results were obtained with 1% EBS at 110 °C, achieving the fastest crystallization. The XRD analysis showed that at 80 °C, the disordered α'phase predominated, while more stable α phases formed at 110 °C and 140 °C. Combining the 1% nucleating agent and 110 °C promotes densely packed crystalline lamellae. The nucleated PLA exhibited a well-organized spherulitic morphology in agreement with the Avrami modeling of DSC data. Higher nucleating agent concentrations yielded smaller, more evenly distributed crystalline domains. Utilizing OA or EBS in PLA processing could offer enhanced properties, improved processability, and cost-efficiency, making PLA more competitive in various applications.

Keywords: polylactic acid (PLA); organic nucleating agent; crystallization behavior; quiescent conditions; differential scanning calorimetry

Citation: Gao, P.; Alanazi, S.; Masato, D. Crystallization of Polylactic Acid with Organic Nucleating Agents under Quiescent Conditions. *Polymers* **2024**, *16*, 320. https://doi.org/10.3390/polym16030320

Academic Editors: Hsin-Lung Chen, Vishal Gavande, Vasi Shaikh and Won-Ki Lee

Received: 15 November 2023
Revised: 8 January 2024
Accepted: 22 January 2024
Published: 24 January 2024

1. Introduction

Biobased polymer materials have gained increased attention during the last decade with various applications, including medical devices, rigid packaging, and agriculture. Biobased materials are derived from renewable resources such as plants, agricultural residues, and algae. Unlike fossil-based materials, the production and decomposition of biobased materials generally result in lower net greenhouse gas emissions [1–3]. Biobased materials often have lower levels of toxic additives and chemicals than their conventional counterparts. This reduces the pollution and health risks associated with manufacturing, use, and disposal. Additionally, some biobased materials can be part of circular economies, where products are designed for reuse, recycling, or composting. This minimizes waste and extends the lifespan of materials, contributing to a more sustainable consumption pattern. Among these materials, polylactic acid (PLA) is the most promising.

PLA is a biobased and biodegradable material from renewable resources such as starch and wheat. PLA refers to a family of materials that share slightly different characteristics. PLA has two stereoisomers, namely the L-lactic and the D-lactic acid. The three forms of PLA that are commercially available include the following: pure L-lactide, pure D-lactide, and a mix of L and D-lactide (meso-lactides). Relatively pure L-feed and D-feed PLA are referred to as PLLA and PDLA, respectively [4]. Typical commercial grade PLA with high crystallinity contains a majority of L-feed mixed with a minimum of 1–2% D-feed content, whereas the amorphous grades may contain up to 20% D-feed.

A significant drawback limiting PLA applications is the slow and uncontrollable crystallization kinetics. Indeed, high and controlled crystallization in the final product can significantly impact the mechanical behavior of this product. Harris et al. showed that PLA with 20% crystallinity could achieve 20% higher flexural modulus than the same amorphous PLA. As the crystallinity further increased to 40%, the flexural modulus increased by another 25% [5].

Four crystalline forms of PLA can be developed in PLA depending on the composition and crystallization conditions. PLA's most common crystalline structure is α form [6,7], which crystallizes from the melt or solvents under quiescent conditions. The disordered α form, known as the α' crystal, is formed from a melt or solvent but at a lower temperature (<120 °C) [8–10]. The β form is usually formed under high shear forces and temperatures [11,12]. The γ crystal is mainly obtained from the hexamethylbenzene substrate via epitaxial growth and is rarely observed in other conditions [13]. Recently, a unique crystal form, the stereocomplex crystal (SC-crystal), received significant attention. The melting point of SC-crystal is 230 °C, which is 50–70 °C higher than the other PLA crystal forms [13,14]. Recent research by Han et al. and Ma et al. shows that the formation of SC crystals is an effective method to enhance the physical and mechanical properties of PLA products [15,16].

The crystallinity of PLA can be enhanced by employing methods such as isothermal annealing [17–21], polymer blending [4,22–26], and strain-induced crystallization during processing [27–35]. Isothermal annealing at temperatures of 85–115 °C for an extended time has been reported to initiate and develop the crystalline domains for injection-molded amorphous PLA samples [21,28].

Nucleating agents have been used to increase the nuclei density and crystallization rate. Mineral-based inorganic nucleating agents have been used to enhance the crystallization behaviors of PLA. It was demonstrated that the nucleation density of PLA can be increased by 600% with the addition of 6% talc [36]. Carbon nanotubes (CNTs) were also identified as efficient nucleating agents to promote the crystallization behaviors of PLA. Through polymer grafting, 5–10 wt.% PLA-g-CNT can increase the degree of crystallinity of PLA by 12–14% and reduce the half-crystallization time ($t_{1/2}$) from 4.2 min to 1.9 min [37–39].

Compared to inorganic nucleating agents, the focus on organic nucleating agents has increased more recently. Many organic additives are derived from renewable sources and are more likely to be biodegradable than inorganic ones. Organic additives can also enhance the biocompatibility of PLA products. Qiu et al. observed that the crystallization density of PLLA increased by >200% by adding 0.3 wt.% of orotic acid (OA) [40]. Gao et al. observed that 1 wt.% of OA can reduce the crystallization time of PLA from 80 min to <5 min and reduce the energy barrier required for crystallization from 90 °C to 70 °C during the injection molding process [41].

EBS (ethylene bis-stearamide) is a synthetic wax-like compound commonly used as a processing aid and lubricant in various industries. It is derived from the reaction of ethylenediamine with two molecules of stearic acid, resulting in a long-chain amide compound. However, EBS has not been widely recognized as nucleating agents for PLA and other biopolymers [42,43]. Harris et al. reported that 2% of EBS increased the crystallization of PLA3001D from NatureWorks to 18% compared to 10% for neat PLA. In total, 2% of EBS was also reported to decrease the annealing time required by 50% for PLA3001D to reach 40% crystallinity [5]. Limited research has focused on the effect of different concentrations of organic nucleating agents. Moreover, studies used different polymer processing techniques, which can affect crystallization through shear and are, thus, not directly comparable.

In this research, the efficiency of two organic nucleating agents, orotic acid (OA) and ethylene bis-stearamide (EBS), are investigated using differential scanning calorimetry (DSC) isotherm studies at different isotherm temperatures. The objective of this work is the analysis of crystallization dynamics and the morphology obtained under quiescent conditions (i.e., without the introduction of any shear effects). A pure commercial-grade PLA was selected for this research. The degree of crystallinity, crystallization rate, and incubation time are investigated on neat PLA, the PLA-OA blend, and the PLA-EBS blend with different concentration levels. The phases and morphology of the crystalline structures obtained in the PLA samples were examined using X-ray diffraction and directly observed via scanning electron microscope (SEM).

2. Materials and Methods

2.1. Material Selection

A PLA crystallization investigation was conducted using Ingeo 2500HP obtained from NatureWorks LLC (Plymouth, MN, USA). The extrusion grade of PLA contains PLLA blended with <2% PDLA and can be fabricated into semi-crystalline samples. The material, provided in pellet form, was dried at 40 °C for 12h before any characterization. Table 1 summarizes the main properties of the commercial PLA.

Table 1. Main PLA properties. (* molded crystalline with a 120 °C mold temperature; the formula included 1 wt% of the nucleating agent (LAK-301, Takemoto Oil & Fat, Gamagori-shi Japan)).

Properties	Ingeo 2500HP	ASTM Standard
Specific gravity	1.24	D792 [44]
MFR, g/10 min (210 °C, 2.16 kg)	8	D1238 [45]
Relative viscosity (in 1.0 g/dL chloroform, 30 °C)	4.0	D5225 [46]
Highest crystallization melting point *, °C	160–180	D3418 [47]

Two nucleating agents were considered to promote crystallization as follows: (1) 97% anhydrous orotic acid (OA) from MilliporeSigma (Burlington, MA, USA), and (2) *N,N'*-ethylene bis stearamide (EBS) from Acme-Hardesty Co. (Blue Bell, PA, USA). Both nucleating agents were received as white powders and dried separately at 100 °C for 2h. The nucleating agents were mixed with the PLA according to the concentrations shown in Table 2. The batch names indicate the selected nucleating agent and the weight concentration.

Table 2. Definition of different batch samples for the experimental investigation.

Batch Name	Nucleating Agent	Concentration (wt.%)
Neat PLA	N/A	N/A
PLA-0.3OA	Orotic acid	0.3
PLA-1OA	Orotic acid	1
PLA-2OA	Orotic acid	2
PLA-0.3EBS	EBS	0.3
PLA-1EBS	EBS	1
PLA-2EBS	EBS	2

2.2. Sample Preparation

The neat PLA was compounded with additives and then quiescently crystallized into the samples. These samples were directly characterized by their crystallization kinetics and morphology. The cryo-fracturing technique and chemical etching were used to create a fractured surface for the direct observation of crystalline structures. Details about each step of the sample preparation procedure are introduced below.

2.2.1. Compounding

The PLA and nucleating agents were mixed using a static batch mixer (C.W. Brabender Intelli-Torque Plasti-corder, C.W. Brabender Inc., South Hackensack, NJ, USA). For each batch (cf. Table 3), 50 g of neat PLA and a specific amount of the nucleating agent were dry-mixed and then fed into the mixing chamber. A flat temperature profile of 200 °C was used. Each sample was processed at a rotational speed of 100 rpm for 5 min after stabilizing the mixing torque to ensure uniform mixing. The mixed samples were collected and stored in sealed bags at the end of the run. In between runs, a purging compound (Dyna-purge D2, Dyna-purge, Buffalo, NY, USA) was used to clean the mixing chamber.

Table 3. Degree of crystallinity of PLA samples measured from the DSC isotherm.

PLA Batch	Degree of Crystallinity (%)		
	Isotherm Temperature		
	80 °C	110 °C	140 °C
Neat PLA	2.5	50.5	9.1
PLA-0.3OA	7.9	43.0	62.2
PLA-1OA	14.7	36.4	51.6
PLA-2OA	35.5	37.4	53.2
PLA-0.3EBS	7.8	41.2	3.9
PLA-1EBS	8.8	40.4	44.6
PLA-2EBS	6.9	41.4	55.3

2.2.2. Quiescent Crystallization

The compounded samples were shaped into disks under quiescent isothermal conditions to allow morphological characterization. The samples were cut into smaller pieces (~10 g) and then arranged in aluminum pans (diameter 70 mm, height 10 mm) to avoid any shear effect on the material. The samples were placed between two hot plates (4394, Carver Compression Molder, Carver, Inc., Wabash, IN, USA), and the temperature was initially stabilized at 220 °C. No pressure was applied. After 60 min, water cooling was used to cool the samples to an isotherm temperature value, which kept below the crystallization point. Three different isotherm temperatures (i.e., 80 °C, 110 °C, and 140 °C) were investigated. The isotherm time was selected to be 60 min to ensure sufficient crystallization. Overall, a total of 21 disk-shaped samples were fabricated and further prepared for characterization.

2.2.3. Sample Etching Procedure

The samples were cryo-fractured to create a flat cross-sectional area to allow the observation of crystal morphology. Each sample was first notched (depth: 1–2 mm) using a bench saw, then kept in a freezer at −20 °C for more than two hours before cracking using pliers. The fractured samples were then chemically etched to expose the crystal structures according to the following procedure:

- In total, 5 g of sodium hydroxide (NaOH), purchased from Sigma-Aldrich (Millipore Sigma, Saint Louis, MO, USA), was dissolved in 250 mL of water to achieve a 0.5 mol/L concentration. Twenty-one clean glass bottles with lids were prepared to etch and hold the PLA samples. The liquid prepared was distributed into 21 bottles.
- Etching: the 21 PLA samples prepared with quiescent crystallization were immersed individually into the solution in glass bottles for 12 h.
- Cleaning: After etching, samples were kept in the bottles for 20 min at 25 °C in an ultrasonic bath (Branson CPX 2800H, Brookfield, CT, USA) to remove residual particles. After cleaning, the samples were removed from the etching solvent and dried with compressed air. The samples were then kept in sealed bags individually for further characterization.

2.3. *Characterization Techniques*

2.3.1. Isothermal Differential Scanning Calorimetry

Differential scanning calorimetry (DSC, 3+ system, Mettler Toledo, Columbus, OH, USA) was used to gain a fundamental understanding of the phase transitions and the melting behavior of the different samples. Fresh samples weighing 5.0–8.0 mg were prepared from the edges of the disk-shaped samples for each DSC run at all isotherm temperatures. Of particular interest was the effect of different concentrations of nucleating agents on the crystallization of PLA. A DSC isothermal test protocol (cf. Figure 1) was defined according to the following steps:

- Segment 1: Heat the sample from 25 °C to 240 °C at 20 °C/min, followed by an isotherm at 240 °C for 3 min. This segment was intended to melt and remove all thermal history of the pellets. The 3 min isotherm ensured the complete melting of the sample.
- Segment 2: Cool the sample rapidly using the maximum cooling rate at 60 °C/min to various isothermal temperatures (80 °C, 110 °C, and 140 °C). This segment is intended to quench the PLA polymer melt to the designed isotherm temperature using the maximum cooling rate, thus minimizing the crystallization behavior during the cooling period.
- Segment 3: The abovementioned isotherm temperatures are held for 60 min. This segment was intended to capture the crystallization process of PLA even with low concentrations of additives and at low temperatures. The 60 min holding time ensures no additional crystallization at the current temperatures and concentrations.
- Segment 4: Heat the sample at 10 °C/min to 240 °C. The degree of crystallinity achieved from the isotherm was quantified from the melting peak observed during this heating segment. The degree of crystallinity (X_C) was calculated using the following:

$$X_C = \frac{\Delta H_m - \Delta H_c}{\Delta H_M} \times 100 \tag{1}$$

where ΔH_m is the melting enthalpy [J/g], ΔH_c is the cold crystallization enthalpy [J/g], and ΔH_M is the melting enthalpy of a PLA crystal of infinite size. The latter was assumed to be 93 J/g, as obtained from the literature [48].

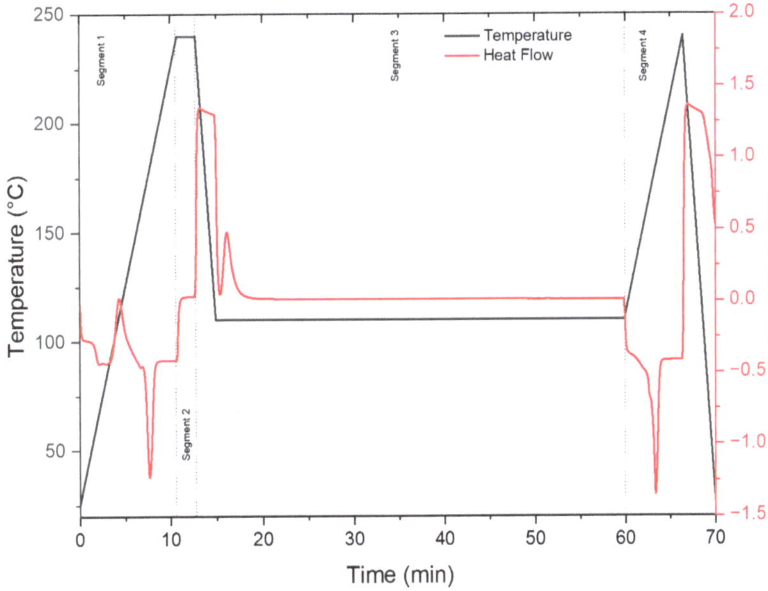

Figure 1. DSC procedure for the characterization of isotherm crystallization.

All DSC experiments were conducted on the PLA-OA, and PLA-EBS compounded using the static batch mixer. Different samples obtained from each batch were tested using preliminary DSC tests with only one heating cycle from 20 to 240 °C, and each batch presented similar melting points and melting peaks, hence suggesting homogeneous mixing. The isotherm DSC experiments were conducted for each batch without replication.

The kinetics of the crystallization process, observed from the DSC experiments, were quantified using the following Avrami equation:

$$v_c = 1 - \exp\left[-K(t)^n\right] \tag{2}$$

where v_c is the volumetric fraction of the converted phase at time t; K is the crystallization rate constant; and n is the Avrami index. However, the Avrami equation only describes the crystallization process after it has been initiated and does not account for the incubation time. The incubation time highly depends on the temperature and the polymer material [49–51]. In this work, the effect of the incubation time on PLA crystallization was quantified using the time derivative of the modified Avrami equation [41]:

$$\frac{dQ}{dt} = \Delta H_c \cdot K \cdot \exp\left[-K(t-t_0)^n\right] \cdot n \cdot (t-t_0)^{n-1} \tag{3}$$

where $\frac{dQ}{dt}$ is the heat flow obtained from the DSC isotherm test, ΔH_c is the enthalpy of crystallization per unit mass, K is the crystallization rate constant, t_0 is the incubation time, and n is the Avrami index for isotherm crystallization.

2.3.2. X-ray Diffractometry

The crystalline phase morphology and the distance between adjacent crystal planes were characterized via X-ray diffractometry (XRD, Rigaku SmartLab II, Rigaku USA, The Woodlands, TX, USA). The XRD experiments were performed using Bragg–Brentano geometry (Cu-Kα source, 1.54184 Å, 40 kV, 50 mA). A step scan protocol with a step size of $2\theta = 0.002°$ and a scanning speed of $2\theta = 5.0°/\text{min}$ was defined. An area of the 5 mm × 5 mm section at the center of each sample was scanned. The Cu-Kα2 signal was removed from the raw signal numerically and corrected for any shift in the diffraction angle. A Pearson VII peak function was utilized to fit the peaks and extract quantitative peak data, such as the peak center, full width at half maximum (FWHM), etc. [52].

$$f(x) = a\left[1 + \frac{(x-d)^2}{b^2}\right]^{-m} \tag{4}$$

where a is the maximum height of the peak, d is the peak's center, b is proportional to the full width at half-maximum (FWHM), and m is the shape factor. When the exponent $m = 1$, the shape becomes Cauchy; $m = 2$, modified Lorentzian; $m = \infty$, Gaussian [53].

The distance between adjacent crystal planes was calculated using Bragg's Law:

$$n\lambda = 2d\, sin(\theta) \tag{5}$$

where n is an integer representing the order of the diffraction peak, λ is the wavelength of the incident radiation, d is the lattice distance spacing between adjacent crystal planes, also known as d-spacing, and θ is the angle between the incident X-ray beam and the crystal plane.

2.3.3. Scanning Electron Microscopy

The morphology of the PLA samples was characterized using scanning electron microscopy (SEM, JEOL JSM 6390, JEOL USA, Inc., Peabody, MA, USA). Before imaging, the etched samples were sputter-coated (Denton Vacuum Desk IV Sputter Coater, Denton North America, Moorestown, NJ, USA) using gold with a 3–4 nanometer thickness. An acceleration voltage of 10 kV and a working distance of 15 mm was selected for all SEM experiments. For each sample, multiple spots were scanned to ensure consistency. For each spot, micrographs were taken at 750×, 1500×, and 3000× magnifications to capture different levels of detail on the crystalline domains. The micrographs were used to observe crystal structures on fractured surfaces. The crystalline domain sizes were measured on the 750× and 1500× micrographs using the image processing software ImageJ (National Institute of Health, Bethesda, MD, USA).

3. Results and Discussion

3.1. Thermodynamics of Crystallization

Degree of Crystallinity

The results for the degree of crystallization, calculated from Segment 4 in the isotherm DSC, for the different samples are summarized in Table 3. The second heating behaviors for PLA-OA and PLA-EBS blends are presented in Figure 2.

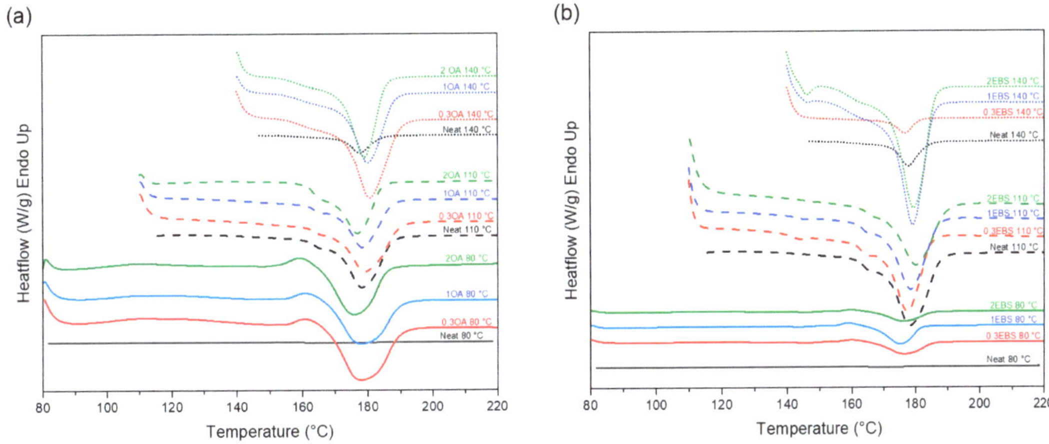

Figure 2. DSC curve for PLA-OA and PLA-EBS samples; the curves show the second heating process (**a**) neat PLA and PLA-OA blends, and (**b**) neat PLA and PLA-EBS blends. Detailed plots for PLA-OA and PLA-EBS can be found in Figures S1 and S2 in the Supplementary Materials.

At 80 °C, the isotherm temperature is below the optimal crystallization temperature for the selected PLA grade (i.e., 103–140 °C) [41]. However, it can be observed that with the aid of OA, the degree of crystallization increased from 7.9% to 35.5% as the concentration of the nucleating agent increased from 0.3% to 2%. This indicates that the degree of crystallinity of PLA is highly affected by the concentration of OA, even at the isotherm temperature below the optimal range. For PLA-EBS blends, the nucleating agent concentration seemed to have a smaller effect on the degree of crystallinity at 80 °C (i.e., maintained at 6.9–8.8%).

When the isotherm temperature increased to 110 °C, the degree of crystallinity increased significantly for both OA and EBS blends compared to the degree of crystallinity achieved at 80°C. A further increase to 140 °C resulted in a higher degree of crystallinity. However, an exception was observed for the PLA-0.3EBS at 140 °C, for which the degree of crystallinity decreased significantly to 4.6%.

Figure 3 presents the effect of the isotherm temperature and concentration of the nucleating agent on the degree of crystallinity. For OA blends, the degree of crystallinity followed an increasing trend as the isotherm temperature increased from 80 °C to 140 °C, and the concentration affected the 80 °C samples significantly. The degree of crystallinity increased as the concentration increased. However, for EBS blends, a low nucleating agent concentration negatively affected the samples prepared at higher isotherm temperatures.

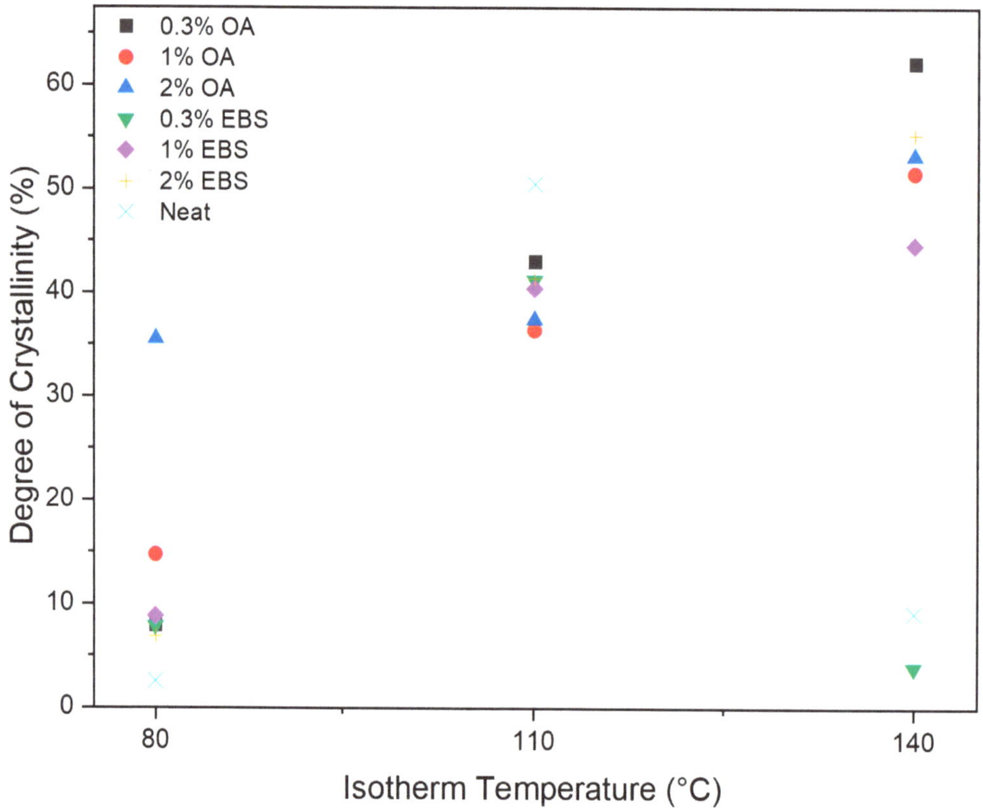

Figure 3. Effect of isotherm temperature and concentration on degree of crystallinity.

The neat PLA offered the highest degree of crystallinity (i.e., 50.5%) at 110 °C. At the same temperature, EBS and OA blends could not reach the same degree of crystallinity. In fact, nucleating agents enhanced the polymer's crystallization kinetics and promoted the formation of smaller and more numerous crystalline structures, which could result in more amorphous appearances due to a lack of growth of crystalline domains [21]. However, at lower temperatures (i.e., 80 °C), the degree of crystallinity increased from 6.5% for neat PLA to a maximum of 35.5% for PLA-2OA. Overall, the nucleating agents expanded the processing window for the PLA, which has relevant implications for manufacturing. Indeed, the ability to increase the crystallization at lower temperatures significantly facilitates processing by allowing the use of a water-heating system. Moreover, the energy consumption used for heating is reduced.

When comparing batches, the melt temperature of the PLA-EBS samples showed a ~4 °C decrease compared to the PLA-OA ones at the same isothermal temperatures. The change in melt temperature was minimal but might affect the crystalline morphology and adhesion between crystalline structures. It was also observed that at 80 °C, all DSC curves showed a unique endothermic peak at ~160 °C. For all PLA-OA blends, the intensity of this exothermic peak remained similar. However, for the PLA-EBS samples, the intensity decreased as the concentration increased from 0.3% to 2%. The same endothermic behavior was observed by other researchers [22,54,55] and could be related to the melting of α′ PLA crystals. In general, α′ crystals are a less stable form of crystalline PLA. They have a less ordered and less dense molecular arrangement than α crystals. α′ crystals have a lower melting temperature and mechanical strength than the α form. These crystals are formed when PLA chains have less packing density and exhibit some chain dislocations or imperfections. The non-ideal crystallization temperature, low cooling rate, and the absence of nucleating agents promote the formation of α' crystals. The melting of α' crystals is also observed in all DSC curves, represented by the endothermic peak in the 142–155 °C range. When the DSC scanning temperature surpasses their melting temperature, the α' crystals start to dissolve or transform into the more stable α phase, represented by the exothermic peaks observed at ~160 °C. The absence of this unique exothermic peak indicates that there is no α' crystal existing at 160 °C that can be transformed into α-phase crystals. All α' crystals formed during the isothermal segment were melted at 142–155 °C. The subsequent XRD experiments further investigated the formation of α' crystals and α crystals.

The crystallization kinetics were calculated according to the exothermic peaks in Segment 3 (cf. Figure 1). The heat flow values were baseline corrected using a tangential baseline before being treated, and Equation (3) was fitted to obtain the crystallization rate, the Avrami index, and the incubation time. The relative degree of crystallization values was obtained from integrating the heat flow data. The crystallization kinetics for PLA-OA and PLA-EBS blends are summarized in Table 4.

Table 4. Crystallization kinetics of PLA-OA and PLA-EBS blends. Detailed plots can be found in Figures S3–S6 in the Supplementary Materials.

Blends	Isotherm Temperature (°C)	Crystallization Rate (/min^{-n})	Avrami Index	Incubation Time (min)	$t_{1/2}$ (min)
Neat PLA	80	$0.0001 \pm 2.92 \times 10^{-6}$	$1.0 \pm 1.2 \times 10^{-4}$	$5.86 \pm 2.2 \times 10^{-4}$	34.2
Neat PLA	110	$0.0012 \pm 4.13 \times 10^{-4}$	3.3 ± 0.02	0.5 ± 0.02	2.14
Neat PLA	140	$0.002 \pm 1.63 \times 10^{-4}$	1.2 ± 0.002	0.32 ± 0.07	3.9
PLA-0.3OA	80	$3.6 \times 10^{-4} \pm 3.7 \times 10^{-6}$	2.0 ± 0.03	1.6 ± 0.2	23.6
PLA-1OA	80	$0.0029 \pm 3.4 \times 10^{-5}$	1.9 ± 0.004	1.4 ± 0.03	19.2
PLA-2OA	80	$0.00814 \pm 6.5 \times 10^{-5}$	1.7 ± 0.003	0.8 ± 0.01	16.9
PLA-0.3OA	110	1.149 ± 0.015	1.7 ± 0.02	0.8 ± 0.008	1.61
PLA-1OA	110	1.099 ± 0.029	1.8 ± 0.05	0.4 ± 0.01	1.45
PLA-2OA	110	1.482 ± 0.026	2.2 ± 0.04	0.1 ± 0.01	1.4
PLA-0.3OA	140	0.152 ± 0.006	2.2 ± 0.03	2.1 ± 0.02	4.6
PLA-1OA	140	0.179 ± 0.007	2.0 ± 0.03	2.1 ± 0.02	4.2
PLA-2OA	140	0.211 ± 0.017	2.3 ± 0.06	1.8 ± 0.04	3.7
PLA-0.3EBS	80	$8.7 \times 10^{-4} \pm 1.1 \times 10^{-5}$	2.7 ± 0.3	1.6 ± 0.2	17.8
PLA-1EBS	80	$0.0084 \pm 7.7 \times 10^{-4}$	1.7 ± 0.01	1.2 ± 0.1	21.3
PLA-2EBS	80	0.0047 ± 0.0024	2.0 ± 0.1	1.0 ± 0.1	17.3
PLA-0.3EBS	110	1.285 ± 0.018	2.4 ± 0.02	0.2 ± 0.01	1.1
PLA-1EBS	110	2.190 ± 0.059	2.1 ± 0.1	0.2 ± 0.02	0.9
PLA-2EBS	110	1.294 ± 0.129	2.4 ± 0.1	0.2 ± 0.05	1.4
PLA-0.3EBS	140	$0.050 \pm 3.45 \times 10^{-4}$	2.1 ± 0.2	1.6 ± 0.02	6.7
PLA-1EBS	140	0.017 ± 0.004	1.9 ± 0.02	1.6 ± 0.02	11.15
PLA-2EBS	140	0.193 ± 0.001	1.8 ± 0.01	1.6 ± 0.01	4.2

The crystallization peaks for both PLA-OA and PLA-EBS blends indicated that the crystallization behavior is more intense at 110 °C than at 80 °C and 140 °C. For PLA-OA blends, the initial crystallization rate was found to be the lowest at 80 °C for PLA-0.3OA, and the highest crystallization rate was 1.482 min^{-n} at 110 °C for PLA-2OA. The crystallization rate increased by two or more orders of magnitude at 110 °C. At 110 °C, there were no significant thermal behaviors after ~500 s during the isotherm segment. However, the crystallization behavior was not completed after 40 min at 80 °C. The initial crystallization rate also increased with a higher OA concentration.

The crystallization half-time ($t_{1/2}$), the time a sample takes to reach 50% relative crystallinity, provides critical crystallization information for process optimization and predictive modeling. Indeed, it captures the effect of both initial nucleation and the growth of crystalline domains, which occur in the later stage of crystallization behavior. It also provides more general information about the crystallization process and is easier to obtain than the crystallization rate (k) (cf. Equation (3)). The crystallization half-time indicated that PLA-2OA and the isotherm temperature of 110 °C provided the fastest nucleation and growth. However, the differences between different concentrations were minimal. The crystallization half-time changed from 1.4–1.61 min at 110 °C to 3.7–4.6 min at 140 °C.

The isotherm DSC results showed that EBS blends were characterized by a higher degree of crystallinity, a faster initial crystallization rate, lower incubation time, and shorter crystallization half-time. The PLA-EBS blends also showed sharper and narrower crystallization peaks at 110 °C compared to 80 °C and 140 °C (cf. Figures S3 and S5 in Supplementary Materials). Unlike PLA-OA blends, the initial crystallization rates were the highest at 1% EBS at 80 °C and 110 °C, while at 140 °C, the crystallization rate was the lowest at 1%. However, the PLA-1EBS blend showed a wider crystallization peak, indicating that crystallization growth was enhanced at this condition. The degree of crystallization was 44.6% for PLA-1EBS at 140 °C. This also confirmed that even though the initial crystallization rate was lower, the intensity of molecular movement was significant enough for the nucleation sites to grow into crystalline structures.

The effects of isotherm temperature and the concentration of nucleating agents on the crystallization kinetics are shown in Figure 4. At the 110 °C isotherm temperature, the crystallization rate was the fastest, while the incubation time and crystallization half-time were the lowest.

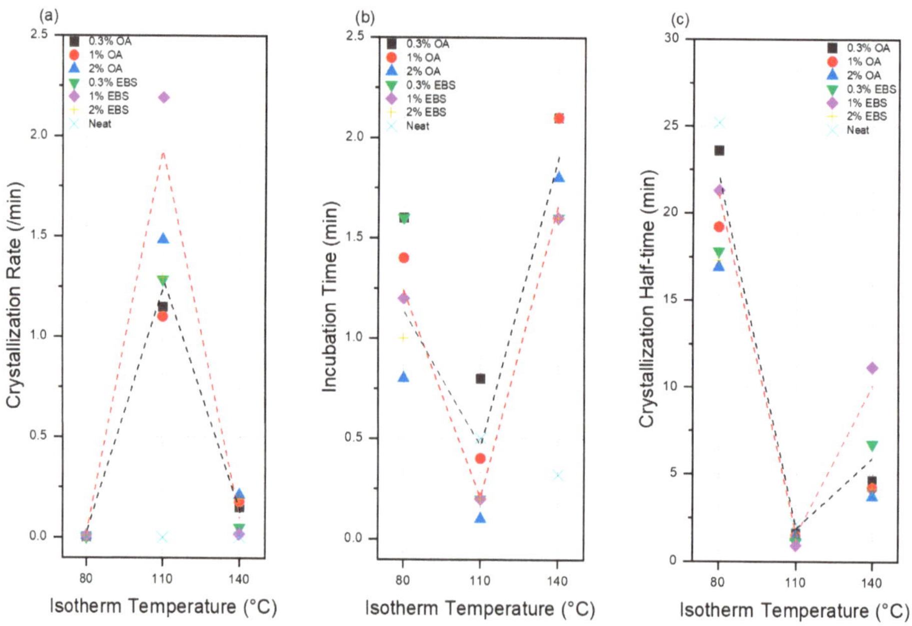

Figure 4. Effects of isotherm temperature and concentration on different crystallization kinetics, (**a**) crystallization rate, (**b**) incubation time, and (**c**) crystallization half-time. The black curve indicates the trend line for PLA-OA samples, and the red curve indicates the trend line for PLA-EBS samples.

The Avrami index (n) provides information about the nucleation and growth mechanism during crystallization. The value of n can vary, and different values suggest different crystallization mechanisms as follows:

$n = 1$: Indicates that the crystallization occurs through one-dimensional growth. This suggests that the growth of crystalline structures is linear (e.g., fibric growth, shish-kebab structures).

$n = 2$: Suggests that crystallization occurs through three-dimensional growth. This indicates that the crystalline structure growth is volumetric (e.g., disc-shaped structures, spherulites).

$n > 2$: Suggests heterogeneous nucleation and/or diffusion-controlled growth mechanisms.

The Avrami index for all tested samples was between 1.7 and 2.4, indicating that major crystalline structures were either disc-shaped crystals or spherulites. This observation is also supported by other research [30,56] and following SEM imaging analysis.

3.2. Polymorphic Analysis of Crystalline Domains

The DSC data in Figure 2 suggest the formation of both α and α′ crystals. The XRD curves of all samples are presented in Figure 5 and can be compared with the XRD patterns for the α and α′phase according to standard files to further prove the phase differences. Neat PLA samples crystallized at the same conditions were characterized as baselines for the raw material. The peak centers of the most significant peaks (200) are summarized in Table 5. The interplanar spacing of the crystal planes was calculated based on the (200) peaks.

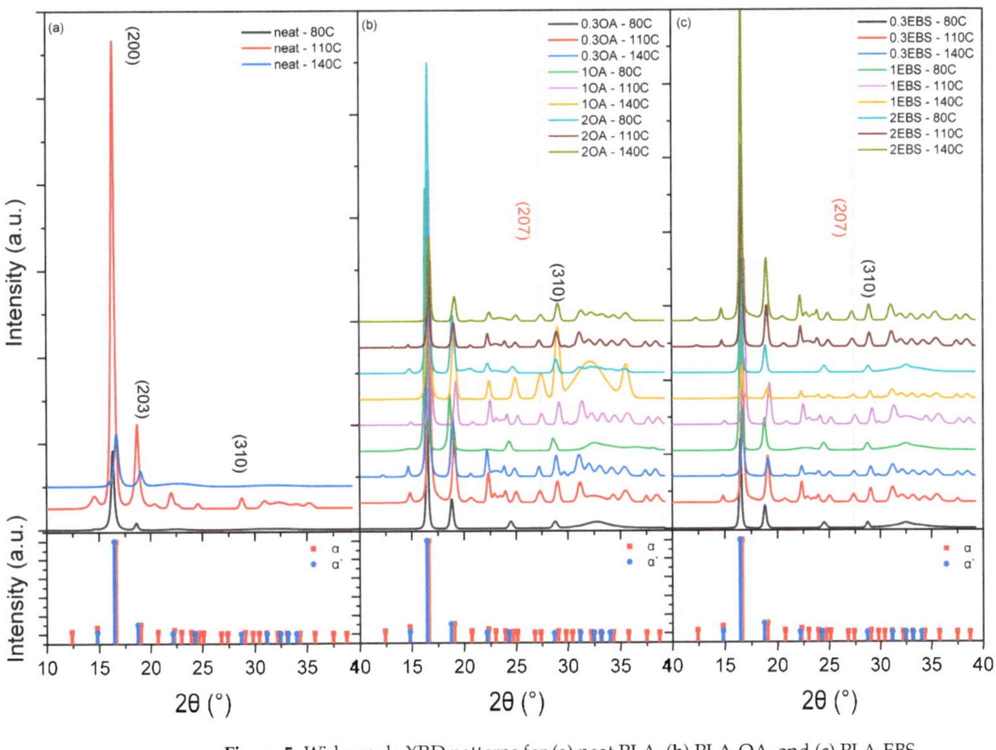

Figure 5. Wide-angle XRD patterns for (**a**) neat PLA, (**b**) PLA-OA, and (**c**) PLA-EBS.

Table 5. Peak centers and Basal distance spacing (d-spacing) obtained from XRD measurements of all PLA-OA and PLA-EBS samples according to (200) peaks.

Sample	Isotherm Temperature (°C)	2θ (°)	D-Spacing (Å)
Neat PLA	80	16.31	5.48
Neat PLA	110	16.34	5.47
Neat PLA	140	16.67	5.36
PLA-0.3OA	80	16.49	5.43
PLA-0.3OA	110	16.70	5.36
PLA-0.3OA	140	16.55	5.41
PLA-1OA	80	16.34	5.47
PLA-1OA	110	16.85	5.31
PLA-1OA	140	16.61	5.39
PLA-2OA	80	16.45	5.44
PLA-2OA	110	16.63	5.38
PLA-2OA	140	16.72	5.35
PLA-0.3EBS	80	16.47	5.43
PLA-0.3EBS	110	16.69	5.36
PLA-0.3EBS	140	16.70	5.36
PLA-1EBS	80	16.46	5.43
PLA-1EBS	110	16.86	5.31
PLA-1EBS	140	16.64	5.38
PLA-2EBS	80	16.55	5.41
PLA-2EBS	110	16.61	5.39
PLA-2EBS	140	16.57	5.40

The peak positions were compared using the JCPDS standards for α (JCPDS#00-064-1624) and α′phases (JCPDS#00-064-1624). The (200) peak for the α phase is centered at 2θ = 16.62° and 2θ = 16.44° for the α′phase. The (200) peak positions of all samples fabricated were between 2θ = 16.34° and 2θ = 16.86°. Despite the systematic errors in the XRD measurements, the crystalline structures in all samples were a combination of the α′phase and α phase. Lower peak positions indicated a more significant amount of α′phase, while higher peak positions suggested more α phase. Additionally, the α phase crystals exhibit unique peaks at the following positions: (204) peak at 2θ = 20.71°, (213) peak at 2θ = 23.92° and (207) peak at 2θ = 27.36°. The peak center positions of the common peaks for α and α′phases are very close and hard to separate numerically. The existence of the α phase was determined by the (207) peak and is indicated by the red dotted lines in Figure 6. The samples showed a significant peak at the (207) peak location, indicating the existence of phase crystals. Otherwise, most crystal structures were α′phase if no peak was observed at the (207) peak position. The XRD scans suggested that the α′phase was the majority phase of the crystals in all neat PLA samples, while PLA-OA and PLA-EBS blends were crystallized at 80 °C. This observation confirms the DSC analysis, where significant exothermic peaks were observed at around 160 °C, representing the α′ to α phase transition during heating.

The d-spacing data, describing the distance between adjacent crystal lattice planes within a crystalline material, of the (200) crystals are presented in Table 5. The standard d-spacing for the (200) crystal in the α′phase is 5.39 Å, and 5.33 Å in the α phase, according to ICDD files. The lowest d-spacing values (5.31 Å) were found on PLA-1OA and PLA-1EBS samples crystallized at 110 °C. The value was smaller than the ICDD standard value, indicating that nucleating agents improved the molecular orientation compared to ICDD conditions, thus resulting in densely packed molecular chains and crystalline structures. Decreasing or increasing the nucleating agent concentration disrupted the well-oriented molecular chains and led to larger d-spacing values.

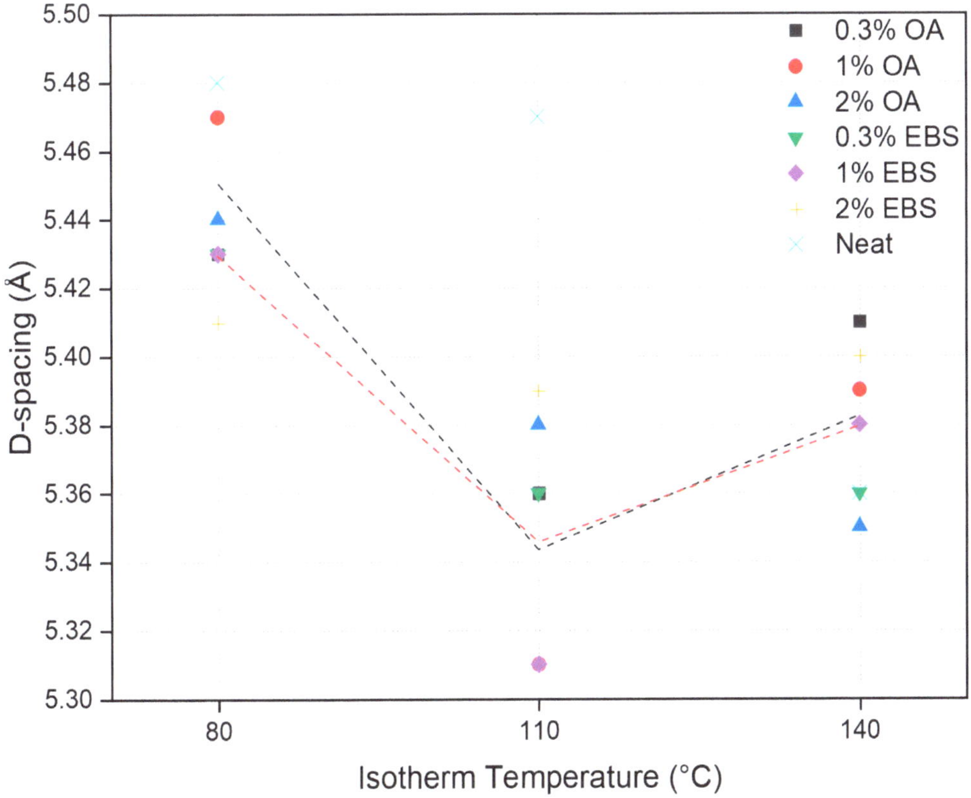

Figure 6. Effects of isotherm temperature and concentration on Basal distance spacing. The black curve indicates the trend line for PLA-OA samples, and the red curve indicates the trend line for PLA-EBS samples.

The isotherm temperature mostly dominated the d-spacing values. Figure 6 presents the effect of the isotherm temperature and concentration on the d-spacing of crystalline domains. This trend showed the smallest d-spacing values at a 110 °C isotherm temperature with 1% of nucleating agents adopted. This indicated that at 110 °C and a 1% nucleating agent concentration, the distance between the PLA crystalline lamellae was the lowest. Hence, the most perfect crystals were formed under this condition.

3.3. Crystallization Morphology

The crystallization morphology was characterized using SEM images taken from the same conditions. Table 6 summarizes the SEM images obtained at a magnification of 750×. A micrograph of neat crystallized PLA is shown in Table 7 as a reference.

Table 6. SEM images (750×) of PLA-OA and PLA-EBS samples.

OA	0.3%	1%	2%
80 °C	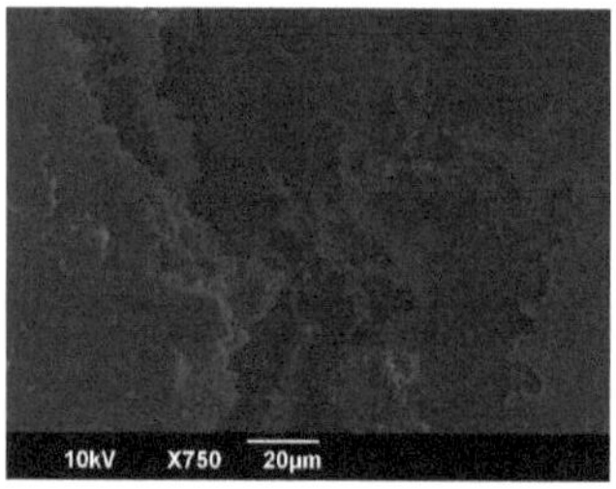	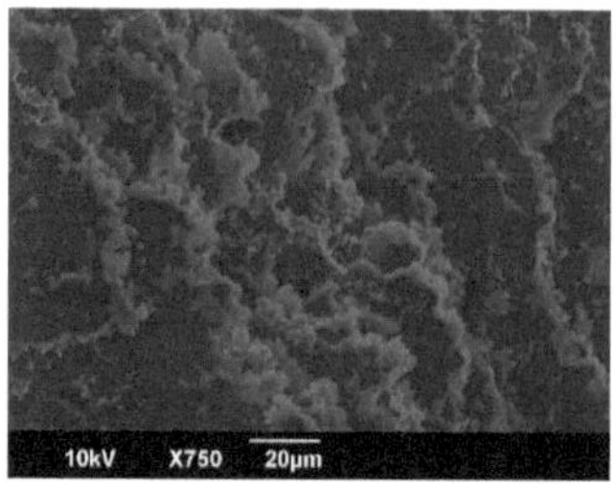	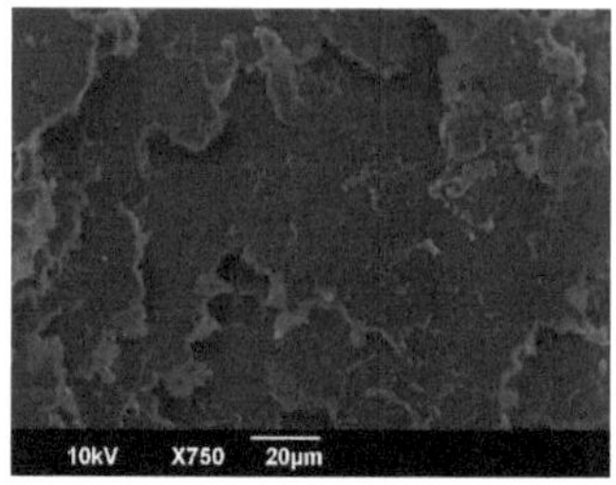
110 °C	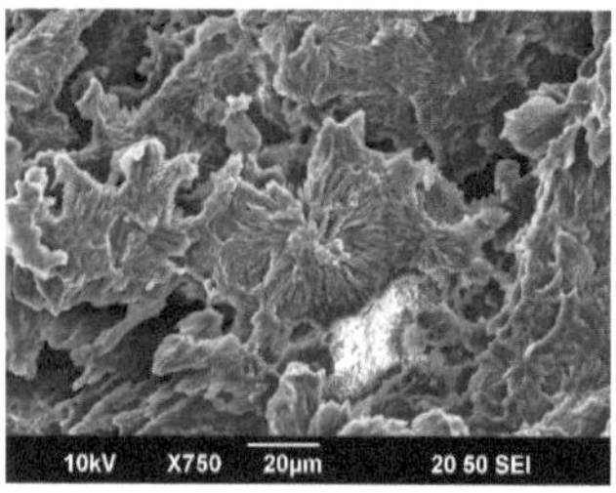		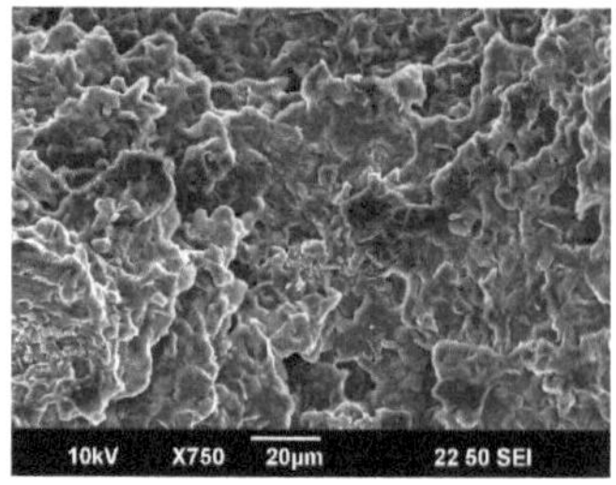
140 °C	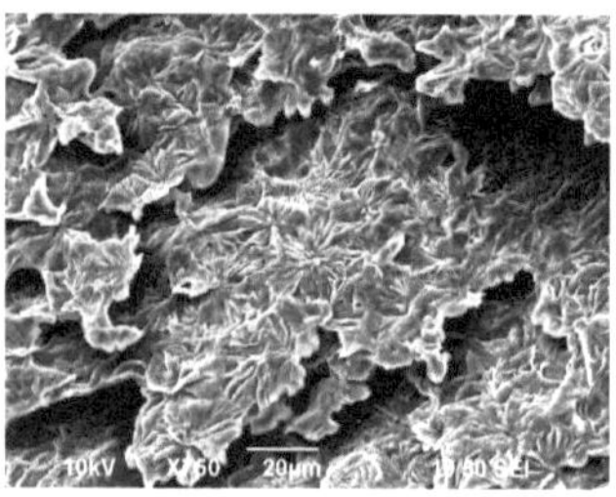	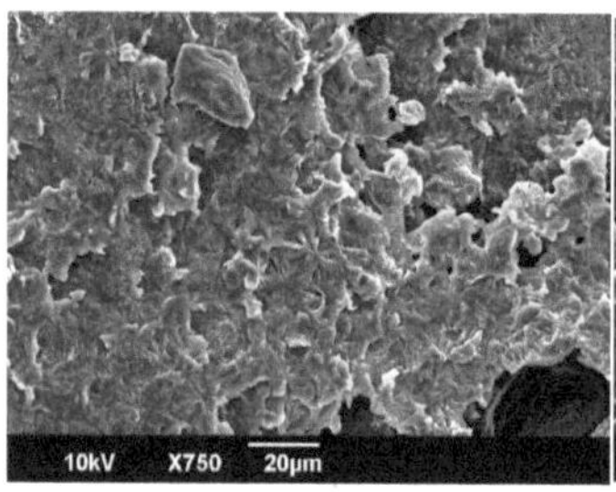	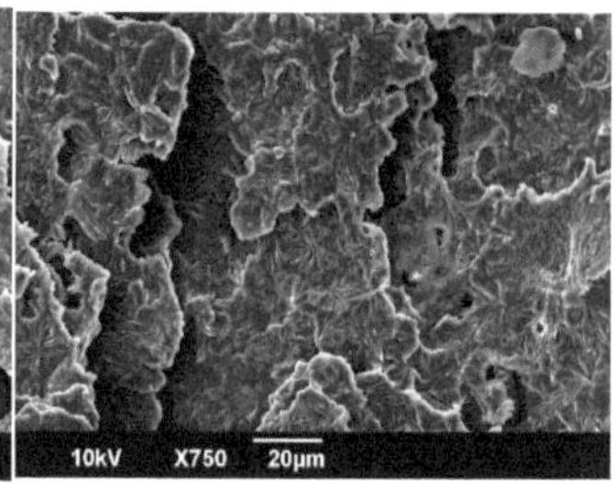
EBS	**0.3%**	**1%**	**2%**
80 °C	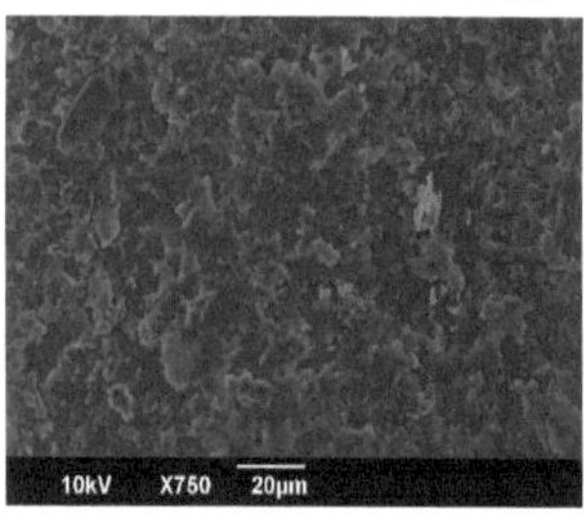	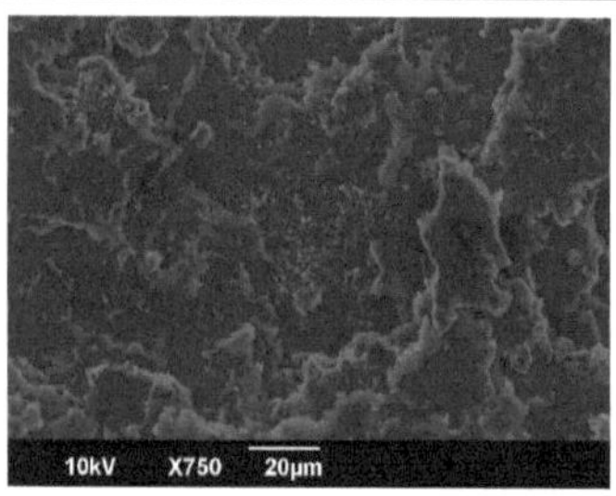	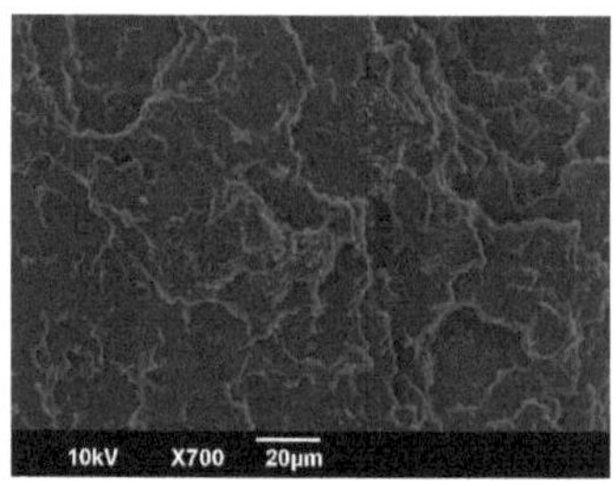
110 °C	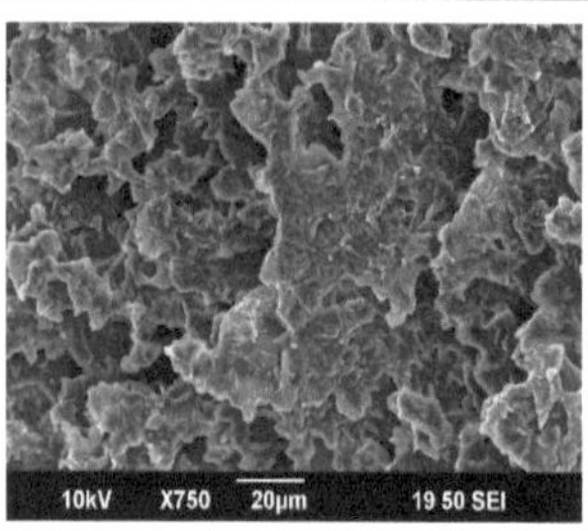	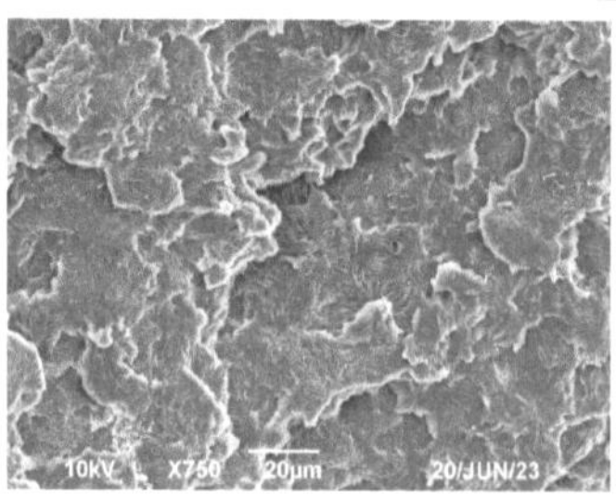	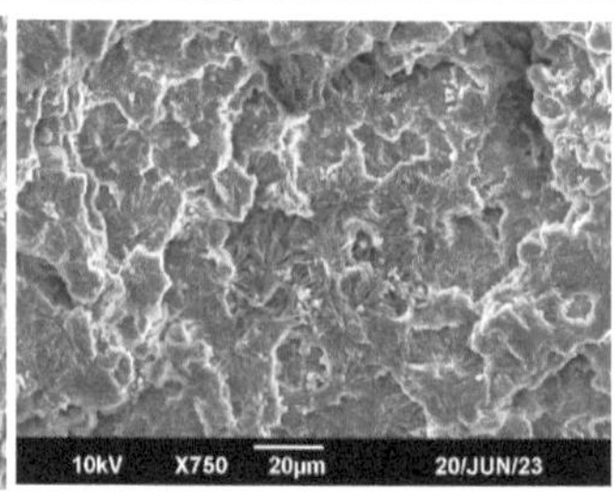

Table 6. *Cont.*

140 °C	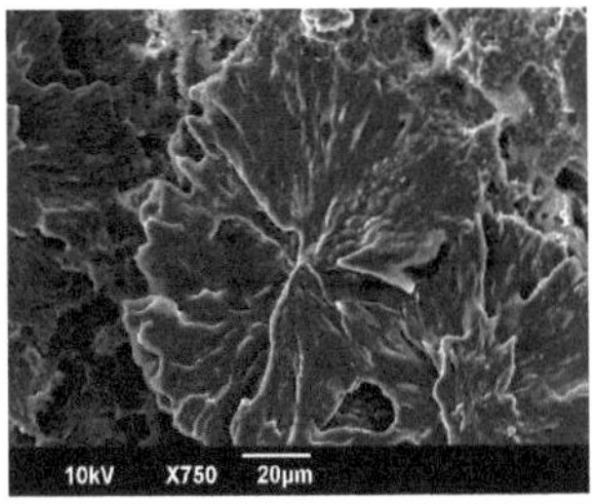	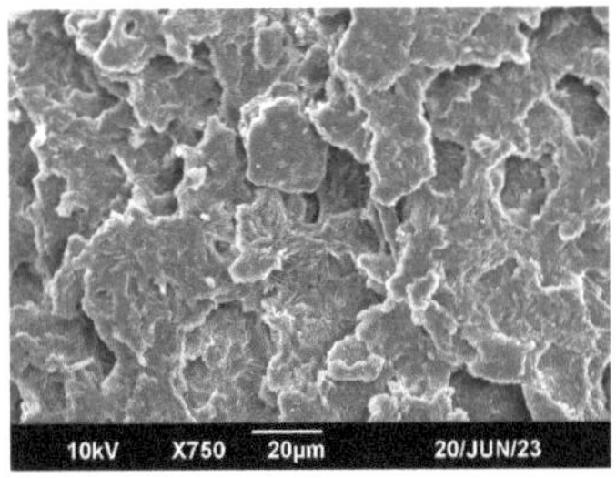	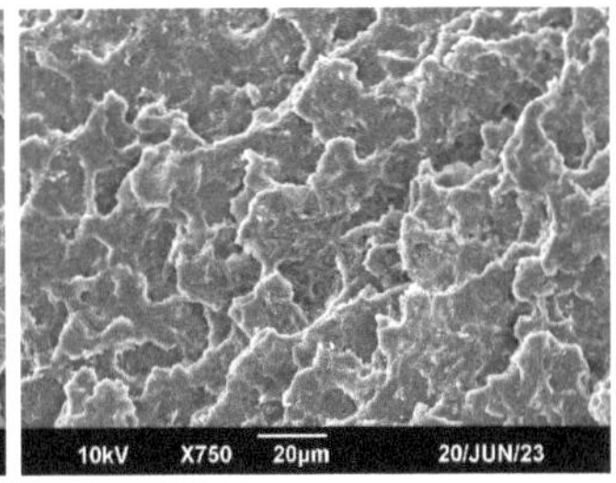

Table 7. SEM images (750×) of neat PLA crystallized at 80 °C, 110 °C, and 140 °C for 45 min.

Neat PLA, 80 °C	Neat PLA, 110 °C	Neat PLA, 140 °C
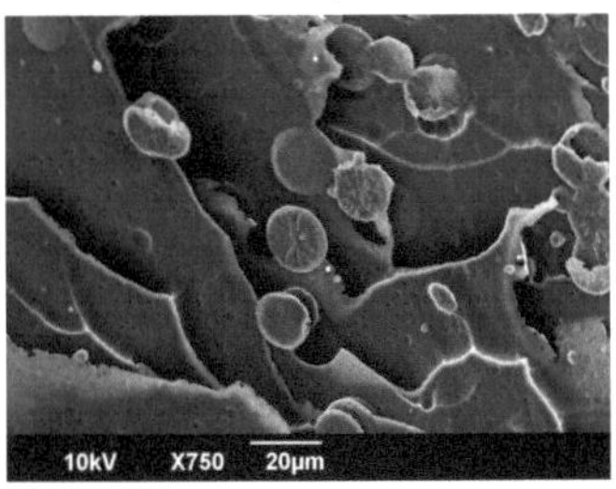		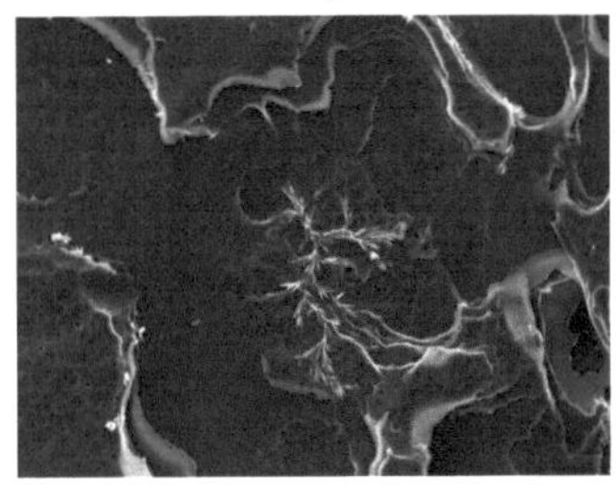

The radius of individual crystal structures was measured directly from the SEM images at 750× magnification, and the results are summarized in Table 8. All crystalline domains showed a disc/circle shape, indicating that the crystalline structures are spherulites/lamellae in growth. The results confirmed the Avrami index values obtained from the isotherm DSC analysis.

Table 8. Radius of crystalline domains in PLA-OA and PLA-EBS samples.

	Isotherm Temperature		
	80 °C	**110 °C**	**140 °C**
PLA-0.3OA	7.8 ± 1.0 μm	22.6 ± 5.2 μm	11.7 ± 1.4 μm
PLA-1OA	9.2 ± 0.6 μm	15.6 ± 2.4 μm	7.8 ± 0.7 μm
PLA-2OA	8.8 ± 1.0 μm	10.8 ± 2.0 μm	6.5 ± 1.0 μm
PLA-0.3EBS	7.2 ± 2.4 μm	8.0 ± 2.0 μm	40.0 ± 5.2 μm
PLA-1EBS	9.3 ± 0.5 μm	9.0 ± 1.5 μm	11.5 ± 1.2 μm
PLA-2EBS	10.0 ± 1.5 μm	9.0 ± 1.2 μm	8.0 ± 0.5 μm

Compared to neat PLA crystals (i.e., radius of 54.5 ± 2.0 μm), all samples fabricated with the nucleating agents had smaller crystalline domain sizes. Additionally, relatively higher nucleation densities were found in all PLA-OA and PLA-EBS samples, indicating that both OA and EBS acted as nucleating agents to provide heterogeneous nucleation sites for PLA spherulitic growth. In all 80 °C samples, the crystalline domain density was lower than 110 °C and 140 °C samples, and only a small amount of crystal structures could be observed. This supported the observation of a low degree of crystalline in the 80 °C samples. In samples crystallized at 110 °C, nucleation density and crystalline domain sizes were relatively high, and the interfaces between crystalline domains were observed on the fractured surfaces. In samples crystallized at 140 °C, the mean crystalline sizes were 6.5 ± 1.0 μm to 11.7 ± 1.4 μm, except for the PLA-0.3EBS sample. The radius of the crystal structure of PLA-0.3EBS was measured as 40.4 ± 5.2 μm, which is closer to the value for neat PLA samples. Additionally, the nucleation density, indicated by the number of individual crystalline domains observed in the SEM images with 750× magnification, was much lower than the PLA-1EBS and PLA-2EBS samples. This also supports the low degree of crystallinity in the PLA-0.3EBS sample. The crystallization condition at 140 °C and low concentration of EBS provided a low crystallization density, and the high crystallization temperature allowed more intense molecular movement, resulting in better molecular alignment. The nucleation density was much higher in PLA-1EBS and PLA-2EBS samples at 140 °C as more nucleation sites were provided. As a result, the sizes of the individual crystalline domains were smaller since there were not enough free volumes for the crystalline domains to grow.

Additionally, the variation in the crystal sizes was significantly lower with PLA-2OA and PLA-2EBS samples compared to PLA-0.3OA and PLA-0.3EBS samples. This also confirmed that the material achieved a state where most of the crystallites reached their thermodynamically favored size at higher concentrations. Under such conditions, the energy barriers for crystal growth and nucleation are balanced, leading to a distribution of relatively similar crystallite sizes. Achieving a state of more stable crystal sizes can improve material properties, such as mechanical strength, thermal stability, and degradation stability. Controlled crystallization processes at higher isotherm temperatures during PLA processing can influence the crystallites' size and arrangement, impacting the final products' performances.

Overall, the nucleation domain to initialize the crystallization of neat PLA is (a) meso-lactide and (b) SC-crystals [36,57]. Both nucleation sites are rare in neat PLA due to the high l-lactide in polymerization for the PLA 2500HP and the high-temperature requirement for forming sc-crystals. Orotic acid and EBS serve as nucleating agents for PLA via introducing nucleation sites. As a result, the number density increased, and the sizes of nucleation domains were smaller. Additionally, the hydrogen bonding between the nucleating agent and PLA altered the chemical structures when blended [58]. This also affected the mobility of the molecular chains and the thermodynamics of the crystallization behavior.

4. Conclusions

The experiments and analysis reported in this work suggest that both OA and EBS can be effectively used as nucleating agents for PLA. The nucleating agents enhanced PLA quiescent crystallization from 80 °C to 140 °C. OA and EBS significantly increased the initial crystallization rate during quiescent isotherm annealing and reduced the incubation time and crystallization half-time. The fastest crystallization rate and smallest crystallization half-time were achieved in the case of blends containing 1% EBS at 110 °C.

Both nucleating agents efficiently reduced the energy barrier of nucleation by achieving high crystallinity at 80 °C, which is 40 °C lower than the recommended processing temperature for PLA 2500HP. At 80 °C, the disordered α'phase was the primary phase in all samples, while the more stable α phase was obtained in samples annealed at 110 °C and 140 °C. Combining the 1wt.% nucleating agent and isotherm temperature at 110 °C promoted the smallest Basal distance spacing, indicating densely packed crystalline lamellae.

The well-organized and evenly distributed spherulitic morphology of the nucleated PLA crystalline domains was observed on the cryo-fractured surfaces. The SEM micrographs confirmed the calculated Avrami index values from the DSC experiments. It was also observed that increasing the concentration of nucleating agents resulted in the formation of smaller and more evenly distributed crystalline domains.

Overall, the use of OA and EBS led to a higher degree of crystallinity at lower temperatures, smaller and more evenly distributed crystalline structures, a faster crystallization rate, lower crystallization temperature, and lower crystallization half-time when compared to neat PLA. Hence, the blends are expected to improve the processability, reduce energy consumption, and enhance the performance of semi-crystalline PLA products. Future studies should consider the shear-induced effects of manufacturing conditions on the blend morphology.

Supplementary Materials: The following supporting information can be downloaded at: https://www.mdpi.com/article/10.3390/polym16030320/s1, Figure S1: DSC curves for PLA-OA samples. Curves showed melting behavior during the second heating cycle, (**a**) overall, (**b**) 80 °C, (**c**) 110 °C, and (**d**) 140 °C; Figure S2: DSC curves for PLA-EBS samples. Curves showed melting behavior during the second heating cycle, (**a**) overall, (**b**) 80 °C, (**c**) 110 °C, and (**d**) 140 °C; Figure S3: DSC curves for PLA-OA samples. Curves showed heat flow data obtained during isotherm cycle, (**a**) 80 °C, (**b**) 110 °C, and (**c**) 140 °C; Figure S4: Curves for relative degree of crystallinity for PLA-OA samples, (**a**) overall, (**b**) 80 °C, (**c**) 110 °C, and (**d**) 140 °C; Figure S5: DSC curves for PLA-EBS samples. Curves showed heat flow data obtained during isotherm cycle, (**a**) 80 °C, (**b**) 110 °C, and (**c**) 140 °C; Figure S6: Curves for relative degree of crystallinity for PLA-EBS samples, (**a**) overall, (**b**) 80 °C, (**c**) 110 °C, and (**d**) 140 °C.

Author Contributions: Conceptualization, P.G., S.A. and D.M.; methodology, P.G. and D.M.; validation, P.G. and S.A.; formal analysis, P.G. and S.A.; investigation, P.G., S.A. and D.M.; data curation, P.G. and S.A.; writing—original draft preparation, P.G. and S.A.; writing—review and editing, P.G., S.A. and D.M.; supervision, P.G. and D.M.; project administration, P.G. and D.M. All authors have read and agreed to the published version of the manuscript.

Funding: This work was supported by the University of Massachusetts Lowell Provost Office Start-Up funds to Prof. Davide Masato. APC funding, in the amount of $2200 was funded by Western Washington University—Office of Research and Sponsored Programs, Manuscript Preparation Grant #MS0149. The remainder of the APC funding is covered also by Western Washington University—Department startup funds to Prof. Peng Gao, grant number 12563.

Institutional Review Board Statement: Not applicable.

Data Availability Statement: The raw data supporting the conclusions of this article will be made available by the authors on request.

Conflicts of Interest: The authors declare no conflict of interest.

References

1. Cywar, R.M.; Rorrer, N.A.; Hoyt, C.B.; Beckham, G.T.; Chen, E.Y.-X. Bio-based polymers with performance-advantaged properties. *Nat. Rev. Mater.* **2021**, *7*, 83–103. [CrossRef]
2. Balla, E.; Daniilidis, V.; Karlioti, G.; Kalamas, T.; Stefanidou, M.; Bikiaris, N.D.; Vlachopoulos, A.; Koumentakou, I.; Bikiaris, D.N. Poly(lactic Acid): A Versatile Biobased Polymer for the Future with Multifunctional Properties—From Monomer Synthesis, Polymerization Techniques and Molecular Weight Increase to PLA Applications. *Polymers* **2021**, *13*, 1822. [CrossRef] [PubMed]
3. Zhang, C.; Xue, J.; Yang, X.; Ke, Y.; Ou, R.; Wang, Y.; Madbouly, S.A.; Wang, Q. From plant phenols to novel bio-based polymers. *Prog. Polym. Sci.* **2021**, *125*, 101473. [CrossRef]
4. Saeidlou, S.; Huneault, M.A.; Li, H.; Park, C.B. Poly(lactic acid) crystallization. *Prog. Polym. Sci.* **2012**, *37*, 1657–1677. [CrossRef]
5. Harris, A.M.; Lee, E.C. Improving mechanical performance of injection molded PLA by controlling crystallinity. *J. Appl. Polym. Sci.* **2008**, *107*, 2246–2255. [CrossRef]
6. Kobayashi, J.; Asahi, T.; Ichiki, M.; Oikawa, A.; Suzuki, H.; Watanabe, T.; Fukada, E.; Shikinami, Y. Structural and optical properties of poly lactic acids. *J. Appl. Phys.* **1995**, *77*, 2957–2973. [CrossRef]
7. De Santis, P.; Kovacs, A.J. Molecular conformation of poly(S-lactic acid). *Biopolymers* **1968**, *6*, 299–306. [CrossRef]
8. Zhang, J.; Tashiro, K.; Tsuji, H.; Domb, A.J. Investigation of phase transitional behavior of poly(l-lactide)/poly(d-lactide) blend used to prepare the highly-oriented stereocomplex. *Macromolecules* **2007**, *40*, 1049–1054. [CrossRef]
9. Zhang, J.; Tashiro, K.; Tsuji, H.; Domb, A.J. Disorder-to-order phase transition and multiple melting behavior of poly(l-lactide) investigated by simultaneous measurements of WAXD and DSC. *Macromolecules* **2008**, *41*, 1352–1357. [CrossRef]
10. Pan, P.; Kai, W.; Zhu, B.; Dong, T.; Inoue, Y. Polymorphous crystallization and multiple melting behavior of poly(l-lactide): Molecular weight dependence. *Macromolecules* **2007**, *40*, 6898–6905. [CrossRef]
11. Hoogsteen, W.; Postema, A.R.; Pennings, A.J.; Brinke, G.T.; Zugenmaier, P. Crystal structure, conformation and morphology of solution-spun poly(L-lactide) fibers. *Macromolecules* **1990**, *23*, 634–642. [CrossRef]
12. Puiggali, J.; Ikada, Y.; Tsuji, H.; Cartier, L.; Okihara, T.; Lotz, B. The frustrated structure of poly(l-lactide). *Polymer* **2000**, *41*, 8921–8930. [CrossRef]

13. Cartier, L.; Okihara, T.; Lotz, B. Triangular Polymer Single Crystals: Stereocomplexes, Twins, and Frustrated Structures. *Macromolecules* **1997**, *30*, 6313–6322. [CrossRef]
14. Okihara, T.; Tsuji, M.; Kawaguchi, A.; Katayama, K.-I.; Tsuji, H.; Hyon, S.-H.; Ikada, Y. Crystal structure of stereocomplex of poly(L-lactide) and poly(D-lactide). *J. Macromol. Sci. Part B* **1991**, *30*, 119–140. [CrossRef]
15. Han, L.; Pan, P.; Shan, G.; Bao, Y. Stereocomplex crystallization of high-molecular-weight poly(l-lactic acid)/poly(d-lactic acid) racemic blends promoted by a selective nucleator. *Polymer* **2015**, *63*, 144–153. [CrossRef]
16. Ma, P.; Shen, T.; Xu, P.; Dong, W.; Lemstra, P.J.; Chen, M. Superior Performance of Fully Biobased Poly(lactide) via Stereocomplexation-Induced Phase Separation: Structure versus Property. *ACS Sustain. Chem. Eng.* **2015**, *3*, 1470–1478. [CrossRef]
17. Zisopol, D.G.; Portoaca, A.I.; Nae, I.; Ramadan, I. A Comparative Analysis of the Mechanical Properties of Annealed PLA. *Eng. Technol. Appl. Sci. Res.* **2022**, *12*, 8978–8981. [CrossRef]
18. Simmons, H.; Tiwary, P.; Colwell, J.E.; Kontopoulou, M. Improvements in the crystallinity and mechanical properties of PLA by nucleation and annealing. *Polym. Degrad. Stab.* **2019**, *166*, 248–257. [CrossRef]
19. Bhandari, S.; Lopez-Anido, R.A.; Gardner, D.J. Enhancing the interlayer tensile strength of 3D printed short carbon fiber reinforced PETG and PLA composites via annealing. *Addit. Manuf.* **2019**, *30*, 100922. [CrossRef]
20. Barkhad, M.S.; Abu-Jdayil, B.; Mourad, A.H.I.; Iqbal, M.Z. Thermal Insulation and Mechanical Properties of Polylactic Acid (PLA) at Different Processing Conditions. *Polymers* **2020**, *12*, 2091. [CrossRef]
21. Di Lorenzo, M.L. The crystallization and melting processes of poly(l-lactic acid). *Macromol. Symp.* **2006**, *234*, 176–183. [CrossRef]
22. Gao, P.; Kundu, A.; Alqosaibi, K.; Coulter, J. Enhanced crystallinity development of poly-lactic acid by dynamic melt manipulation. In Proceedings of the ASME 2021 International Mechanical Engineering Congress and Exposition, Online, 1–5 November 2021; pp. 1–9. [CrossRef]
23. Refaa, Z.; Boutaous, M.; Xin, S.; Siginer, D.A. Thermophysical analysis and modeling of the crystallization and melting behavior of PLA with talc: Kinetics and crystalline structures. *J. Therm. Anal. Calorim.* **2017**, *128*, 687–698. [CrossRef]
24. As'habi, L.; Jafari, S.H.; Khonakdar, H.A.; Häussler, L.; Wagenknecht, U.; Heinrich, G. Non-isothermal crystallization behavior of PLA/LLDPE/nanoclay hybrid: Synergistic role of LLDPE and clay. *Thermochim. Acta* **2013**, *565*, 102–113. [CrossRef]
25. Xu, Z.; Su, L.; Jiang, S.; Gu, W.; Peng, M.; Wang, P. Crystallization behavior and water vapor permeability of poly(lactic acid) nanocomposite under oscillatory shear. *J. Appl. Polym. Sci.* **2015**, *132*, 42321. [CrossRef]
26. Nagarajan, V.; Mohanty, A.K.; Misra, M. Crystallization behavior and morphology of polylactic acid (PLA) with aromatic sulfonate derivative. *J. Appl. Polym. Sci.* **2016**, *133*. [CrossRef]
27. Alzahrani, F.J.; Duhduh, A.A.; Gao, P.; Coulter, J.P. The influence of cooling rate and mold temperature on polymers crystallization kinetics in injection molding. In Proceedings of the ASME 2021 International Mechanical Engineering Congress and Exposition, Online, 1–5 November 2021; pp. 1–7. [CrossRef]
28. Gao, P.; Kundu, A.; Coulter, J. Vibration-assisted injection molding: An efficient process for enhanced crystallinity development and mechanical characteristics for poly lactic acid. *Int. J. Adv. Manuf. Technol.* **2022**, *121*, 3111–3124. [CrossRef]
29. Dorgan, J.R.; Williams, J.S.; Lewis, D.N. Melt rheology of poly(lactic acid): Entanglement and chain architecture effects. *J. Rheol.* **1999**, *43*, 1141–1155. [CrossRef]
30. Refaa, Z.; Boutaous, M.; Siginer, D.A. PLA crystallization kinetics and morphology development. *Int. Polym. Process.* **2018**, *33*, 336–344. [CrossRef]
31. Xu, H.; Zhong, G.-J.; Fu, Q.; Lei, J.; Jiang, W.; Hsiao, B.S.; Li, Z.-M. Formation of Shish-Kebabs in Injection-Molded Poly(l-lactic acid) by Application of an Intense Flow Field. *ACS Appl. Mater. Interfaces* **2012**, *4*, 6774–6784. [CrossRef]
32. Ghosh, S.; Viana, J.; Reis, R.; Mano, J. Effect of processing conditions on morphology and mechanical properties of injection-molded poly(l-lactic acid). *Polym. Eng. Sci.* **2007**, *47*, 1141–1147. [CrossRef]
33. Wang, Y.; Zhang, Q.; Na, B.; Du, R.; Fu, Q.; Shen, K. Dependence of impact strength on the fracture propagation direction in dynamic packing injection molded PP/EPDM blends. *Polymer* **2003**, *44*, 4261–4271. [CrossRef]
34. Volpe, V.; Foglia, F.; Pantani, R. Effect of the application of low shear rates on the crystallization kinetics of PLA. *Polym. Cryst.* **2020**, *3*, e10139. [CrossRef]
35. Song, Y.; Zhao, Q.; Yang, S.; Ru, J.; Lin, J.; Xu, J.; Lei, J.; Li, Z. Flow-induced crystallization of polylactide stereocomplex under pressure. *J. Appl. Polym. Sci.* **2018**, *135*, 46378. [CrossRef]
36. Kolstad, J.J. Crystallization kinetics of poly(L-lactide-co-meso-lactide). *J. Appl. Polym. Sci.* **1996**, *62*, 1079–1091. [CrossRef]
37. Shieh, Y.; Twu, Y.; Su, C.; Lin, R.; Liu, G. Crystallization kinetics study of poly(L-lactic acid)/carbon nanotubes nanocomposites. *J. Polym. Sci. Part B* **2010**, *48*, 983–989. [CrossRef]
38. Tang, H.; Chen, J.-B.; Wang, Y.; Xu, J.-Z.; Hsiao, B.S.; Zhong, G.-J.; Li, Z.-M. Shear flow and carbon nanotubes synergistically induced nonisothermal crystallization of poly(lactic acid) and its application in injection molding. *Biomacromolecules* **2012**, *13*, 3858–3867. [CrossRef]
39. Xu, J.-Z.; Chen, T.; Yang, C.-L.; Li, Z.-M.; Mao, Y.-M.; Zeng, B.-Q.; Hsiao, B.S. Isothermal crystallization of poly(l-lactide) induced by graphene nanosheets and carbon nanotubes: A comparative study. *Macromolecules* **2010**, *43*, 5000–5008. [CrossRef]
40. Qiu, Z.; Li, Z. Effect of Orotic Acid on the Crystallization Kinetics and Morphology of Biodegradable Poly(L-lactide) as an Efficient Nucleating Agent. *Ind. Eng. Chem. Res.* **2011**, *50*, 12299–12303. [CrossRef]
41. Gao, P.; Masato, D.; Kundu, A.; Coulter, J.P. An Investigation on the Efficacy of Orotic Acid as a Bio-Nucleating Agent for Poly-Lactic Acid under Quiescent Condition and Injection Molding. *Micromachines* **2022**, *13*, 2186. [CrossRef] [PubMed]

42. Nouri, A.; Hodgson, P.D.; Wen, C. Study on the Role of Stearic Acid and Ethylene-bis-stearamide on the Mechanical Alloying of a Biomedical Titanium Based Alloy. *Met. Mater. Trans. A* **2010**, *41*, 1409–1420. [CrossRef]
43. Zhang, L.; Yin, H.; Xiong, Z.; Xiong, Y.; Xu, W. Flowability and Mechanical and Thermal Properties of Nylon 6/Ethylene bis-Stearamide/Carboxylic Silica Composites. *J. Macromol. Sci. Part B* **2011**, *50*, 2255–2270. [CrossRef]
44. *ASTM D792-20*; Standard Test Methods for Density and Specific Gravity (Relative Density) of Plastics by Displacement. ASTM: West Conshohocken, PA, USA, 2020.
45. *ASTM D1238-10*; Standard Test Method for Melt Flow Rates of Thermoplastics by Extrusion Plastometer. ASTM International: West Conshohocken, PA, USA, 2010.
46. *ASTM D5225-22*; Standard Test Method for Measuring Solution Viscosity of Polymers with a Differential Viscometer. ASTM International: West Conshohocken, PA, USA, 2022.
47. *ASTM D3418-21*; Standard Test Method for Transition Temperatures and Enthalpies of Fusion and Crystallization of Polymers by Differential Scanning Calorimetry. ASTM International: West Conshohocken, PA, USA, 2021.
48. Mikešová, J.; Hašek, J.; Tishchenko, G.; Morganti, P. Rheological study of chitosan acetate solutions containing chitin nanofibrils. *Carbohydr. Polym.* **2014**, *112*, 753–757. [CrossRef] [PubMed]
49. Wang, M.; Lai, Z.; Jin, X.; Sun, T.; Liu, H.; Qi, H. Multifunctional Liquid-Free Ionic Conductive Elastomer Fabricated by Liquid Metal Induced Polymerization. *Adv. Funct. Mater.* **2021**, *31*, 2101957. [CrossRef]
50. Al Hosni, A.S.; Pittman, J.K.; Robson, G.D. Microbial degradation of four biodegradable polymers in soil and compost demonstrating polycaprolactone as an ideal compostable plastic. *Waste Manag.* **2019**, *97*, 105–114. [CrossRef] [PubMed]
51. Wei, R.; Breite, D.; Song, C.; Gräsing, D.; Ploss, T.; Hille, P.; Schwerdtfeger, R.; Matysik, J.; Schulze, A.; Zimmermann, W. Biocatalytic Degradation Efficiency of Postconsumer Polyethylene Terephthalate Packaging Determined by Their Polymer Microstructure. *Adv. Sci.* **2019**, *6*, 1900491. [CrossRef] [PubMed]
52. Hall, M.M.; Veeraraghavan, V.G.; Rubin, H.; Winchell, P.G. The approximation of symmetric X-ray peaks by Pearson type VII distributions. *J. Appl. Crystallogr.* **1977**, *10*, 66–68. [CrossRef]
53. Prevey, P.S. The Use of Person VII Distribution Functions in X-Ray Diffraction Residual Stress Measurement. *Adv. X-ray Anal.* **1985**, *29*, 103–111. [CrossRef]
54. Zhuo, R.; Zhang, Y.; Li, G.; Shao, C.; Wang, Y.; Liu, C.; Li, Q.; Cao, W.; Shen, C. Structural evolution of poly(lactic acid) upon uniaxial stretching investigated by in situ infrared spectroscopy. *Vib. Spectrosc.* **2016**, *86*, 262–269. [CrossRef]
55. Salač, J.; Šerá, J.; Jurča, M.; Verney, V.; Marek, A.A.; Koutný, M. Photodegradation and biodegradation of poly(lactic) acid containing orotic acid as a nucleation agent. *Materials* **2019**, *12*, 481. [CrossRef]
56. Wen, H.; Wang, Y.; Wang, D.; de Claville Christiansen, J.; Yu, D.; Jiang, S.; Chen, C. Evaluation of Relationship Between Crystallization Structure and Thermal-Mechanical Performance of PLA with MCC Addition. *ChemistrySelect* **2019**, *4*, 10174–10180. [CrossRef]
57. Cunha, B.L.C.; Bahú, J.O.; Xavier, L.F.; Crivellin, S.; de Souza, S.D.A.; Lodi, L.; Jardini, A.L.; Filho, R.M.; Schiavon, M.I.R.B.; Concha, V.O.C.; et al. Lactide: Production Routes, Properties, and Applications. *Bioengineering* **2022**, *9*, 164. [CrossRef] [PubMed]
58. Kowalewska, A.; Nowacka, M. Supramolecular Interactions in Hybrid Polylactide Blends—The Structures, Mechanisms and Properties. *Molecules* **2020**, *25*, 3351. [CrossRef] [PubMed]

polymers

MDPI

Article

Ternary Blends from Biological Poly(3-hydroxybutyrate-*co*-3-hydroxyhexanoate), Poly(propylene carbonate) and Poly(vinyl acetate) with Balanced Properties

Yujie Jin [1], Changyu Han [2], Yi Li [1,*], Hongda Cheng [2], Dongdong Li [1] and Huan Wang [1]

[1] School of Materials Science and Engineering, Jilin Jianzhu University, Changchun 130118, China
[2] Key Laboratory of Polymer Ecomaterials, Changchun Institute of Applied Chemistry, Chinese Academy of Sciences, Changchun 130022, China
* Correspondence: liyi@jlju.edu.cn

Abstract: Poly(3-hydroxybutyrate-*co*-3-hydroxyhexanoate) (PHBH) has gained significant attention because of its biodegradability and sustainability. However, its expanded application in some fields is limited by the brittleness and low melt viscoelasticity. In this work, poly(vinyl acetate) (PVAc) was introduced into PHBH/poly(propylene carbonate) (PPC) blends via melt compounding with the aim of obtaining a good balance of properties. Dynamic mechanical analysis results suggested that PPC and PHBH were immiscible. PVAc was miscible with both a PHBH matrix and PPC phase, while it showed better miscibility with PHBH than with PPC. Therefore, PVAc was selectively localized in a PHBH matrix, reducing interfacial tension and refining dispersed phase morphology. The crystallization rate of PHBH slowed down, and the degree of crystallinity decreased with the introduction of PPC and PVAc. Moreover, the PVAc phase significantly improved the melt viscoelasticity of ternary blends. The most interesting result was that the remarkable enhancement of toughness for PHBH/PPC blends was obtained by adding PVAc without sacrificing the strength markedly. Compared with the PHBH/PPC blend, the elongation at the break and yield strength of the PHBH/PPC/10PVAc blend increased by 1145% and 7.9%, respectively. The combination of high melt viscoelasticity, toughness and strength is important for the promotion of the practical application of biological PHBH.

Keywords: biodegradable polymers; poly(3-hydroxybutyrate-*co*-3-hydroxyhexanoate); poly(propylene carbonate); poly(vinyl acetate); mechanical properties; polymer blends

Citation: Jin, Y.; Han, C.; Li, Y.; Cheng, H.; Li, D.; Wang, H. Ternary Blends from Biological Poly(3-hydroxybutyrate-*co*-3-hydroxyhexanoate), Poly(propylene carbonate) and Poly(vinyl acetate) with Balanced Properties. *Polymers* **2023**, *15*, 4281. https://doi.org/10.3390/polym15214281

Academic Editors: Vishal Gavande, Vasi Shaikh and Won-Ki Lee

Received: 13 September 2023
Revised: 26 October 2023
Accepted: 27 October 2023
Published: 31 October 2023

1. Introduction

Interest focusing on biodegradable biopolymers has exponentially increased over the past few decades on account of the growing plastic accumulation and depletion of fossil resources. Poly(3-hydroxyalkanoate) (PHA) is one of aliphatic bio-polyesters produced from the microorganism fermentation of sugar or plant-based oils, which has attracted much industrial and scientific attention due to its biodegradability, biocompatibility and renewability of raw materials [1,2]. So far, over 150 kinds of PHAs composed of different co-monomers have been synthesized, but only a few of them have the potential to be produced commercially, such as polyhydroxybutyrate (PHB), poly(hydroxybutyrate-*co*-hydroxyhexanoate) (PHBH) and poly(hydroxybutyrate-*co*-hydroxyvalerate) (PHBV) [3].

PHBH is a thermoplastic copolymer containing two different monomers, namely the short-chain 3-hydroxybutyrate (3HB) units and the medium-long-chain 3-hydroxyhexanoate (3HHx) units [4]. The 3HB/3HHx copolymerization ratio can affect the physical properties of PHBH with a range from amorphous elastomer to semicrystalline plastic. Therefore, PHBH is expected to play a role in multiple fields, such as food packaging, disposable plastics and agricultural mulch films [5,6]. However, PHBH with low 3HHx content is brittle, which originated from its high crystallinity, which limits its practical applications.

Therefore, it is necessary to improve the flexibility of PHBH while maintaining its elastic modulus and mechanical strength. Polymer blending is an effective approach with a low production cost to overcome polymer drawbacks and combine characteristics of blend components. Blending PHBH with other biodegradable polymers can improve the performance and is in line with the sustainable development strategies and eco-friendliness concepts. Many biodegradable PHBH-based blends have been reported. For example, Lim et al. [7] prepared immiscible PHBH/polycaprolactone (PCL) blends and found that the PCL phase improved the yield strength and modulus of PHBH/PCL blends. Blending poly(butylene adipate-*co*-terephthalate) (PBAT) with PHBH inhibited the crystallization of PHBH and improved the processability and mechanical properties [8]. Yu et al. [9] found that the miscibility of PHBH/poly(ethylene oxide) (PEO) blends was dependent on the blend composition, and the mechanical properties of PHBH were improved by the introduction of 5–17.5 wt% PEO. PHBH was miscible with poly(3-hydroxybutyrate-*co*-4-hydroxybutyrate) (P34HB), and the PHBH/P34HB blends displayed higher thermal stability, flexibility and tensile strength compared with neat PHBH [10].

Poly(propylene carbonate) (PPC), as an aliphatic polycarbonate, is made by alternating copolymerization of carbon dioxide (CO_2) with propylene oxide, whose synthesis process mitigates CO_2 accumulation in the environment [11]. The outstanding properties of PPC, including good processability, biodegradability and ductility, can render it of much industrial interest for applications in packaging, foams and agricultural mulch films. More recently, the utilization of rare earth catalysts allowed for the large-scale synthesis of PPC, which made it easy to utilize PPC at low cost [12,13]. PPC had been blended with a number of biodegradable polymers, for instance, poly(lactic acid) (PLA) [14,15], PHBV [16], PHB [17] and poly(butylene succinate) (PBS) [18], to enhance their mechanical properties. Zhang et al. [19] investigated the PHBH/PPC solution blend films with different weight ratios. It was observed that PHBH matrix and PPC phase were immiscible in the melted state. As PPC content was increased from 0 to 50 wt%, the elastic modulus decreased from 92.9 to 69.6 MPa, and the elongation at the break increased from 16% to 20%.

For thermodynamically immiscible polymer blends, lower interfacial interactions and poorer phase morphology limit the combination of the advantages of blend components. Therefore, compatibilization is required for an immiscible PHBH/PPC blend to improve the phase morphology and give the blends desired properties. The most common approaches to enhance the compatibility of immiscible polymer blends are the addition of block or graft copolymers, or "reactive blending" [20]. The main limitations of block or graft copolymers are the limited availability and high cost. Reactive blending is possible when the polymer components being blended can be modified by mutually reactive functional groups and remain structurally stable under processing conditions. Therefore, another method of introducing the third polymer component into immiscible blend systems to improve their compatibility has attracted attention. Moussaif et al. [21] reported that the introduction of polymethylmethacrylate (PMMA) into immiscible polycarbonate (PC)/polyvinylidenefluoride (PVDF) polymer blends decreased interfacial tension and improved tensile properties. Zhang et al. [22] prepared ternary PLA/PHBV/PBS blends with a good balance of toughness and stiffness and found that PHBV and PLA were partially miscible, while PBS was immiscible with PLA or PHBV.

Poly(vinyl acetate) (PVAc) with amorphous characteristics was reported to be miscible with PHB [23], PCL [24] and PLA [25]. Gao et al. [26] revealed that in the PPC/PLA/PVAc ternary blends, PVAc was selectively localized at the interface between the PLA dispersed phase and PPC matrix, which served as compatibilizers to increase interfacial adhesion and improve the mechanical properties of the blends. Therefore, in the present study, in view of the complementary property between PHBH and PPC, coupled with the miscibility between PVAc, PHBH and PPC, PVAc was selected as the third polymer phase to be introduced into the PHBH/PPC blends. Furthermore, the effect of PVAc concentration on the dynamic mechanical property, phase morphologies, melting and crystallization, tensile properties and rheological properties was investigated in detail. It was expected that PVAc phase could improve the compatibility of the PHBH/PPC blends and refine the phase structure, which resulted in an improvement in the mechanical and rheological properties of PHBH/PPC/PVAc blends.

2. Materials and Methods

2.1. Materials

PHBH was provided by Jiangsu Lansu Biomaterial Co., Ltd. (Yancheng, China). It had a weight-average molecular weight (M_w) and polydispersity (PDI) of 754 kg mol^{-1} and 4.89, respectively. PHBH copolymer had 6 mol% 3HH unit based on ^{1}H NMR. PPC was provided by Changchun Institute of Applied Chemistry. It had an M_w of 130 kg mol^{-1} and PDI of 2.6. PVAc with M_w of 315 kg mol^{-1} and PDI of 1.9 was bought from Nuoda New Materials Company (Yantai, China). Chemical structures of the three polymer components are provided in Scheme 1.

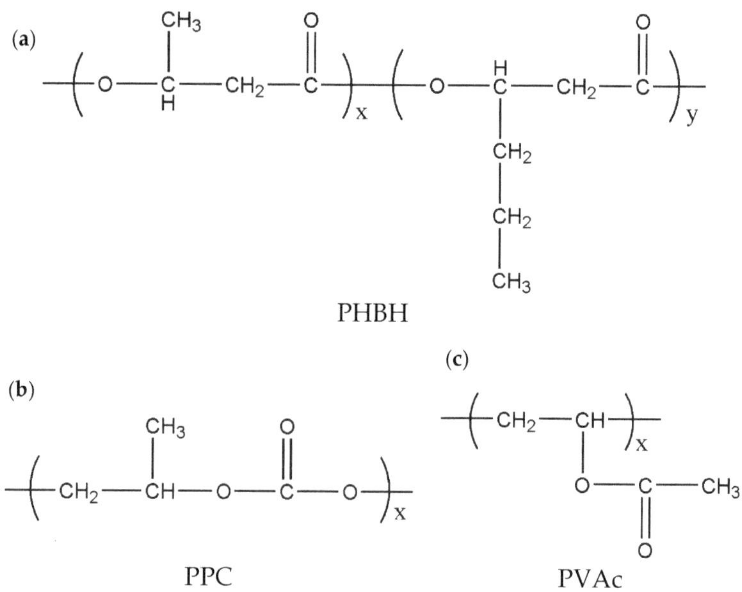

Scheme 1. Chemical structures of (**a**) PHBH, (**b**) PPC and (**c**) PVAc.

2.2. Sample Preparation

To remove moisture, before processing, PHBH, PPC and PVAc were dried at 80 °C under vacuum for 10 h. Melt blending of samples with different PVAc content was carried out with an internal mixer (XSS300, Shanghai Kechuang Rubber Plastic Mechanical Equipment Co., Ltd., Shanghai, China). Blending parameters included temperature of 160 °C, screw speed of 60 rpm and residence time of 7 min. The blended samples were hot-pressed at 170 °C and under a fixed pressure of 10 MPa for 3 min, then cold-pressed at room temperature and 10 MPa for 3 min to obtain 1-mm thick samples for tests. The weight composition of PHBH/PPC = 70 wt%/30 wt% was fixed in binary and ternary blends. For the PHBH/PPC/PVAc blends, PVAc contents were 5, 10 and 20 wt% of the total ternary blends. It was worth noting that neat PHBH, neat PPC, PHBH/PPC binary, PPC/PVAc binary and PHBH/PVAc binary blends undergoing the same thermal processing were also prepared as control samples. The PVAc weight content was set as 30 and 50 wt% in the PPC/PVAc binary blends and as 10, 20 and 30 wt% in the PHBH/PVAc binary blends.

2.3. Characterizations

2.3.1. Dynamic Mechanical Analysis (DMA)

Dynamic mechanical property tests were conducted on a DMA850 from a TA Instruments (New Castle, DE, USA) at oscillating amplitude of 5 μm and frequency of 1 Hz. The tests were carried out at temperatures ranging from −30 to 100 °C with a heating rate of 3 °C min^{-1}. The rectangle sample dimension was 20 × 10 × 1.0 mm^3.

2.3.2. Rheological Measurements

A rheometer (AR2000EX, TA Instrument, USA) with a parallel-plate geometry (diameter = 25 mm) was performed to investigated the melt viscoelastic properties of neat PHBH and its blends. The dynamic frequency sweep was carried out at 155 °C and swept from 100 to 0.05 rad s^{-1} under dry nitrogen. The fixed strain of 1.25% ensured that the response was within the linear viscoelastic range.

2.3.3. Scanning Electronic Microscopy (SEM)

The phase morphologies of neat PHBH and all blends were observed using a field emission scanning electron microscopy (XL30 ESEM FEG, FEI Co., Hillsboro, OR, USA) with an accelerating voltage of 10 kV. All sample films were cooled in liquid nitrogen for 20 min, then were cryogenically fractured. The smooth fractured surface of neat PHBH was taken for direct SEM observation, and all blends were etched by immersing them in acetone solution for 120 min, ensuring the dissolution of PPC phase but the preservation of PHBH matrix. Prior to SEM observation, the surfaces of all samples were sputter-coated with gold. The average size of dispersed phase (D) of the blends was calculated using Nano Measurer 1.2 software and analyzing 50–100 particles of SEM images for each sample.

2.3.4. Differential Scanning Calorimetry (DSC)

Melting and crystallization behaviors of neat PHBH and all its blends were probed by TA Instruments DSC (Q20, USA) under N_2 atmosphere. Around 5−8 mg of samples were heated from 40 to 190 °C at a heating rate of 100 °C min^{-1} and left at 190 °C for 3 min to erase the thermal history. The first cooling scan was monitored from 190 to −30 °C at a rate of 5 °C min^{-1} for determining the crystallization temperature (T_c) and crystallization enthalpy (ΔH_c). After that, the second heating scan was performed between −30 and 190 °C at a heating rate of 10 °C min^{-1}, from which the glass transition temperature (T_g), cold crystallization temperature (T_{cc}), cold crystallization enthalpy (ΔH_{cc}), melting temperature (T_m) and melting enthalpy (ΔH_m) could be obtained. The degree of crystallinity (X_c) of PHBH in all blends was calculated with the following formula:

$$X_c = \frac{\Delta H_m - \Delta H_{cc}}{\Delta H_m^0 \alpha} \times 100\%, \quad (1)$$

where ΔH_m^0 represents the fusion enthalpy of 100% crystalline PHBH (146 J g^{-1}) [27], and α represents the weight fraction of PHBH in the blends. It was worth noting that all enthalpy values including ΔH_c, ΔH_{cc} and ΔH_m, were corrected based on the weight content of PHBH in the blends.

For the isothermal crystallization analysis, all samples were first heated at a rate of 100 °C min^{-1} from 40 to 190 °C and held for 3 min, then quickly cooled at a rate of 45 °C min^{-1} to the desired crystallization temperature. Finally, the samples were further isothermally crystallized at the crystallization temperature for the sufficient time.

2.3.5. Tensile Tests

The tensile properties of neat PHBH and all its blends were measured at room temperature by using an Instron-1121 tensile tester (Canton, MA, USA). The gauge length of samples was 20 mm, and crosshead speed was set to 10 mm min^{-1}. At least 5 bars for each specimen were measured under the same conditions to obtain an average value.

3. Results and Discussion

3.1. Dynamic Mechanical Analysis

In order to investigate the mutual miscibility of the three polymer components, DMA tests were performed. This is because the phase structure and morphology of immiscible blends depend on the interfacial interactions and miscibility between polymer components. The damping factor (tan δ) curves of neat PHBH, PPC, PVAc, PHBH/PPC binary blends and PHBH/PPC/PVAc ternary blends are shown in Figure 1a. In order to study the miscibility between components more clearly, tan δ curves of PPC/PVAc binary and PHBH/PVAc binary blends are given in Figure 1b and 1c, respectively.

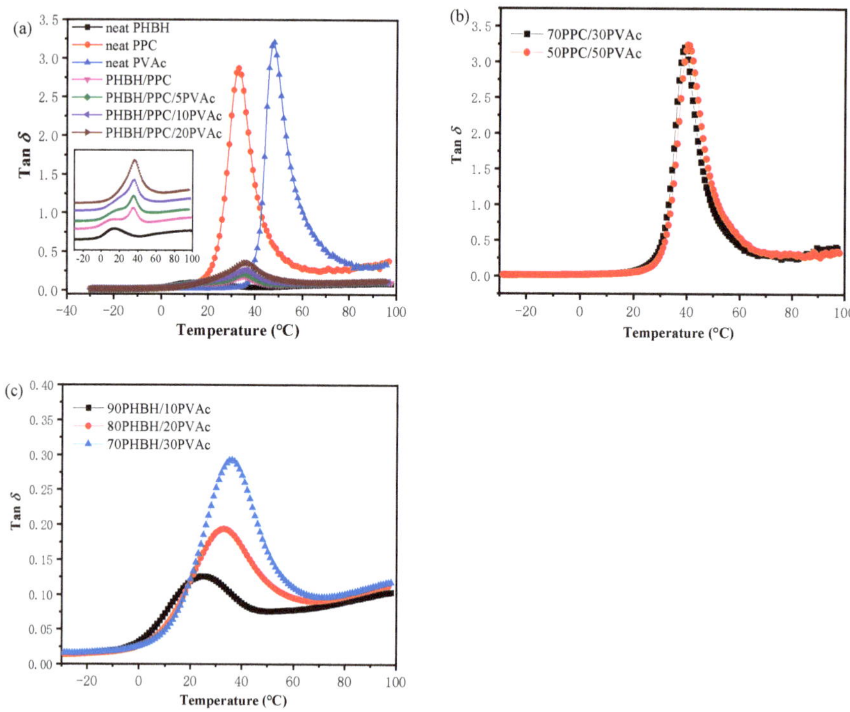

Figure 1. Damping factor (tan δ) curves of (**a**) neat PHBH, PPC, PVAc, PHBH/PPC binary and PHBH/PPC/PVAc ternary; (**b**) PPC/PVAc binary and (**c**) PHBH/PVAc binary blends.

The tan δ peak is directly related to the transition in molecular mobility, which represents the glass transitions of polymer. It is well known that the glass transition temperature (T_g) of the polymer blend system is an important criteria to assess the miscibility of polymer blends. If the blend has only one T_g, between the two neat components, it means that the two components of the blend are completely miscible. If the blend shows two composition-independent T_gs, which are close to the T_gs of neat components, it suggests that the blend is completely immiscible. Two composition-dependent T_gs, which are between those of neat components, indicates a partially miscible polymer blend [28]. Table 1 shows T_g values of neat components and all blends. As shown in Figure 1a and Table 1, sharp tan δ peaks were observed at about 13.9 °C for neat PHBH, 33.0 °C for neat PPC and 47.9 °C for neat PVAc, corresponding to their glass transitions. For the PHBH/PPC binary blend, two T_gs, one at around 12.5 °C, close to that of neat PHBH, and the other at around 34.7 °C, close to that of neat PPC, were observed, suggesting that PHBH and PPC were immiscible. Moreover, such a phenomenon was also in accordance with the literature report [19]. For the PHBH/PPC/PVAc blends, the damping peaks of PPC were located at about 35 °C and did not show significant changes with increasing PVAc content. The increase in the T_g of the PHBH matrix was observed for PHBH/PPC/5PVAc and PHBH/PPC/10PVAc blends, and the T_g of PHBH in the PHBH/PPC/10PVAc blend was around 6 °C higher than that of PHBH in the PHBH/PPC binary blend. The T_g of PHBH in the PHBH/PPC/20PVAc blend was hardly detected due to the fact that the T_g of PHBH was close to that of the PPC phase. As can be seen from Figure 1b, PPC/PVAc binary blends showed only one tan δ peak, which was between those of neat PPC and PVAc and increased with PVAc content. A variation of the T_g of PHBH/PVAc binary blends was similar to that of PPC/PVAc blends, as shown in Figure 1c. These results demonstrated that PVAc was miscible with both PHBH and PPC and displayed better miscibility with PHBH compared with PPC. Therefore, it can be inferred that PVAc was probably selectively localized in the PHBH matrix, increasing the T_g of PHBH in the PHBH/PPC/PVAc blends.

Table 1. Glass transition temperatures (T_gs) of all samples from DMA tests.

Samples	$T_{g,PHBH}$ (°C)	$T_{g,PPC}$ (°C)	$T_{g,PVAc}$ (°C)
neat PHBH	13.9	-	-
neat PPC	-	33.0	-
neat PVAc	-	-	47.9
PHBH/PPC	12.5	34.7	-
PHBH/PPC/5PVAc	16.8	35.2	-
PHBH/PPC/10PVAc	18.9	35.3	-
PHBH/PPC/20PVAc	Not detected	35.8	-
70PPC/30PVAc	-	38.9	-
50PPC/50PVAc	-	40.6	-
90PHBH/10PVAc	25.6	-	-
80PHBH/20PVAc	32.9	-	-
70PHBH/30PVAc	35.6	-	-

Figure 2 shows the temperature dependence of storage modulus (E') of neat components, the PHBH/PPC blends and PHBH/PPC/PVAc blends. Of the three neat glassy components at −10 °C, neat PHBH exhibited the highest E' of about 5.2 GPa, and neat PPC had the smallest E' of about 2.9 GPa. Due to the glass transition, E' of PHBH, PPC and PVAc decreased sharply at about 10, 30 and 50 °C, respectively. In case of PHBH/PPC binary blend, E' was between that of neat PHBH and that of neat PPC. For ternary blends, PVAc phase had little influence on E' of the PHBH/PPC/PVAc blends at temperature below T_g of PVAc (about 50 °C). However, E' of the PHBH/PPC/PVAc blends decreased significantly with the increase in PVAc content above the T_g of PVAc. Such a result was possibly owing to the dilution effect of PVAc on the PHBH matrix.

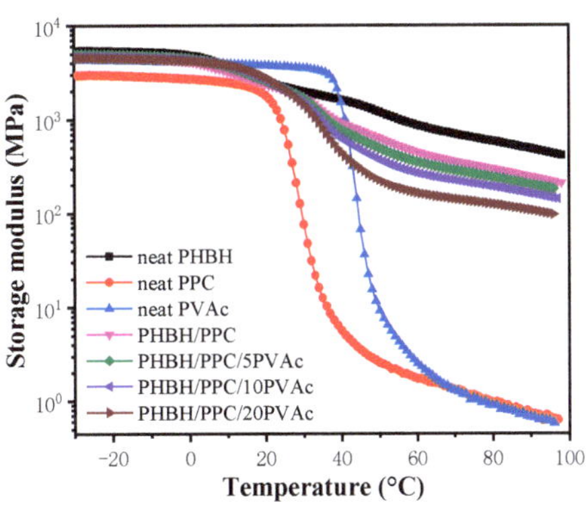

Figure 2. Storage modulus (E') curves of neat PHBH, PPC, PVAc, PHBH/PPC blends and PHBH/PPC/PVAc blends.

3.2. Rheological Properties

Rheology is often used to investigate the melt behavior of non-Newtonian fluids, such as polymers. The viscosity of a Newtonian fluid changes with temperature and does not change with the strain rate. However, a non-Newtonian liquid displays a change in viscosity with the strain rate. Therefore, the investigation of rheology is particularly important for determining the fluid mechanics of polymer [29]. The rheological properties of polymer blends are essential for understanding the relationships between the processibility and structure-property of polymer blends. Figure 3 shows the curves of the storage modulus (G'), loss modulus (G''), complex viscosity ($|\eta^*|$) and damping factor ($\tan\delta$) of neat PHBH, PPC, PVAc, the PHBH/PPC blends and the PHBH/PPC/PVAc blends at 155 °C.

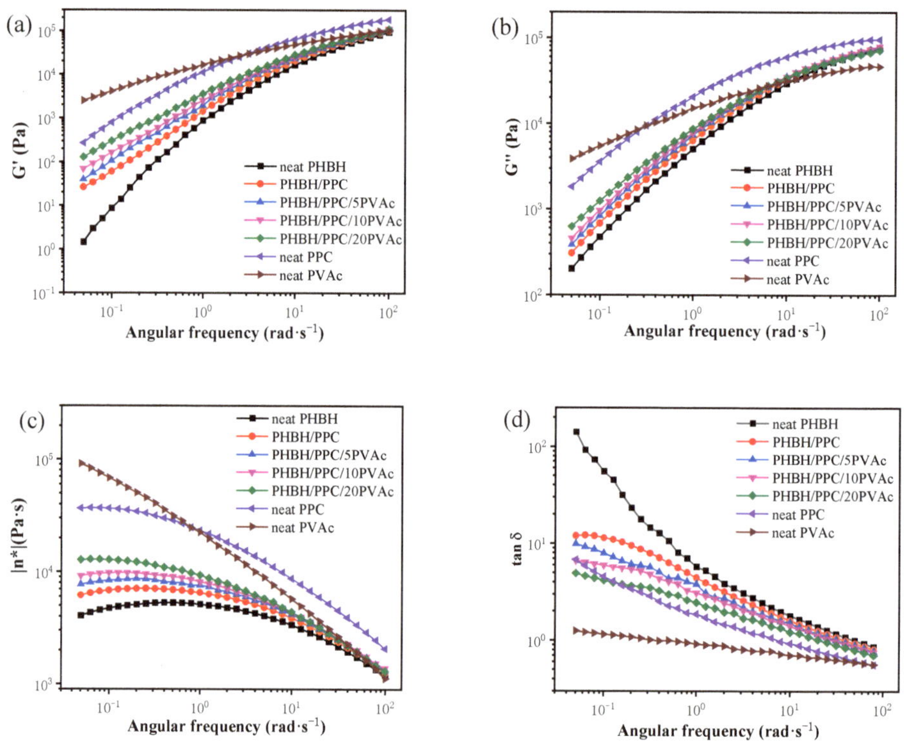

Figure 3. Dynamic viscoelastic properties of neat PHBH, PPC, PVAc, PHBH/PPC blends and PHBH/PPC/PVAc blends at 155 °C: (**a**) storage modulus (G'), (**b**) loss modulus (G''), (**c**) complex viscosity ($|\eta^*|$) and (**d**) loss tangent ($\tan\delta$).

According to G' in Figure 3a, neat PHBH had much lower melt elasticity than those of neat PVAc and PPC. Neat PVAc showed higher melt elasticity at low and intermediate frequency regions and lower melt elasticity at high frequency regions compared to that of neat PPC. The binary and ternary blends displayed a lower G' compared to neat PPC and a higher G' compared to neat PHBH. The increase in the PVAc content gradually increased the melt elasticity of ternary blends at low frequency regions due to the fact that PVAc had higher melt elasticity compared with the neat PHBH. Similar to G', the gradual introduction of the PVAc component into the PHBH/PPC blend also raised the G'' of the blends (Figure 3b). As shown in Figure 3c, neat PHBH showed a Newtonian plateau at intermediate frequency regions with the smallest viscosity of the three neat components. In the low-frequency region, the decreasing complex viscosity with decreasing frequency might be due to the slight degradation of PHBH during testing at 155 °C. Neat PVAc showed significant shear-thinning behaviour, suggesting that its complex viscosity was a function of frequency. Clearly, with the introduction of PVAc, the Newtonian plateau of the PHBH/PPC/PVAc blends at low frequency regions became progressively smaller, and the complex viscosity increased significantly. The damping factor (tan δ) is the ratio of dissipated energy (G'') to stored energy (G'). The tan δ data in Figure 3d demonstrated that PHBH had the largest damping factor among the three neat components and displayed a frequency dependence of tan δ, suggesting its viscous liquid characteristic. The tan δ values of PVAc were smaller than those of PHBH and PPC and were not very dependent on frequency, suggesting its elastic liquid characteristic. The tan δ curves of binary and ternary blends were between those of PHBH and PPC and decreased with an increase in the PVAc content at a low frequency due to the lower tan δ values of PVAc compared to those of PHBH.

As stated earlier, the dynamic viscoelastic properties of the blends implied that the presence of PVAc affected the rheological properties of PHBH/PPC blends. The influence of PVAc content on the rheological properties was due to its localization in the continuous phase of PHBH, which improved the interfacial interactions between the PHBH matrix and PPC dispersed phase, thereby promoting the dispersion of PPC domains in the PHBH matrix. By considering that the melt viscosity and elasticity of PVAc was higher than that of PHBH, adding PVAc into the PHBH/PPC blend resulted in an increase in the entanglement density in the continuous phase and at the interface between the PPC dispersion phase and the PHBH/PVAc continuous phase, as PVAc was mixed with PHBH macromolecular chains at their molecular-scale [30]. The increase in the degree of entanglement density was directly connected with the PVAc content. The enhancement of the entanglement density of the PHBH matrix and between the matrix and dispersion phase resulted in the increased viscosity and elasticity of the melt. This was due to that the increased physical interactions through chain entanglement restricted macromolecular flow and slip under shear deformations. Additionally, the entanglement density in the ternary blends could strongly retard the macromolecular chain relaxations, which might lead to an increase in the melt strength of the PHBH matrix [31]. The improved melt strength was very meaningful for PHBH processing due to one of its drawbacks of a low melt strength.

3.3. Phase Morphology

The phase morphology of blends is a crucial key affecting the macroscopic properties of blend systems. Therefore, the investigation of the size of dispersed phases and morphology types of the PHBH/PPC/PVAc ternary blends in this work was important and helpful for determining the relation between microstructure and resulting performance. Figure 4 gives the SEM micrographs of neat PHBH and all blends.

As shown in Figure 4(a1,a2), neat PHBH showed featureless and smooth cryo-fractured surfaces. For the PHBH/PPC and PHBH/PPC/PVAc blends (Figure 4(b1–e2)), the dispersed PPC phase etched away by acetone left homogeneous and spherical voids, implying that the PHBH matrix and PPC phase were immiscible, which was in line with the observation of DMA. It was worth noting that the acetone solution was a solvent for neat PVAc. From the previous discussion, PVAc was miscible with PHBH at the molecular level and was selectively localized in the PHBH matrix. Therefore, the small amount of PVAc dispersed in the PHBH matrix could not be solubilized in the acetone solution. It was of great interest that a much finer and more uniform phase structure was observed with the addition of the PVAc phase. Figure 5 clearly demonstrates the dispersed phase size obtained using an analysis of the Nano Measurer 1.2 software. As can be seen from Figure 5, the average diameter (D) of the dispersed phase decreased from 1.68 μm for the PHBH/PPC binary blend to a half value for the PHBH/PPC/20PVAc ternary blend (0.84 μm).

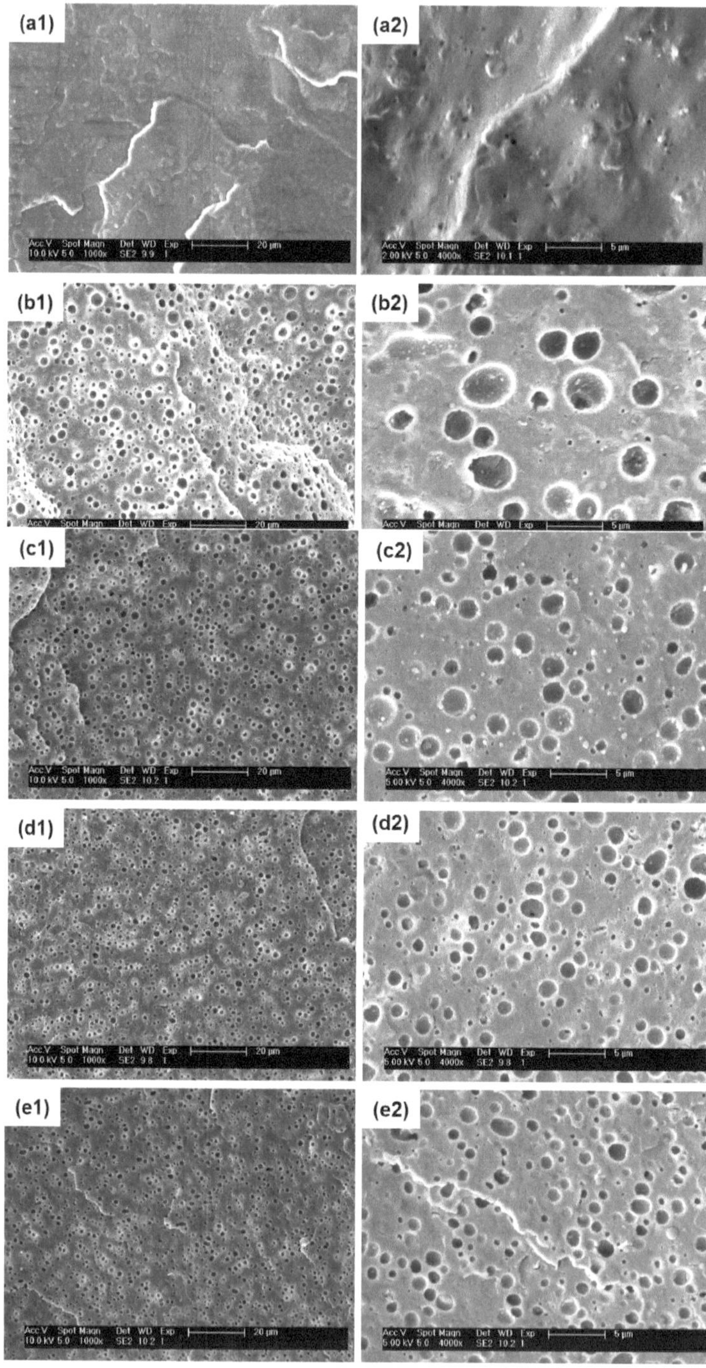

Figure 4. SEM images of surfaces for: (**a1**,**a2**) neat PHBH, (**b1**,**b2**) PHBH/PPC, (**c1**,**c2**) PHBH/PPC/5PVAc, (**d1**,**d2**) PHBH/PPC/10PVAc and (**e1**,**e2**) PHBH/PPC/20PVAc at different magnifications.

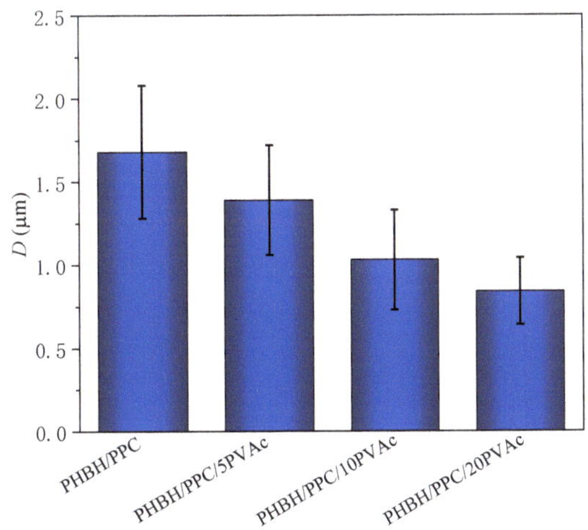

Figure 5. The average size of dispersed phase (D) of PHBH/PPC and PHBH/PPC/PVAc blends.

The phase morphologies of immiscible polymer blends was dependent on interfacial tension between the phases, blend compositions, processing parameters and viscosity ratio of the components [32,33]. The final morphology of immiscible blends was controlled by the competition between the droplet coalescence and break-up during melt processing [34]. It is well known that for the polymer blends with "sea-island" phase structure, phase coarsening frequently occurs with the coalescence of droplets, which reduces the interfacial area and decreases the free energy of blend systems [35]. The addition of compatibilizers to immiscible polymer blends can control phase morphology and enhance the miscibility of the blends by reducing interfacial tension [36]. Therefore, in the present work, it can be inferred that the PVAc phase in the PHBH/PPC/PVAc blends was fully miscible with the PHBH matrix and PPC phase and was selectively localized in the PHBH matrix, acting as compatibilizers to reduce the interfacial tensions between the PPC phase and PHBH matrix and refine the phase structure. In addition, PVAc as a third phase was selectively located in the PHBH matrix, forming a continuous phase together with the PHBH, which also led to a smaller size of the dispersed phase with the increasing PVAc content. Studies on the influence of the viscosity ratio (the ratio of dispersed phase viscosity to matrix phase viscosity) on the morphology of immiscible polymer blends indicated that the larger the viscosity ratio, the coarser the phase structure, while a viscosity ratio of approximately equal to 1 led to much finer morphology [37,38]. In the present work, as shown in Figure 3c, the viscosity of PPC was significantly greater than that of PHBH, indicating that the PHBH/PPC binary blend had a high viscosity ratio. For the PHBH/PPC/PVAc ternary blends, PVAc with a higher viscosity was selectively dispersed in the PHBH matrix, which could increase the matrix viscosity and reduce the viscosity ratio, resulting in a refined morphology. Similar results were observed in partially miscible PHBH/PLA blends with a reactive epoxy compatibilizer (REC) to enhance the compatibility between PHBH and PLA phases [39].

3.4. Thermal and Crystallization Behaviors

A differential scanning calorimetry (DSC) test was performed to study the thermal behaviors of neat components and their blends, as recorded in Figure 6. The thermal parameters are summarized in Table 2. Figure 6a clearly indicated that neat PHBH exhibited a significant crystallization exothermic peak (T_c) at about 50 °C. Neat PPC and PVAc had no crystallization peak, suggesting their amorphous characteristics. For the PHBH/PPC binary blends, it was observed that the T_c of PHBH shifted to a low temperature, suggesting that the crystallization of PHBH was suppressed by adding the PPC phase. Similarly, for the immiscible PHBH/PBAT and PHBH/PBS blends, the presence of PBAT and PBS inhibited the crystallization of PHBH [8,40]. In contrast, PCL was found to promote the crystallization of PHBH in immiscible PHBH/PCL blends [40]. However, for the PHBH/PPC/PVAc blends, no crystallization peaks were observed, suggesting that the presence of PVAc inhibited the crystallization of PHBH.

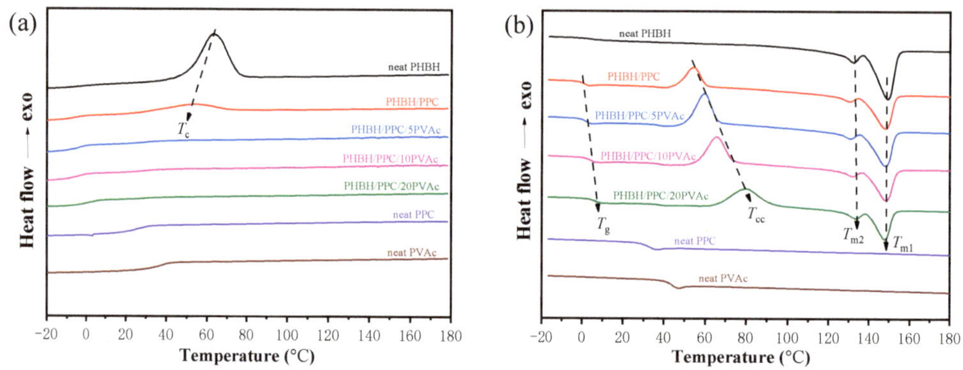

Figure 6. DSC thermograms of (**a**) first cooling at a cooling rate of 5 °C min^{-1} and (**b**) second heating at a heating rate of 10 °C min^{-1} for neat PHBH, PPC, PVAc, PHBH/PPC blends and PHBH/PPC/PVAc blends.

Table 2. Thermal properties of PHBH, PPC, PVAc, PHBH/PPC blends and PHBH/PPC/PVAc blends.

Sample	First Cooling		Second Heating						
	T_c (°C)	ΔH_c (J/g)	$T_{g,PHBH}$ (°C)	T_{cc} (°C)	ΔH_{cc} (J/g)	T_{m1} (°C)	T_{m2} (°C)	ΔH_m (J/g)	X_c (%)
neat PHBH	62.8	49.7	2.8	-	-	132.5	149.3	55.9	38.3
neat PPC	-	-	33.3	-	-	-	-	-	
neat PVAc	-	-	44.7	-	-	-	-	-	
PHBH/PPC	53.0	17.6	1.6	54.4	24.7	130.5	148.4	49.0	16.6
PHBH/PPC/5PVAc	-	-	2.4	59.6	40.8	130.7	148.5	52.9	8.3
PHBH/PPC/10PVAc	-	-	4.3	65.8	43.7	132.8	148.9	49.4	3.9
PHBH/PPC/20PVAc	-	-	6.5	80.4	47.5	133.5	148.5	49.3	1.2

As shown in Figure 6b, neat PHBH, PPC and PVAc showed the T_gs at 2.8, 33.3 and 44.7 °C, respectively. The T_g of PPC was not apparent for all blends, probably because the glass transition peaks overlapped with the onset of cold crystallization exothermic peaks. The T_g of PHBH in the blends was increased with increasing PVAc content, which was consistent with DMA observations, suggesting that PVAc was miscible with the PHBH matrix.

From Figure 6b, neat PHBH had no cold crystallization (T_{cc}), indicating its relatively complete crystallization during the cooling process. Furthermore, the T_{cc}s of the PHBH/PPC/PVAc blends increased significantly with increasing PVAc content. For instance, the T_{cc} of the PHBH/PPC blend was 54.4 °C, and it increased to 80.4 °C for the PHBH/PPC/20PVAc blend. The obvious increase in the T_{cc} of PHBH in the ternary blends once again illustrated that the introduction of PVAc into the PHBH/PPC blend inhibited the crystallization of PHBH. First, PVAc, which was located in the PHBH matrix, would restrict the stacking ability and mobility of PHBH molecular chains due to its high viscosity. Second, the presence of amorphous PVAc increased the entanglement density between PHBH molecular chains, which also inhibited the chains' mobility [30].

Neat PHBH and all blends showed typical double melting peaks during the second heating process. The melting peaks on the low-temperature side and high-temperature side are labeled as T_{m1} and T_{m2}, which correspond to the melting endothermic peaks of primary and recrystallized PHBH crystals, respectively [41]. The T_{m1} of PHBH in the PHBH/PPC blend was lower than that of neat PHBH, which was probably owing to the fact that the PPC phase affected the thickness of the lamellae of the primary crystals of PHBH. The T_{m1}s of PHBH in the PHBH/PPC/PVAc ternary blends were found to increase gradually with increasing PVAc content. This might be due to the fact that the incorporation of PVAc raised the T_{cc} of PHBH in the ternary blends, leading to the formation of more stable and perfect primary crystals of PHBH. The T_{m2}s of the blends did not change significantly compared to that of neat PHBH.

From Table 2, it was observed that the degree of crystallinity (X_c) of neat PHBH was 38.3%, and the X_c of the binary blend decreased to 16.6%, which was due to the incorporation of PPC inhibiting the crystallization of PHBH. For the PHBH/PPC/PVAc blends, the X_c continued to decrease with the introduction of PVAc. This was due to a decrease in the number of PHBH segments and chains growing toward the crystal growth front caused by the dilution of PVAc. In addition, the complete miscibility of PHBH and PVAc led to a migration of heterogeneities from the PHBH to the PVAc phase. Consequently, the number of heterogeneous primary nuclei of PHBH was decreased.

3.5. Isothermal Melt Crystallization Behavior and Kinetics

In this section, the influence of PVAc content on the isothermal melt crystallization kinetics of the PHBH/PPC blend was further investigated with DSC. DSC scans of isothermal crystallization of neat PHBH and all blends at 80 and 85 °C are presented in Figure 7. The relative crystallinity (X_t) at crystallization time (t) was determined with the following equation [42]:

$$X_t = \frac{X_c(t)}{X_c(\infty)} = \int_0^t \frac{dH(t)}{dt}dt \Big/ \int_0^\infty \frac{dH(t)}{dt}dt, \tag{2}$$

where X_c (t) is defined as the crystallization enthalpy at time t, and X_c (∞) is defined as the enthalpy when crystallization is finished. The $dH(t)/dt$ is the rate of heat flow during isothermal crystallization at time t. The development of X_t versus crystallization time t is showed in Figure 8. Obviously, all plots presented the shape of "S", and the crystallization time of the PHBH/PPC/PVAc blends was prolonged with an increase in the PVAc content. Furthermore, the Avrami equation was used to analyze isothermal crystallization kinetics of PHBH and its binary and ternary blends as follows [43]:

$$X_t = 1 - \exp(-kt^n), \tag{3}$$

where n ($n = n_1 + n_2$) is the Avrami exponent depending on both nucleation type (n_1) and growth dimension (n_2), while n_1 equals 1 for homogeneous nucleation and n_1 equals 0 for heterogeneous nucleation. The variable k is the crystallization rate constant, which is associated with both nucleation and growth. Figure 9 presents the Avrami plots of neat PHBH and all blends, and n and k values obtained from the linear portion of Avrami plots are listed in Table 3. An R^2 value greater than 0.99 indicated that the Avrami equation could describe the isothermal crystallization kinetics of the binary and ternary blend system well. In general, neat polymers have an n value of 4 with sporadic nucleation mode, while polymers containing heterogeneous nucleating agents show an n value of 3 with instantaneous nucleation [44]. In practice, impurities in the neat polymers result in the heterogeneous nucleation with a three-dimensional spherulite growth at n values of 3. From Table 3, regardless of the crystallization temperature and PVAc content, neat PHBH and its blends displayed similar values of n in a range of 2.4 to 2.8, indicating heterogeneous nucleation with two-dimensional to spherulitic crystal growth (n_1 = 0, an d n_2 = 2 or 3) [45]. Moreover, the crystallization temperature and blending with PPC and PVAc did not change the crystallization mechanisms of PHBH.

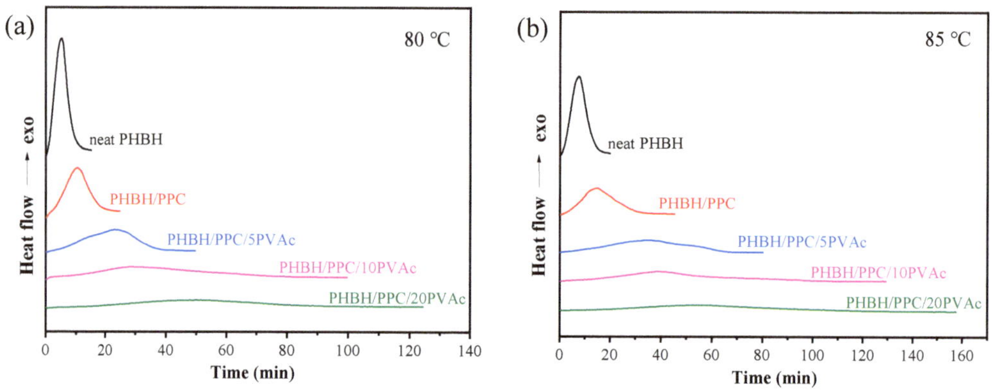

Figure 7. DSC curves for isothermal crystallization of neat PHBH and its blends at (**a**) 80 and (**b**) 85 °C.

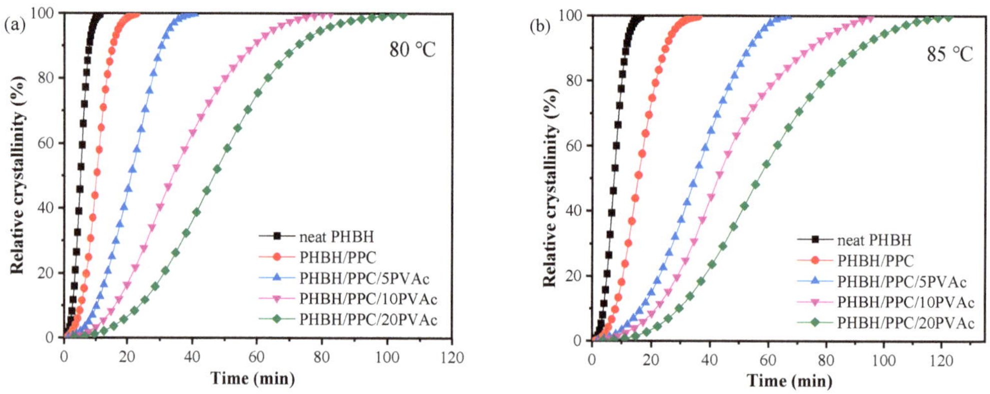

Figure 8. Relative crystallinity curves of neat PHBH and its blends at (**a**) 80 and (**b**) 85 °C.

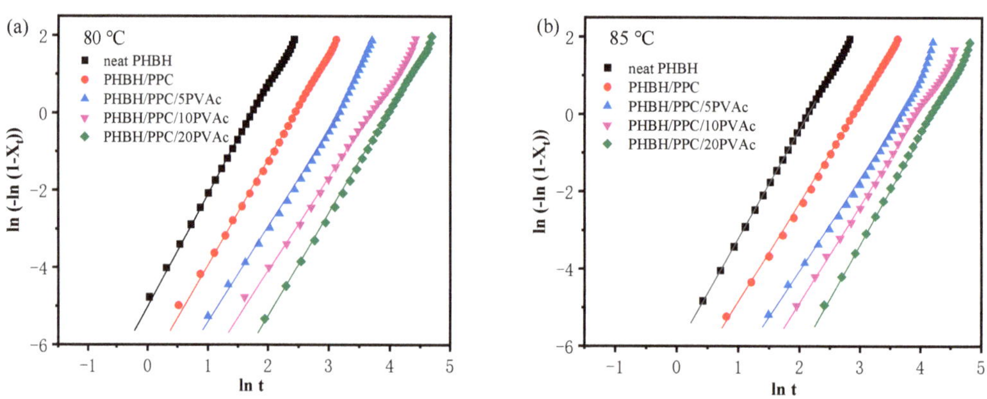

Figure 9. Avrami curves of neat PHBH and its blends at (**a**) 80 and (**b**) 85 °C.

Table 3. Isothermal crystallization kinetic parameters of neat PHBH and its blends based on the Avrami equation.

Sample	Crystallization Temperature of 80 °C				Crystallization Temperature of 85 °C			
	$t_{1/2}$ (min)	n	k (min^{-n})	R^2	$t_{1/2}$ (min)	n	k (min^{-n})	R^2
PHBH	4.8	2.8	8.23×10^{-3}	0.9998	7.2	2.8	2.36×10^{-3}	0.9998
PHBH/PPC	10.2	2.7	1.39×10^{-3}	0.9982	15.5	2.6	5.26×10^{-4}	0.9986
PHBH/PPC/5PVAc	20.9	2.6	3.16×10^{-4}	0.9940	35.3	2.4	1.20×10^{-4}	0.9963
PHBH/PPC/10PVAc	34.1	2.4	1.63×10^{-4}	0.9986	43.1	2.5	5.54×10^{-5}	0.9982
PHBH/PPC/20PVAc	46.8	2.7	2.30×10^{-5}	0.9994	56.9	2.8	8.81×10^{-6}	0.9992

Crystallization half time ($t_{1/2}$), defined as the time to achieve 50% X_t, could be used to evaluate the crystallization rate and was calculated with the following equation:

$$t_{1/2} = \left(\frac{\ln 2}{k}\right)^{1/n} \quad (4)$$

The $t_{1/2}$ values of neat PHBH and its blends are summarized in Table 3. As can be seen in Table 3, when the crystallization temperature was increased from 80 to 85 °C, the crystallization time of the samples was increased. For example, the $t_{1/2}$ of PHBH/PPC/5PVAc at 80 °C was 20.9 min, which was increased to 35.3 min at 85 °C. This was due to the fact that an increase in the crystallization temperature led to a decrease in supercooling, which reduced the driving force and resulted in difficulties in crystallization nucleation [46]. At crystallization temperature of 80 or 85 °C, the $t_{1/2}$ of the blends was greater than that of neat PHBH and increased with the increasing PVAc content. These results indicated that the incorporation of PVAc reduced the isothermal crystallization rate of the PHBH matrix, which could be due to the dilution influence of the amorphous PVAc melt on the PHBH crystal region. Another possible reason for the decrease in the crystallization rate of PHBH in the PHBH/PPC/PVAc blends was that the high viscosity of PVAc as well as the increased entanglement density inhibited the diffusion and stacking of PHBH molecular chains [22].

3.6. Tensile Mechanical Properties

The stress–strain curves of neat PHBH, the PHBH/PPC blends and the PHBH/PPC/PVAc blends are presented in Figure 10, and the mechanical properties of all samples are summarized in Table 4. Based on Figure 10a, neat PHBH was a rigid polymer and exhibited a brittle fracture fashion with much lower elongation at the break of 4.0%. It could be seen that the PHBH/PPC binary blend also failed in brittle mode, while it showed improved elongation at the break compared to neat PHBH. It could be expected that the modulus and yield strength of the binary blend were smaller than those of neat PHBH. This could be mainly due to the fact that the phase-separated morphology of the PHBH/PPC blend led to the interfacial debonding.

Table 4. Mechanical properties of neat PHBH and its blends.

Sample	Yield Strength (MPa)	Breaking Strength (MPa)	Young's Modulus (MPa)	Elongation at Break (%)
neat PHBH	31.6 ± 3.1	31.6 ± 3.1	705 ± 47	4.0 ± 0.1
PHBH/PPC	27.8 ± 2.4	27.8 ± 2.4	650 ± 41	6.8 ± 0.2
PHBH/PPC/5PVAc	34.3 ± 1.6	34.3 ± 1.6	634 ± 40	19.2 ± 1.6
PHBH/PPC/10PVAc	30.0 ± 0.4	18.5 ± 1.2	648 ± 13	84.7 ± 4.4
PHBH/PPC/20PVAc	27.9 ± 0.5	20.8 ± 0.5	386 ± 27	636 ± 31.7

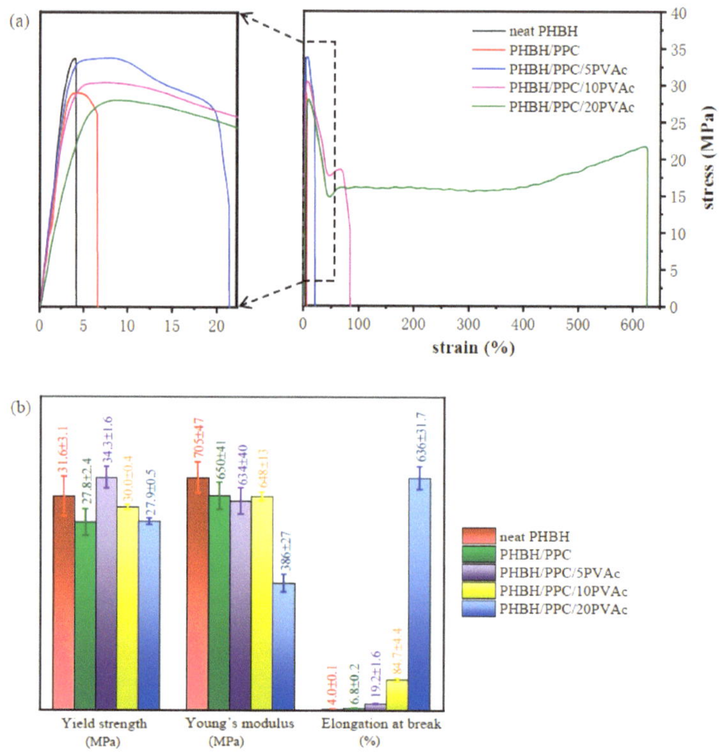

Figure 10. (**a**) Typical stress–strain curves of neat PHBH and its blends and (**b**) detailed tensile results about yield strength, Young's modulus and elongation at break.

For the PHBH/PPC/PVAc ternary blends, the elongation at the break was increased with increasing PVAc content. The ternary blends with 10 and 20 wt% PVAc underwent distinct yielding, suggesting the occurrence of brittle-to-ductile fracture transition. The ternary blend with 20 wt% PVAc especially showed a considerable and stable cold drawing after yielding, and the elongation at the break increased to 636%, which was more than 150 times higher than that of neat PHBH. It is well known that strength and toughness are important requirements for most structural materials. It is a pity that strength and toughness are usually mutually exclusive. Typically, the toughening of polymers would be accompanied by a sharp decrease in the strength of binary blends. The blends of PHBH also showed such trends [8,19,47]. For example, the elastic modulus and tensile strength of PHBH/PPC (70 wt%/30 wt%) decreased by 11.2% and 0.05%, respectively, and the elongation at the break increased by 23.9% compared to neat PHBH [19]. Katsumata et al. [47] observed that the introduction of a small amount of PCL could increase the toughness of the P(3HB-*co*-7 mol% 3HH) cast film, where the Young's modulus and the maximum stress descended greatly. However, for the blend of PHBH/PPC/5PVAc in this work, the presence of the PVAc phase increased the strength of PHBH from 31.6 to 34.3 MPa, and the elongation at the break was enhanced from 4 to 19%. With the addition of 10 wt% PVAc in the PHBH/PPC blend, the yield strength of the blend decreased slightly from 31.6 to 30.0 MPa, and a more significant increase in the elongation at the break, up to 84.7%, was observed.

The significantly greater elongation at the break of the PHBH/PPC/PVAc blends compared to neat PHBH was derived from the presence of the PVAc phase in the PHBH matrix, which confined the crystallization of PHBH and significantly reduced the degree of crystallinity due to the entanglement density and the dilution effect of PVAc, as shown in Figure 11. As a result, more amorphous PHBH regions were deformed and oriented when being stretched. On the other hand, the dilution effect of the PVAc phase weakened the intermolecular interactions of PHBH in the PHBH/PPC/PVAc blends, leading to an easier flow of PHBH molecular chains. The PVAc phase selectively localized in the PHBH matrix acted as a compatibilizer, reducing the interfacial tensions and refining the phase morphology, leading to an increase in yield strength.

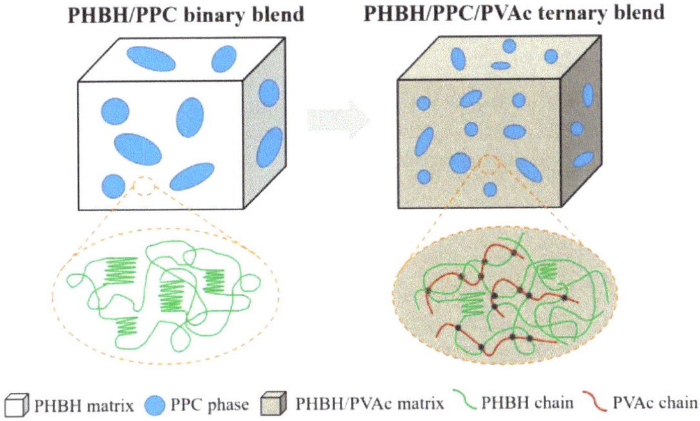

Figure 11. Schematic illustration of selective dispersion of PVAc in the ternary blends.

Figure 12 shows the comprehensive properties of neat PHBH, the PHBH/PPC blends and the PHBH/PPC/PVAc blends, including the yield strength, Young's modulus, elongation at the break, melt viscosity and storage modulus at an angular frequency of 0.05 rad s^{-1}. As can be seen from Figure 12, the incorporation of PVAc into the PHBH/PPC blends significantly improved the flexibility and melt viscoelasticity of ternary blends. Compared to neat PHBH, the melt viscosity, melt elasticity and elongation at the break of PHBH/PPC/10PVAc blend were increased by 126%, 4636% and 2000%, respectively, while the yield strength and Young's modulus decreased by only 8% and 5%, respectively. Therefore, the significant increase in flexibility and melt viscoelasticity with no deterioration in strength compared to neat PHBH were achieved by adding small amounts of amorphous PVAc into the PHBH/PPC binary blend, which was positive for expanding the applications of PHBH blends.

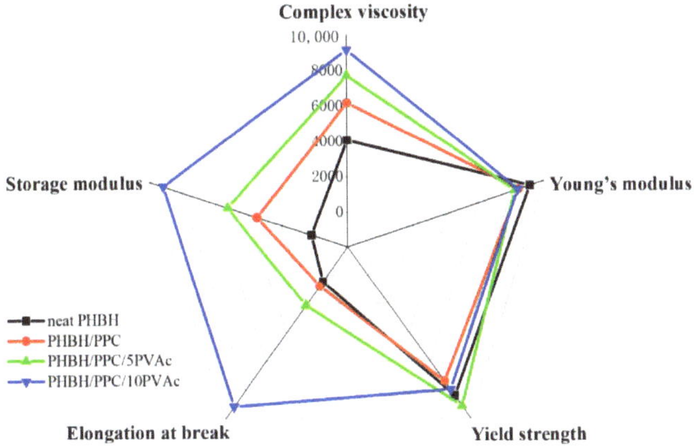

Figure 12. Comprehensive properties of neat PHBH, PHBH/PPC blends and PHBH/PPC/PVAc blends.

4. Conclusions

PHBH/PPC/PVAc ternary blends were prepared via a melt blending method with the aim of obtaining multiple good performances. A DMA analysis revealed that PHBH and PPC were immiscible. PVAc was miscible with both the PHBH matrix and PPC dispersed phase and displayed better miscibility with PHBH compared with PPC. Rheological property studies demonstrated that the elasticity and viscosity of blend melts increased significantly with the increasing PVAc content due to the high viscosity and elasticity of the PVAc melt and the increased entanglement density. The SEM results of all blends exhibited phase-separated morphology, and dispersed phase diameter gradually decreased with the increasing PVAc content. The PVAc phase was selectively localized in the PHBH matrix. Based on the DSC results, the PVAc phase suppressed the cold crystallization and reduced the degree of crystallinity of the PHBH matrix. The isothermal crystallization rate of the PHBH/PPC/PVAc blends was significantly decreased with adding PVAc, while the crystallization mechanism did not change. It was of great interest to observe that the introduction of a certain amount of PVAc significantly improved the toughness of the ternary blends and kept the strength without deterioration. This work provided a simple method to tailor the strength, ductility, viscosity and elasticity of the melt and the crystallization rate of the PHBH blends, which represented promise for the realization of specific properties, thus further expanding the application of PHBH.

Author Contributions: Conceptualization, C.H.; methodology, Y.J. and C.H.; validation, H.C., Y.J. and D.L.; formal analysis, Y.L. and H.C.; investigation, Y.J. and C.H.; resources, Y.J., Y.L. and H.W.; writing—original draft preparation, Y.L.; writing—review and editing, C.H.; supervision, Y.L. All authors have read and agreed to the published version of the manuscript.

Funding: This research was funded by Jilin Province science and technology development planning project (grant number 20220203087SF, 20230508042RC, and 20230508059RC) and the Science and technology research project of Education Department of Jilin Province (grant number JJKH20230325KJ).

Institutional Review Board Statement: Not applicable.

Data Availability Statement: The raw/processed data required to reproduce these findings are available upon request from the authors.

Conflicts of Interest: The authors declare no conflict of interest.

References

1. Tamiya, T.; Hsu, Y.-I.; Asoh, T.-A.; Uyama, H. Improvement of interfacial adhesion between poly(3-hydroxybutyrate-*co*-3-hydroxyhexanoate) and silica particles. *Ind. Eng. Chem. Res.* **2020**, *59*, 13595–13602. [CrossRef]
2. Frank, C.; Emmerstorfer-Augustin, A.; Rath, T.; Trimmel, G.; Nachtnebel, M.; Stelzer, F. Bio-polyester/rubber compounds: Fabrication, characterization, and biodegradation. *Polymers* **2023**, *15*, 2593. [CrossRef] [PubMed]
3. Che, X.; Ye, H.; Chen, G. Effects of uracil on crystallization and rheological property of poly(R-3-hydroxybutyrate-*co*-4-hydroxybutyrate). *Compos. Part A* **2018**, *109*, 141–150. [CrossRef]
4. Giubilini, A.; Siqueira, G.; Clemens, F.J.; Sciancalepore, C.; Messori, M.G.; Nyström, G.; Bondioli, F. 3D-printing nanocellulose-poly(3-hydroxybutyrate-*co*-3- hydroxyhexanoate) biodegradable composites by fused deposition modeling. *ACS Sustain. Chem. Eng.* **2020**, *8*, 10292–10302. [CrossRef]
5. Volova, T.G.; Uspenskaya, M.V.; Kiselev, E.G.; Sukovatyi, A.G.; Zhila, N.O.; Vasiliev, A.D.; Shishatskaya, E.I. Effect of monomers of 3-hydroxyhexanoate on properties of copolymers poly(3-hydroxybutyrate-*co* 3-hydroxyhexanoate). *Polymers* **2023**, *15*, 2890. [CrossRef]
6. Oyama, T.; Kobayashi, S.; Okura, T.; Sato, S.; Tajima, K.; Isono, T.; Satoh, T. Biodegradable compatibilizers for poly(hydroxyalkanoate)/poly(ε-caprolactone) blends through click reactions with endfunctionalized microbial poly(hydroxyalkanoate)s. *ACS Sustain. Chem. Eng.* **2019**, *7*, 7969–7978. [CrossRef]
7. Lim, J.; Chong, M.S.K.; Teo, E.Y.; Chen, G.; Chan, J.K.Y.; Teoh, S.-H. Biocompatibility studies and characterization of poly(3-hydroxybutyrate-*co*-3-hydroxyhexanoate)/polycaprolactone blends. *J. Biomed. Mater. Res. B* **2013**, *101*, 752–761. [CrossRef]
8. Zhang, M.; Diao, X.; Jin, Y.; Weng, Y. Preparation and characterization of biodegradable blends of poly(3-hydroxybutyrate-*co*-3-hydroxyhexanoate) and poly(butylene adipate-*co*-terephthalate). *J. Polym. Eng.* **2016**, *36*, 473–480. [CrossRef]
9. Yu, F.; Nakayama, T.; Nakamura, N.; Katsumata, K.; Pan, P.; Inoue, Y. Miscibility and physical properties of poly-(3-hydroxybutyrate-*co*-3-hydroxyhexanoate)/poly(ethylene oxide) binary blends. *Macromol. Mater. Eng.* **2009**, *294*, 868–876. [CrossRef]

10. Luo, L.; Wei, X.; Chen, G. Physical properties and biocompatibility of poly(3-hydroxybutyrate-*co*-3-hydroxyhexanoate) blended with poly(3-hydroxybutyrate-*co*-4-hydroxybutyrate). *J. Biomater. Sci. Polym. Ed.* **2009**, *20*, 1537–1553. [CrossRef]
11. Muthuraj, R.; Mekonnen, T. Recent progress in carbon dioxide (CO_2) as feedstock for sustainable materials development: Co-polymers and polymer blends. *Polymer* **2018**, *145*, 348–373. [CrossRef]
12. Wang, D.; Yu, J.; Zhang, J.; He, J.; Zhang, J. Transparent bionanocomposites with improved properties from poly(propylene carbonate) (PPC) and cellulose nanowhiskers (CNWs). *Compos. Sci. Technol.* **2013**, *85*, 83–89. [CrossRef]
13. Han, D.; Guo, Z.; Chen, S.; Xiao, M.; Peng, X.; Wang, S.; Meng, Y. Enhanced properties of biodegradable poly(propylene carbonate)/polyvinyl formal blends by melting compounding. *Polymers* **2018**, *10*, 771. [CrossRef]
14. Haneef, I.N.H.M.; Buys, Y.F.; Shaffiar, N.M.; Shaharuddin, S.I.S.; Nor Khairusshima, M.K. Miscibility, mechanical, and thermal properties of polylactic acid/polypropylene carbonate (PLA/PPC) blends prepared by melt-mixing method. *Mater. Today* **2019**, *17*, 534–542. [CrossRef]
15. Wang, Z.; Zhang, M.; Liu, Z.; Zhang, S.; Cao, Z.; Yang, W.; Yang, M. Compatibilization of the poly(lactic acid)/poly(propylene carbonate) blends through in situ formation of poly(lactic acid)-*b*-poly(propylene carbonate) copolymer. *J. Appl. Polym. Sci.* **2018**, *135*, 46009. [CrossRef]
16. Enriquez, E.; Mohanty, A.K.; Misra, M. Biobased blends of poly(propylene carbonate) and poly(hydroxybutyrate-*co*-hydroxyvalerate): Fabrication and characterization. *J. Appl. Polym. Sci.* **2017**, *134*, 44420. [CrossRef]
17. El-Hadi, A.M. Improvement of the miscibility by combination of poly(3-hydroxy butyrate) PHB and poly(propylene carbonate) PPC with additives. *J. Polym. Environ.* **2017**, *25*, 728–738. [CrossRef]
18. Han, X.; Jin, Y.; Wang, B.; Tian, H.; Weng, Y. Reinforcing and toughening modification of PPC/PBS blends compatibilized with epoxy terminated hyperbranched polymers. *J. Polym. Environ.* **2022**, *30*, 461–471. [CrossRef]
19. Zhang, L.; Zheng, Z.; Xi, J.; Gao, Y.; Ao, Q.; Gong, Y.; Zhao, N.; Zhang, X. Improved mechanical property and biocompatibility of poly(3-hydroxybutyrate-*co*-3-hydroxyhexanoate) for blood vessel tissue engineering by blending with poly(propylene carbonate). *Eur. Polym. J.* **2007**, *43*, 2975–2986. [CrossRef]
20. Chun, S.; Han, C. Morphology of model A/B/(Cblock-D) ternary blends and compatibilization of two immiscible homopolymers A and B with a C-block-D copolymer. *Macromolecules* **2000**, *33*, 3409–3424. [CrossRef]
21. Moussaif, N.; Jérôme, R. Compatibilization of immiscible polymer blends (PV/PVDF) by the addition of a third polymer (PMMA): Analysis of phase morphology and mechanical properties. *Polymer* **1999**, *40*, 3919–3932. [CrossRef]
22. Zhang, K.; Mohanty, A.K.; Misra, M. Fully biodegradable and biorenewable ternary blends from polylactide, poly(3-hydroxybutyrate-*co*-hydroxyvalerate) and poly(butylene succinate) with balanced properties. *ACS Appl. Mater. Interfaces* **2012**, *4*, 3091–3101. [CrossRef] [PubMed]
23. Madbouly, A.A.; Mansour, A.A.; Abdou, N.Y. Crystallization kinetics of PHB/PVAc blends using time resolved dielectric spectroscopy. *Eur. Polym. J.* **2007**, *43*, 3933–3942. [CrossRef]
24. Sivalingam, G.; Karthik, R.; Madras, G. Blends of poly(3-caprolactone) and poly(vinyl acetate): Mechanical properties and thermal degradation. *Poly. Degrad. Stabil.* **2004**, *84*, 345–351. [CrossRef]
25. Gajria, A.M.; Dave, V.; Gross, R.A.; McCarthy, S.P. Miscibility and biodegradability of blends of poly(lactic acid) and poly(vinyl acetate). *Polymer* **1996**, *37*, 437–444. [CrossRef]
26. Gao, J.; Bai, H.; Zhang, Q.; Gao, Y.; Chen, L.; Fu, Q. Effect of homopolymer poly(vinyl acetate) on compatibility and mechanical properties of poly(propylene carbonate)/poly(lactic acid) blends. *Express Polym. Lett.* **2012**, *6*, 860–870. [CrossRef]
27. Lee, J.; Hikima, Y.; Sekiguchi, T.; Ohshima, M. Thermal, rheological, and mechanical properties of cellulose nanofber (CNF) and poly(3-hydroxybutyrate-*co*-3-hydroxyhexanoate) (PHBH) biopolymer nanocomposites. *Cellulose* **2022**, *29*, 3901–3913. [CrossRef]
28. Qiu, Z.; Yang, W.; Ikehara, T.; Nishi, T. Miscibility and crystallization behavior of biodegradable blends of two aliphatic polyesters. Poly(3-hydroxybutyrate-*co*-hydroxyvalerate) and poly(3-caprolactone). *Polymer* **2005**, *46*, 11814–11819. [CrossRef]
29. Ching, Y.C.; Ali, M.E.; Abdullah, L.C.; Choo, K.W.; Kuan, Y.C.; Julaihi, S.J.; Chuah, C.H.; Liou, N.-S. Rheological properties of cellulose nanocrystal-embedded polymer composites: A review. *Cellulose* **2016**, *23*, 1011–1030. [CrossRef]
30. Hao, X.; Kaschta, J.; Liu, X.; Pan, Y.; Schubert, D.W. Entanglement network formed in miscible PLA/PMMA blends and its role in rheological and thermo-mechanical properties of the blends. *Polymer* **2015**, *80*, 38–45. [CrossRef]
31. Hemelrijck, E.V.; Puyvelde, P.V.; Macosko, C.W.; Moldenaers, P. The effect of block copolymer architecture on the coalescence and interfacial elasticity in compatibilized polymer blends. *J. Rheol.* **2005**, *49*, 783–798. [CrossRef]
32. Ardakani, F.; Jahani, Y.; Morshedian, J. The impact of viscoelastic behavior and viscosity ratio on the phase behavior and morphology of polypropylene/polybutene-1 blends. *J. Vinyl Addit. Technol.* **2015**, *21*, 94–101. [CrossRef]
33. Rameshwaram, J.K.; Yang, Y.S.; Jeon, H.S. Structure–property relationships of nanocomposite-like polymer blends with ultrahigh viscosity ratios. *Polymer* **2005**, *46*, 5569–5579. [CrossRef]
34. Liebscher, M.; Tzounis, L.; Pötschke, P.; Heinrich, G. Influence of the viscosity ratio in PC/SAN blends filled with MWCNTs on the morphological, electrical, and melt rheological properties. *Polymer* **2013**, *54*, 6801–6808. [CrossRef]
35. Feng, J.; Liu, X.; Bao, R.; Yang, W.; Xie, B.; Yang, M. Suppressing phase coarsening in immiscible polymer blends using nano-silica particles located at the interface. *RSC Adv.* **2015**, *5*, 74295.
36. Salehiyan, R.; Yoo, Y.; Choi, W.J.; Hyun, K. Characterization of morphologies of compatibilized polypropylene/polystyrene blends with nanoparticles via nonlinear rheological properties from FT-rheology. *Macromolecules* **2014**, *47*, 4066–4076. [CrossRef]

37. Slouf, M.; Radonjic, G.; Hlavata, D.; Sikora, A. Compatibilized iPP/aPS blends: The effect of the viscosity ratio of the components on the blends morphology. *J. Appl. Polym. Sci.* **2006**, *101*, 2236–2249. [CrossRef]
38. Lu, X.; Zhao, J.; Yang, X.; Xiao, P. Morphology and properties of biodegradable poly(lactic acid)/poly (butylene adipate-*co*-terephthalate) blends with different viscosity ratio. *Polym. Test.* **2017**, *60*, 58–67. [CrossRef]
39. Zhou, Y.; Huang, Z.; Diao, X.; Weng, Y.; Wang, Y. Characterization of the effect of REC on the compatibility of PHBH and PLA. *Polym. Test.* **2015**, *42*, 17–25. [CrossRef]
40. Jeepery, I.F.; Sudesh, K.; Abe, H. Miscibility and enzymatic degradability of poly(3-hydroxybutyrate-*co*-3-hydroxyhexanoate)-based polyester blends by PHB depolymerase and lipase. *Polym. Degrad. Stabil.* **2021**, *192*, 109692. [CrossRef]
41. Li, D.; Fu, J.; Ma, X. Improvement in thermal, mechanical, and barrier properties of biocomposite of poly (3-hydroxybutyrate-*co*-3-hydroxyhexanoate)/modified nano-SiO_2. *Polym. Compos.* **2020**, *41*, 381–390. [CrossRef]
42. Cebe, P.; Hong, S.D. Crystallization behaviour of poly(ether-ether-ketone). *Polymer* **1986**, *27*, 1183–1192.
43. Avrami, M. Kinetics of phase change. II transformation–time relations for random distribution of nuclei. *J. Chem. Phys.* **1940**, *8*, 212–224.
44. Piorkowska, E.; Galeski, A.; Haudin, J.-M. Critical assessment of overall crystallization kinetics theories and predictions. *Prog. Polym. Sci.* **2006**, *31*, 549–579.
45. Tsui, A.; Frank, C.W. Comparison of anhydrous and monohydrated forms of orotic acid as crystal nucleating agents for poly(3-hydroxybutyrate-co-3-hydroxyvalerate). *Polymer* **2014**, *55*, 6364–6372.
46. Li, J.; Qiu, Z. Effect of low loadings of cellulose nanocrystals on the significantly enhanced crystallization of biodegradable poly(butylene succinate-*co*-butylene adipate). *Carbohyd. Polym.* **2019**, *205*, 211–216.
47. Katsumata, K.; Saito, T.; Yu, F.; Nakamura, N.; Inoue, Y. The toughening effect of a small amount of poly(ε-caprolactone) on the mechanical properties of the poly(3-hydroxybutyrate-*co*-3-hydroxyhexanoate)/PCL blend. *Polym. J.* **2011**, *43*, 484–492.

polymers

Article

Replacing Harmful Flame Retardants with Biodegradable Starch-Based Materials in Polyethylene Formulations

Bárbara O. Carvalho [1], Luís P. C. Gonçalves [1], Patrícia V. Mendonça [1], João P. Pereira [2], Arménio C. Serra [1] and Jorge F. J. Coelho [1,*]

[1] Centre for Mechanical Engineering, Materials, and Processes, Advanced Production and Intelligent Systems (ARISE), Department of Chemical Engineering, University of Coimbra, Rua Sílvio Lima-Polo II, 3030-790 Coimbra, Portugal
[2] Componit Lda, Estr. Real 3, 2070-621 Vila Chã de Ourique, Portugal
* Correspondence: jcoelho@eq.uc.pt

Abstract: The addition of toxic flame retardants to commercially available polymers is often required for safety reasons due to the high flammability of these materials. In this work, the preparation and incorporation of efficient biodegradable starch-based flame retardants into a low-density polyethylene (LDPE) matrix was investigated. Thermoplastic starch was first obtained by plasticizing starch with glycerol/water or glycerol/water/choline phytate to obtain TPS-G and TPS-G-CPA, respectively. Various LDPE/TPS blends were prepared by means of melt blending using polyethylene graft maleic anhydride as a compatibilizer and by varying the content of TPS and a halogenated commercial flame retardant. By replacing 38% and 76% of the harmful commercial flame retardant with safe TPS-G-CPA and TPS-G, respectively, blends with promising fire behavior were obtained, while the limiting oxygen index (LOI ≈ 28%) remained the same. The presence of choline phytate improved both the charring ability and fire retardancy of starch and resulted in a 43% reduction in fire growth index compared to the blend with commercial flame retardant only, as confirmed by means of cone calorimetry. Standard UL 94 vertical tests showed that blends containing TPS exhibited dripping behavior (rated V2), while those with commercial flame retardant were rated V0. Overall, this work demonstrates the potential of starch as a natural flame retardant that could reduce the cost and increase the safety of polymer-based materials.

Keywords: starch; flame retardant; biodegradable polymer; polyethylene; fire behavior

Citation: Carvalho, B.O.; Gonçalves, L.P.C.; Mendonça, P.V.; Pereira, J.P.; Serra, A.C.; Coelho, J.F.J. Replacing Harmful Flame Retardants with Biodegradable Starch-Based Materials in Polyethylene Formulations. *Polymers* **2023**, *15*, 4078. https://doi.org/10.3390/polym15204078

Academic Editors: Paul Joseph, Vishal Gavande, Vasi Shaikh and Won-Ki Lee

Received: 3 August 2023
Revised: 19 September 2023
Accepted: 11 October 2023
Published: 13 October 2023

1. Introduction

Polyethylene (PE) is one of the most widely used polyolefins in the global plastics market (around 34%) [1] and is available in three main grades: low-density polyethylene (LDPE), linear low-density polyethylene (LLDPE) and high-density polyethylene (HDPE) [2,3]. The popularity and wide range of applications of PE are related to its good mechanical properties, ease of processing, low toxicity, and electrical insulation properties, at an affordable price [4]. LDPE, for example, is one of the most widely used polymers for electrical insulation because it combines suitable mechanical properties and good electrical properties, namely low permittivity, and high electrical breakdown strength [1]. According to the European Commission, the LDPE market was valued at USD 4 billion in 2020 [5]. The market is forecast to grow at a 3% CAGR (compound annual growth rate) from 2023 to 2028 [6]. However, due to its long aliphatic chains, LDPE is highly flammable and cannot tolerate temperatures above 70 °C, which limits its application in many sectors, such as transportation, electronics, among others [7,8]. Therefore, the addition of flame retardants (FRs) to LDPE is a simple strategy to reduce the flammability of this polymer, increase its survival time in the event of fire and ensure the safety levels required for each application. There are several types of commercially available FR, such as compounds based on phosphorus, borate, inorganic hydroxides, silicon, nitrogen, and halogen-containing molecules [9,10], that can be used in LDPE formulations, and these are usually incorporated into the polymer during processing. Information on the mechanism of action and application of these FRs in LDPE can be found in a recent review article [1]. The commercially available FRs used in LDPE are usually halogenated FR in combination with antimony trioxide (ATO, Sb_2O_3), magnesium hydroxide (MDH, $Mg(OH)_2$), or aluminum hydroxide (ATH, $Al(OH)_3$) compounds. Despite their good flame-retardant performance, halogen containing FRs are not good candidates from an environmental point of view due to their high toxicity, and some of them were banned a few decades ago [11]. In view of current environmental concerns, it is therefore highly desirable to develop new halogen-free FRs that are both safe and can be obtained from renewable and sustainable sources [12,13]. Several biobased materials, such as deoxyribonucleic acid (DNA) [14], β-cyclodextrin [14], organic phosphorus compounds [15], phytic acid [7], polydopamine or tannic acid [12], have been proposed as environmentally friendly FRs. The use of various natural polymers, such as starch, wood flour, chitosan, lignin, and others as FRs has also been demonstrated with the aim of improving both the biodegradability and flame retardancy of polymer materials [13].

Using starch as an FR is a wise and promising option because this natural polymer is safe, abundant, and inexpensive. Starch from various sources has been successfully blended with LDPE via extrusion, resulting in blends with inferior mechanical properties in comparison with neat LDPE [16]. In fact, plasticizing starch with different molecules, namely water, polyols, or choline phytate (CPA), to give the so-called thermoplastic starch (TPS) is necessary to provide some workability and mechanical properties to starch-based materials [7,17,18]. However, the incorporation of TPS into LDPE results in materials with poor mechanical properties. This is due to the incompatibility between the highly hydrophilic character and strong intra- and inter-molecular hydrogen bonding of the starch, and the hydrophobic nature of the polyolefin [19]. A straightforward and simple strategy to improve the miscibility of TPS/LDPE blends is to use commercially available compatibilizers, such as PE grafted with glycidyl methacrylate (PE-*g*-GMA) or LDPE-grafted maleic anhydride (LDPE-*g*-MA) [20]. Unfortunately, all reported work uses starch/TPS as a filler for LDPE to improve the biodegradability of this polyolefin. To the best of our knowledge, the use of starch as a flame retardant in LDPE formulations has never been reported.

The aim of this work was to develop a new starch-based flame retardant that can be used in PE formulations. In this way, it will be possible to obtain a low-cost and partially biodegradable product and reduce or eliminate the need for harmful flame retardants. For this purpose, various LDPE/TPS blends with different proportions and compositions of TPS were prepared by means of melt blending. The effect of combining the biobased FRs developed in this work with a commercial halogenated flame retardant in LDPE blends was also investigated. The materials were analyzed using scanning electron microscopy (SEM), tensile tests, thermogravimetric analysis (TGA), the limiting oxygen index (LOI), flammability (UL 94) and cone calorimetry.

2. Materials and Methods

2.1. Materials

Choline chloride (>99.0%, Acros Organics, Geel, Belgium, Germany), ethanol absolute (>99.5%, PanReac AppliChem ITW Reagents, Darmstadt, Germany), deuterated chloroform ($CDCl_3$, Eurisotop, Saarbrücken, Germany), deuterated dimethyl sulfoxide (d_6-DMSO, Eurisotop, Saarbrücken Germany), deuterated water (D_2O, Eurisotop, Saarbrücken Germany), glycerol (>99%, Merck, Darmstadt, Germany), phytic acid (ca. 50% in water, ca. 1.1 mol/L, TCI Europe, Zwijndrecht, Belgium, Germany) and sodium hydroxide (NaOH, José Manuel Gomes dos Santos, LDA, Odivelas, Portugal) were used as received.

IsoAdditive FR L069 flame retardant (brominated-based additives in LDPE carrier) was kindly supplied by Isolago (Aveiras de Baixo, Portugal) and used as received.

Potato starch, low-density polyethylene (LDPE) with a melting temperature of 140 °C (Maxxam™) and polyethylene-*graft*-maleic anhydride (PE-*g*-MA, Orevac OE825, SK-fp) were kindly supplied by Componit (Vila Chã de Ourique, Portugal) and used as received.

2.2. Techniques

TPS and LDPE/TPS blends were prepared by means of melt blending using a laboratory mixer HAAKE™ Polylab™ QC (Thermo Scientific™, Waltham, MA, USA).

We recorded 400 MHz ^{1}H nuclear magnetic resonance (^{1}H NMR) spectra of the compounds using a Bruker Avance II 400 MHz Spectrometer (Bruker Biospin, Wissemboug, France) with a 5 mm TIX triple resonance detection probe.

Fourier transform infrared (FTIR) spectra were obtained in the range of 4000 to 750 cm^{-1}, with 4 cm^{-1} spectral resolution and 64 accumulations, at room temperature using an Agilent Technologies Carey 630 spectrometer (Agilent, Santa Clara, CA, USA) equipped with a golden gate single reflection diamond ATR.

The mechanical properties of the blends were accessed via tensile tests using a Chatillon TCD100 machine (Transcat, Rochester, NY, USA) at 5 mm/min speed and a load cell of 5 kN, with at least 5 specimens (10 $\pm$ 0.5 mm $\times$ 70 $\pm$ 0.5 mm).

Samples for scanning electron microscopy (SEM) analysis were prepared using the fraction surface of the tensile test's specimens. The fracture surface was coated with gold and analyzed in a Field Emission SEM ZEISS MERLIN Compact/VP Compact Gemini II (Zeiss, Madrid, Spain).

Vertical flammability assays were conducted by AIMPLAS (Spain) following the UL 94:2013 standard [21].

LOI assays were conducted by AIMPLAS (Spain) using the Netzsch-Taurus Oxygen (Netzsch, Bobingen, Germany) system following the standard UNE-EN ISO 4589-2:2017 [22]. Average sample thickness (mm): 1.5; average sample width (mm): 10.

Cone calorimeter tests were conducted by AIMPLAS (Spain) using the FTT ICONE CLASSIC machine (Fire Testing Technology, East Grinstead, UK) and following the standard EN ISO 5660-1:2015 [23]. Conditions: exhaust flow rate (L/s): 0.24; irradiance (kW/m^2): 50; orientation: horizontal; separation (mm): 25; wire grid: No; N of specimens: 3; average sample thickness (mm): 1.5; pre-conditioning 48 h at 23 $\pm$ 2 °C, 50 $\pm$ 10%RH; environmental conditions: 23 $\pm$ 2 °C; 50 $\pm$ 10%RH

Thermogravimetric (TG) analyses were carried out using a NETZSCH STA 44F5 (Netzsch, Selb, Germany), from 20 °C to 500 °C, at a heating rate of 10 °C·min^{-1}, under nitrogen purge flow (sample weight in the range of 5 to 10 mg).

2.3. Procedures

2.3.1. Synthesis of Choline Phytate

CPA was prepared according to the procedures described in the literature [24]. First, choline hydroxide ([Chol][OH]) was prepared using the ion exchange method. For this purpose, NaOH (28.0 g, 0.7 mol) and choline chloride ([Chol][Cl]), 97.7 g, 0.7 mol) were dissolved in 375 mL and 250 mL of absolute ethanol, respectively. The solutions were mixed and stirred at room temperature for 1 h. The formed white precipitate (sodium chloride) was removed via vacuum filtration, and the solution of [Chol][OH] in ethanol was obtained. Phytic acid (77.2 g, 0.117 mol) was then dissolved in 100 mL of absolute ethanol. The phytic acid solution was added to the [Chol][OH] solution and stirred for 1 h at 30 °C. The ethanol was removed under reduced pressure. Finally, CPA was obtained after drying under reduced pressure at 80 °C for 24 h. The chemical structure of CPA was confirmed through ^{1}H NMR spectroscopy (Figure S1).

2.3.2. Preparation of TPS

Potato starch was plasticized by two different methods, giving TPS-G (plasticized with glycerol and water) and TPS-G-CPA (plasticized with glycerol, water and CPA). Table 1 shows the composition of each TPS type.

Table 1. Composition of the two types of TPS prepared.

Sample	Potato Starch (wt%)	Glycerol (wt%)	Distilled Water (wt%)	CPA (wt%)
TPS-G	62.5	25.0	12.5	--
TPS-G-CPA	50.0	20.0	10.0	20.0

All compounds were placed in a plastic flask and mixed manually with a spatula. The flask was sealed and the mixture was allowed to stand overnight. The mixture was then processed via melt mixing at 120 °C (TPS-G) or 140 °C (TPS-G-CPA) for 5 min, at 100 rpm.

2.3.3. Preparation of LDPE/TPS Blends

To prepare LDPE/TPS blends, LDPE, TPS, PE-*g*-MA compatibilizer, and a commercial flame retardant (FR L069, if used) were added to a mixing chamber. The mixtures were processed at 140 °C with a rotation speed of 100 rpm for 5 min. The obtained mixtures were cut into small pieces by hand, placed between two silicone squares and pressed in a Carver® hydraulic press under 0.5 bar at 140 °C for 5 min, followed by cooling at room temperature. The compounds were produced in the form of sheets with a thickness of approx. 5 mm.

3. Results and Discussion

3.1. Preparation and Characterization of the Blends

Starch is an inexpensive and naturally occurring polysaccharide that exhibits charring properties, and has been considered as a potential biobased flame retardant for various different applications [25,26]. In this work, we investigated the possibility of replacing a commercial flame retardant (FR L069) used in LDPE formulations with a starch-based flame retardant, thereby reducing production costs and increasing both the safety and biodegradability of the final material. Initially, potato starch was plasticized with water and glycerol to form TPS-G, to obtain starch-based LDPE blends with adequate plasticity. To improve the efficacy of the developed biobased flame retardants, starch was also plasticized in the presence of CPA (TPS-G-CPA), as this compound has been described as both a flame retardant and a starch plasticizer [24]. It is worth noting that CPA is also a biobased compound derived from phytic acid, which is produced by plants to store phosphorus. To increase the compatibility between hydrophobic LDPE and hydrophilic TPS g or TPS-G-CPA [24] and ensure the mechanical properties of the blends, a reactive PE-*g*-MA compatibilizer was also incorporated into the formulation during the melt blending process (Figure 1). This copolymer is commonly used in the preparation of PE/TPS blends because the maleic anhydride groups can react with the hydroxyl groups of starch, while the PE segment has an affinity for the LDPE matrix [19].

Various LDPE/TPS blends compatibilized with 5 wt% PE-*g*-MA (optimized content) were prepared by means of melt blending at 120 °C and 140 °C, respectively, when TPS-G or TPS-G-CPA was used as the biobased flame retardant (Table S1). Considering that the objective of this work was to replace a commercial harmful flame retardant in LDPE formulations with a biodegradable starch-based one, preliminary tests were conducted in the laboratory mimicking the conditions of the UL 94 standard to evaluate the potential of the prepared blends as flame-retardant and self-extinguishing materials. For this purpose, 7 cm × 1 cm samples of LDPE/TPS-G(-CPA) blends were positioned vertically above a Bunsen burner flame and exposed to the flame (approximately 1 cm) for 10 s or until combustion started (if it occurred in less than 10 s). This time was recorded as the ignition time. The flame source was then removed and the time until the flame was extinguished was recorded as the first quenching time. The results (Figure S2) show that all samples did not extinguish the flame on their own, as they burned completely. However, it is worth noting that flame retardancy increased with increasing starch content in the blends, suggesting that this biomolecule indeed has the potential to be used as a flame retardant for LDPE. Interestingly, the combination of starch and CPA improved the flame retardancy of the materials, as shown by the 2.5 times higher quenching time of the blend 47.5 LDPE/47.5 TPS-G-CPA compared to 47.5 LDPE/47.5 TPS-G. Unfortunately, dripping was observed in all samples, which can be considered detrimental in terms of flame resistance, as this event may represent an additional ignition source or a process of flame propagation during a fire [27].

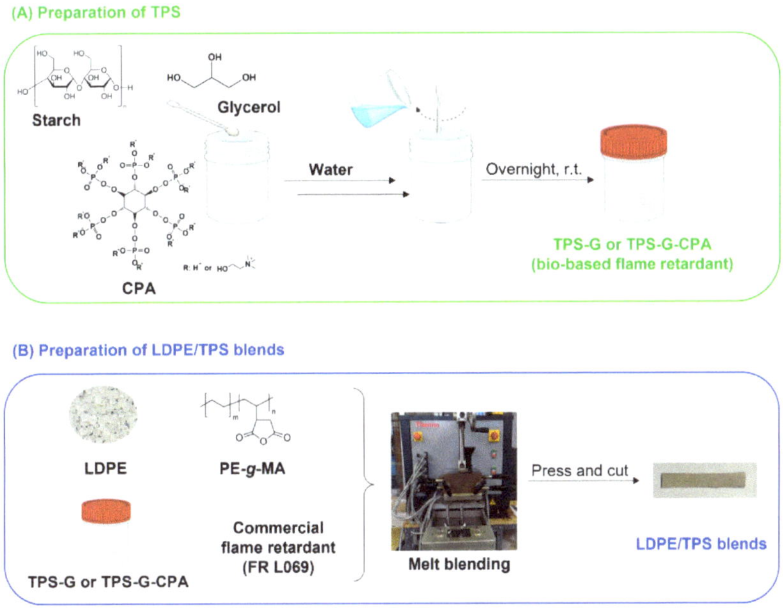

Figure 1. General scheme of the preparation of (**A**) biobased FR, TPS-G and TPS-G-CPA and (**B**) LDPE/TPS blends.

Although not complete, the replacement of any percentage of a harmful FR with those based on TPS in the formulations of PE without adverse effects on the performance of the material is highly desirable from the standpoint of waste management and environmental protection. Therefore, the effect of incorporating a halogenated commercial flame retardant (FR69) into the LDPE/TPS-G(-CPA) blends prepared in this work was investigated. Table 2 shows the composition of the LDPE/TPS-G(-CPA)/FR69 blends studied, where the total flame retardant content was set at 50 wt% and the starch-based flame retardant content ranged from 9.5 to 38 wt%.

Table 2. Composition of LDPE/TPS-G(-CPA)/FR69 blends compatibilized with PE-*g*-MA and prepared via melt blending. LDPE/FR69 was used as reference material.

Sample	LDPE (wt%)	TPS g or TPS-G-CPA (wt%)	FR69 (wt%)	PE-*g*-MA (wt%)
50 LDPE/50 FR69	50	0	50	0
47.5 LDPE/9.5 TPS-G (or TPS-G-CPA)/38 FR69	47.5	9.5	38	5
47.5 LDPE/19 TPS-G (or TPS-G-CPA)/28.5 FR69	47.5	19	28.5	5
47.5 LDPE/28.5 TPS-G (or TPS-G-CPA)/19 FR69	47.5	28.5	19	5
47.5 LDPE/38 TPS-G (or TPS-G-CPA)/9.5 FR69	47.5	38	9.5	5

FTIR analysis (Figure 2) confirmed the chemical structure of LDPE, namely the bands at 2918 cm^{-1}, 2851 cm^{-1}, 1468 cm^{-1} and 718 cm^{-1}, corresponding to CH_2 asymmetrical stretching, CH_2 symmetrical stretching, bending deformation and rocking deformation, respectively [28]. The FTIR spectrum of TPS-G (Figure 2a) was also consistent with the literature and showed the expected signals of $C-O$ stretching at 920 cm^{-1}, 1022 cm^{-1} and 1148 cm^{-1}, bound water at 1648 cm^{-1}, hydroxyl groups at 3277 cm^{-1}, CH stretching at 2914 cm^{-1} and glycerol at 1423 cm^{-1} [29]. TPS-G-CPA showed a similar FTIR spectrum (Figure 2b) to that of TPS-G. The reactive compatibilization between PE-*g*-MA and TPS-G was confirmed by the disappearance of the symmetric and asymmetric maleic anhydride groups of PE-*g*-MA at 1714 cm^{-1} and 1791 cm^{-1}, respectively, in the FTIR spectrum of the blend (Figure 2a). This indicates successful esterification between the anhydride rings of the compatibilizer and the hydroxyl groups of TPS [30]. Compatibilization appeared to be less effective for the blends containing TPS-G-CPA, as indicated by the remaining maleic anhydride groups of PE-*g*-MA at 1714 cm^{-1} in the FTIR spectrum of the blend (Figure 2b).

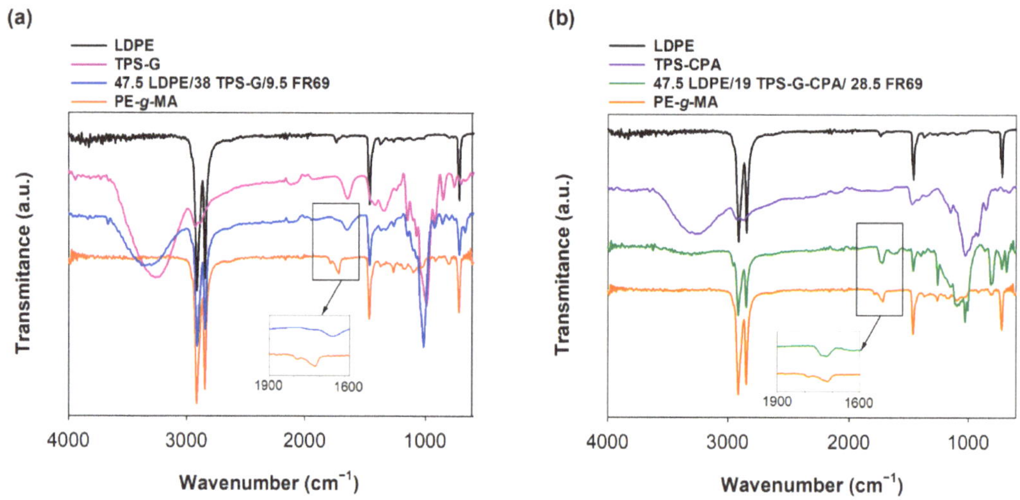

Figure 2. FTIR spectra of LDPE (black line), TPS-G (pink line), 47.5 LDPE/38 TPS-G/9.5 FR69 (blue line), PE-*g*-MA (orange line), TPS-G-CPA (purple line) and 47.5 LDPE/19 TPS-G-CPA/28.5 FR69 (green line). (**a**) LDPE/TPS-G-based blend and (**b**) LDPE/TPS-G-CPA-based blend.

To further evaluate the compatibility of LDPE/TPS/FR69 blends, representative samples were analyzed using SEM. Images of the surface of the blends (Figure 3) showed that both the control sample (LDPE/FR69 blend) and the blend with TPS-CPA had a similar appearance, while the blend with TPS-G had a rougher surface. However, all samples exhibited a smoother surface than that of non-compatibilized LDPE/TPS reported in the literature [31,32], confirming the role of PE-*g*-MA as an enhancer of the interaction between LDPE and TPS. Images of the fractured surfaces (cross section in Figure 3) confirm the FTIR results, indicating that the compatibilization of blends containing TPS-G-CPA was less effective, as indicated by the small gaps between a minority of TPS-G-CPA particles and the LDPE matrix. On the other hand, the sample containing TPS-G appeared to have a continuous phase between the starch particles and the LDPE, indicating better compatibilization, even at 40 wt% TPS-G compared to 19 wt% TPS-CPA.

3.2. Mechanical Analysis

The incorporation of starch into LDPE is expected to result in a loss of mechanical properties of the polyolefin [18]. To investigate the extent of this event, LDPE/TPS-G(-CPA)/FR69 blends with different starch contents were subjected to tensile testing. The results shown in Figure 4 indicate that blends with a TPS-G(-CPA) content of up to 30 wt% generally exhibit similar tensile strength to LDPE/FR69, which was used as the reference material. However, as expected, elongation at break decreased dramatically with increasing

TPS-G(-CPA) content in the blends. Interestingly, the results also confirmed the plasticizing effect of CPA, as the blends containing this molecule (TPS-G-CPA) generally exhibited higher elongation at break than the blends containing TPS plasticized only with glycerol and water (TPS-G) (Figure 4).

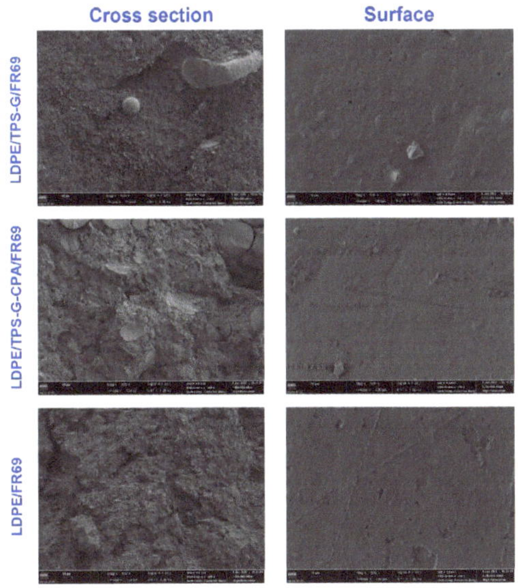

Figure 3. SEM images of the cross section and surface of LDPE/TPS/FR 69 blends, 1 kV and ×500: 47.5 LDPE/38 TPS-G/9.5 FR 69 (**top**); 47.5 LDPE/ 19 TPS-G-CPA/28.5 FR69 (**middle**) and 50LDPE/50FR 69 (**bottom**). Analysis conditions: EHT = 1.00 kV; Magnification = 500×.

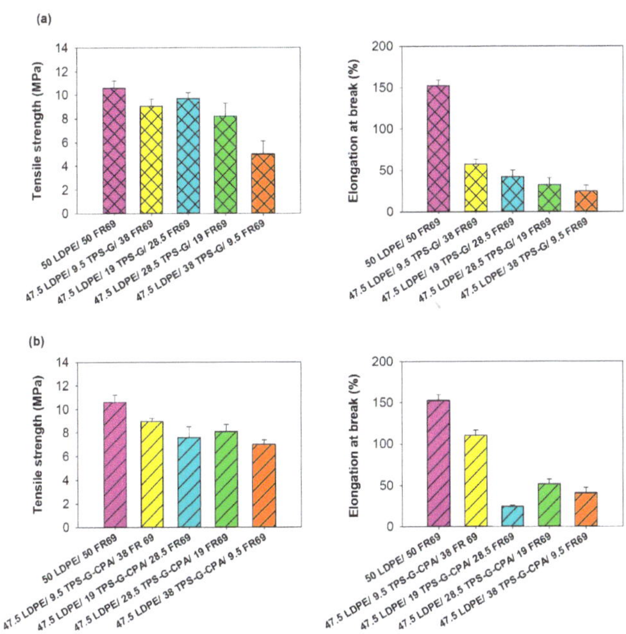

Figure 4. Tensile strength and elongation at break of (**a**) LDPE/TPS-G/FR69 blends and (**b**) LDPE/TPS-G-CPAFR69 blends. LDPE/FR69 was used as a reference material.

3.3. Fire Behavior

Preliminary flammability tests were conducted to evaluate the potential of LDPE/TPS-G(-CPA)/FR69 blends as flame retardant materials. The tests were conducted in the same manner as previously described for LDPE/TSP-G(-CPA) blends, but this time the materials were re-ignited with a flame after the initial quenching, and the time until the flame was extinguished was recorded as the second quenching time. Compared to LDPE/TPS-G(-CPA) blends (Figure S2), LDPE-based blends containing both commercial and starch-based flame retardants exhibited higher flame resistance, as reflected by higher ignition times and lower extinction times (Figure 5). In fact, LDPE/TPS-G(-CPA)-based samples with up to about 30 wt% starch-based flame retardant showed very impressive results compared to the reference (LDPE/FR 69) (Figure 5a). At higher levels of TPS-G, the samples burned longer but still self-extinguished, which is very encouraging. Unexpectedly, the same behavior was not observed with the mixtures containing TPS-G-CPA (Figure 5b), which exhibited a longer first quenching time but were unable to support the second burn cycle. Investigation of this behavior is beyond the scope of this paper, and the results will be published elsewhere.

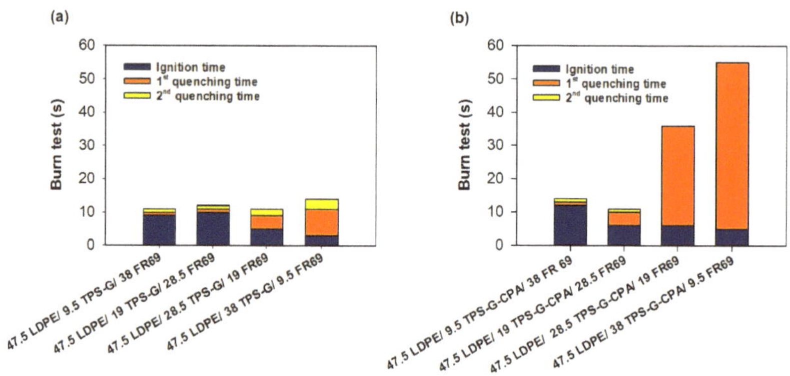

Figure 5. Preliminary burning tests conducted in the laboratory and mimicking the UL94 test for (**a**) LDPE/TPS-G blends and (**b**) LDEP/TPS-G-CPA blends. LDPE with a commercial flame retardant (76 LDPE/24 FR 69) was used as a reference material.

Blends with higher TPS content and higher flame retardancy in preliminary burning tests (47.5 LDPE/38 TPS-G/9.5 FR69 and 47.5 LDPE/19 TPS-G-CPA/28.5 FR69) were selected for further characterization and evaluation of their flame retardancy. LDPE with 50 wt% commercial flame retardant (50 LDPE/50 FR 69) was used as a reference material. The LOI can provide information about the relative flammability of polymers, as it indicates the minimum percentage (vol%) of oxygen in the atmosphere that can sustain a flame on a material. Therefore, the higher the flammability of the material, the lower the LOI value. It is noteworthy that the samples in which a portion of the commercial flame retardant was replaced with the starch-based products developed in this work have a similar LOI value to the reference material that contained only the commercial flame retardant (Figure 6). All tested blends are almost as good as "self-extinguishing" materials (LOI > 28%) [33], i.e., with good flame retardancy. Vertical UL 94 standard tests were conducted to evaluate the flammability of the LDPE-based blends. While LDPE containing the commercial flame retardant was rated V0 (self-extinguishing within 10 s with no dripping), both LDPE blends containing the bio-based and commercial flame retardant showed poorer results, as these samples took longer to self-extinguish and exhibited dripping (V2 rate, Figure 6).

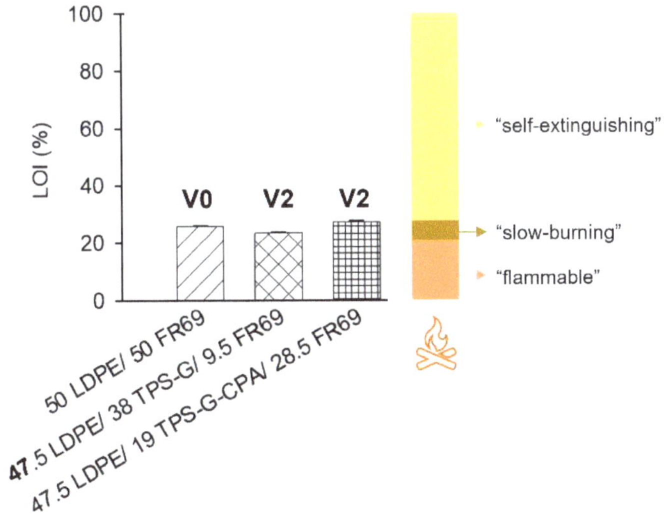

Figure 6. LOI and UL 94 rate (V0, V2) of selected LDPE/flame retardant blends. V0: Burning stops within 10 seconds on a vertical part allowing for drops of plastic that are not inflames; V2: Burning stops within 30 seconds on a part allowing for drops of vertical flammable plastic.

The fire behavior of LDPE/flame retardant blends was evaluated using cone calorimetry, which mimics realistic fire conditions and provides several valuable parameters, particularly the rate of heat release rate during combustion. To increase the flame resistance of LDPE/flame retardant blends and reduce flame spread, it is desirable to have a low total heat release (THR) or a low peak of heat release rate (pHRR) [34]. The results presented in Table 3 show that both blends containing TPS have similar THR values, but they are about 30% higher than those of the reference material. Nevertheless, it is interesting to note that the pHRR value was similar for all LDPE/flame retardant blends (around 650 kW/m^2). The values obtained for time to ignition (TTI) values in Table 3 show that the samples containing TPS started to burn two times faster than the samples containing only the commercial flame retardant. The results also show that the content of effective combustion components in LDPE could be reduced by using the biobased TPS-G-CPA flame retardant, as evidenced by the lower effective heat of combustion (EHC) of the corresponding LDPE blend compared to the reference material and the blend containing TPS-G. The propensity for fire development can be evaluated using the maximum average rate of heat emission (MARHE), which is the cumulative heat emission during the cone calorimetry test divided by time [35]. Both samples with the biobased FR had a higher MARHE (8% and 24%, respectively) than the reference material. However, it is important to note that the fire resistance of materials is affected by several parameters mentioned above. One way to evaluate overall fire safety is to determine both the fire performance index (FPI = TTI/pHRR) and the fire growth index (FGI = pHRR/tpHRR) [36]. For low fire risk and high safety levels, it is desirable to maximize FPI and minimize FGI. The results presented in Table 3 show that the FPI was reduced by 43% when 76% and 38% of the commercial flame retardant was replaced with TPS-G and TPS-G-CPA, respectively. However, it is noteworthy that the blend in which 38% of the commercial flame retardant was replaced with TPS-G-CPA showed a significant reduction (43%) in FGI, indicating that the blends may contribute to lower fire spread than the reference material.

Table 3. Cone calorimetry results of selected LDPE/flame retardant blends.

Sample	50 LDPE/50 FR69	47.5 LDPE/38 TPS-G/9.5 FR69	47.5 LDPE/19 TPS-G-CPA/28.5 FR69
pHRR (kW/m^2)	651.56	621.48	663.01
t_{pHRR} (s)	453.3	325.0	808.0
THR (MJ/m^2)	25.56	34.25	32.27
TTI (s)	46.3	22.3	27.7
EHC (MJ/kg)	14.97	14.12	10.00
MARHE (kW/m^2)	184.83	229.08	200.19
FPI (m^2/kW)	0.07	0.04	0.04
FGI ($kW/m^2 \cdot s$)	1.44	1.91	0.82

pHRR: peak of heat release rate; t_{pHRR}: time to peak of heat release rate; THR: total heat released; TTI: time to ignition; EHC: effective heat of combustion; MARHE: maximum average rate of heat emission, FPI: fire performance index; FGI: fire growth index.

3.4. Thermal Analysis

The thermal stability of the blends was investigated via TGA, and the TG and DTG curves are shown in Figure 7. The weight loss profile (Figure 7a) shows that the decomposition of the blends followed a similar and expected four-step decomposition process, with (i) moisture loss ($T \approx 128$ °C), (ii) degradation of glycerol ($T \approx 270$ °C), (iii) degradation of starch, CPA, and FR69 ($T \approx 350$ °C), and (iv) degradation of LDPE ($T \approx 475$ °C) [24,37]. The blend containing starch plasticized in the presence of CPA exhibited higher thermal stability than the blend containing TPS-G, as judged by both $T_{5wt\%}$ and $T_{10wt\%}$ of the samples (Table 4). In addition, the char residue of the mixture containing 19 wt% TPS-G-CPA was 10.5% at 600 °C, while the residue of the mixture containing about twice the amount (40 wt%) of TPS-G was 6.5% (Table 4). These results indicate that CPA can indeed improve the charring ability and flame retardancy of starch. Similar observations have been made in the literature [24]. As expected, the reference material (50 LDPE/50 FR69) exhibited higher thermal stability than the starch-containing blends.

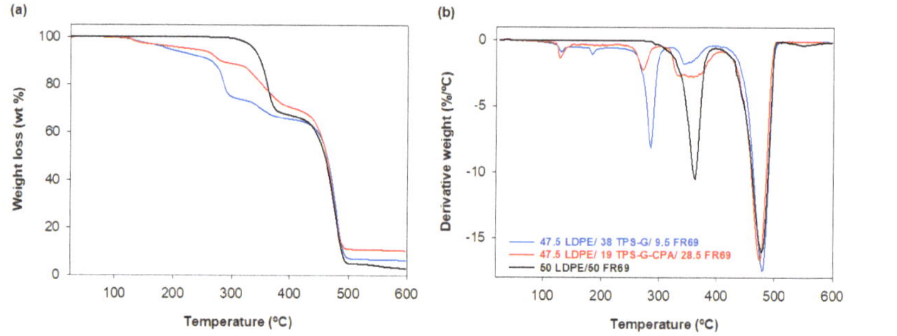

Figure 7. Thermogravimetric curves (**a**) TG and (**b**) DTG of the blends 47.5 LDPE/38 TPS-G/9.5 FR69 and 47.5 LDPE/19 TPS-G-CPA/28.5 FR69 under nitrogen atmosphere.

Table 4. Thermal properties of the blends obtained by means of TG analysis.

Sample	$T_{5wt\%}$ (°C) [a]	$T_{10wt\%}$ (°C) [b]	$CR_{600°C}$ (%) [c]
50 LDPE/50 FR69	337.9	349.4	3.0
47.5 LDPE/38 TPS-G/9.5 FR69	191.7	264.3	6.5
47.5 LDPE/19 TPS-G-CPA/28.5 FR69	230.6	281.2	10.5

[a] $T_{5wt\%}$: temperature at 5% weight loss; [b] $T_{10wt\%}$: temperature at 10% weight loss; [c] $CR_{600°C}$: char residue at 600 °C.

4. Conclusions

In this work, the possibility of using starch as a natural flame retardant for LDPE was investigated. Different blends of LDPE/TPS g and LDPE/TPS-G-CPA were prepared by means of melt blending in the presence of PE-*g*-MA as a compatibilizer. The results of FTIR spectroscopy and SEM analysis showed that there was good compatibilization between starch and LDPE. However, this was less effective when TPS-G-CPA was used. Nevertheless, the mechanical properties of both blends were similar in terms of tensile strength. The plasticizing effect of CPA was confirmed by an increase in elongation at break of the LDPE/TPS-G-CPA blends compared to those with TPS-G. Acceptable fire behavior was achieved by combining the starch-based FR with a commercial flame retardant (FR69), with the most promising blends being 47.5 LDPE/38 TPS-G/9.5 FR69 and 47.5 LDPE/19 TPS-G-CPA/28.5 FR69. These blends had a similar LOI (about 28%) to that of reference blend containing only the commercial flame retardant (50 LDPE/50 FR69). CPA not only imparts higher plasticity to the starch, but also improves the charring ability, flame retardancy, and effective combustion component content of the blends, as well as a 43% reduction in fire growth index compared to the reference blends, suggesting that it may contribute to lower fire spread in the event of a fire. These results indicate that starch plasticized in the presence of water, glycerol and CPA has a promising future as a safe and cost-effective flame retardant for LDPE formulations.

Supplementary Materials: The following supporting information can be downloaded at: https://www.mdpi.com/article/10.3390/polym15204078/s1, Figure S1: The 400 MHz ^{1}H NMR spectrum of CPA in D_2O; Table S1: Composition of LDPE/TPS-G(-CPA) blends compatibilized with PE-*g*-MA and prepared by melt blending. LDPE was used as reference material (100LDPE code); Figure S2: Preliminary burning tests conducted in the laboratory and mimicking the UL94 test for (a) LDPE/TPS-G blends and (b) LDEP/TPS-G-CPA blends. LDPE was used as reference material.

Author Contributions: Conceptualization, J.F.J.C.; methodology, B.O.C., L.P.C.G., P.V.M., J.P.P., A.C.S. and J.F.J.C.; validation, B.O.C. and L.P.C.G.; formal analysis, B.O.C. and L.P.C.G.; investigation, B.O.C. and L.P.C.G.; writing—original draft preparation, P.V.M. and B.O.C.; writing—review and editing, B.O.C., L.P.C.G., P.V.M. and J.F.J.C.; supervision, A.C.S. and J.F.J.C.; project administration, J.F.J.C.; funding acquisition, J.F.J.C. All authors have read and agreed to the published version of the manuscript.

Funding: This research is sponsored by FEDER funds through the program COMPETE—Programa Operacional Factores de Competitividade—and by national funds through FCT—Fundação para a Ciência e Tecnologia, CEMMPRE funding (UID/EMS/00285/2020) and ARISE funding (LA/P/0112/2020). The ^{1}H NMR data were collected at the UC-NMR facility which is supported in part by FEDER—European Regional Development Fund through the COMPETE Programme (Operational Programme for Competitiveness) and by National Funds through FCT (Portuguese Foundation for Science and Technology) through grants REEQ/481/QUI/2006, RECI/QEQ-QFI/0168/2012, CENTRO-07-CT92-FEDER-002012, and Rede Nacional de Ressonância Magnética Nuclear (RNRMN). This research was supported by FEDER funds through the program COMPETE–Programa Operacional Factores de Competitividade—POCI-01-0247-FEDER-069481—Project Starchfire. Patrícia V. Mendonça thanks the FCT—Fundação para a Ciência e a Tecnologia (Portuguese Foundation for Science and Technology) for the financial support through the Scientific Employment Stimulus 2017 (CEECIND/00117/2017).

Institutional Review Board Statement: Not applicable.

Data Availability Statement: Not applicable.

Conflicts of Interest: The authors declare no conflict of interest.

References

1. Ghomi, E.R.; Khosravi, F.; Mossayebi, Z.; Ardahaei, A.S.; Dehaghi, F.M.; Khorasani, M.; Neisiany, R.E.; Das, O.; Marani, A.; Mensah, R.A.; et al. The Flame Retardancy of Polyethylene Composites: From Fundamental Concepts to Nanocomposites. *Molecules* **2020**, *25*, 5157. [CrossRef]
2. Yao, Z.; Seong, H.J.; Jang, Y.S. Environmental Toxicity and Decomposition of Polyethylene. *Ecotoxicol. Environ. Saf.* **2022**, *242*, 113933. [CrossRef]
3. Li, D.; Zhou, L.; Wang, X.; He, L.; Yang, X. Effect of Crystallinity of Polyethylene with Different Densities on Breakdown Strength and Conductance Property. *Materials* **2019**, *12*, 1746. [CrossRef]
4. Shih, Y.F.; Chen, Y.H.; Lai, S.Y.; Chen, Y.X.; Wang, S.C.; Zhang, S.P. The Effect of Thermal, Flammability, and Mechanical Properties of Wood Plastic Composites Made from Recycled Food-Packaging LDPE and Eco-Friendly Phytic Acid. *Int. J. Appl. Sci. Eng.* **2021**, *18*, 1–8. [CrossRef]
5. Crippa, M.; De Wilde, B.; Koopmans, R.; Leyssens, J.; Muncke, J.; Ritschkoff, A.C.; Van Doorsselaer, K.; Velis, C.; Wagner, M. *A Circular Economy for Plastics: Insights from Research and Innovation to Inform Policy and Funding Decisions*; Publications Office of the European Union: Luxembourg, 2019; ISBN 9789279984297.
6. Low Density Polyethylene Market Size & Share Analysis—Growth Trends & Forecasts (2023–2028). Available online: https://www.Mordorintelligence.Com/Industry-Reports/Low-Density-Polyethylene-Market (accessed on 9 May 2023).
7. Feng, X.; Wang, Z.; Ma, H.; Dang, B.; Li, J. A New Bio-Based Flame Retardant Simply Synthesized: Improving the Flame Retardancy and Thermal Stability of Low-Density Polyethylene Together with Melamine Cyanurate. *J. Appl. Polym. Sci.* **2023**, *140*, e53953. [CrossRef]
8. Zhou, Y.; Peng, S.; Hu, J.; He, J. Polymeric Insulation Materials for HVDC Cables: Development, Challenges and Future Perspective. *IEEE Trans. Dielectr. Electr. Insul.* **2017**, *24*, 1308–1318. [CrossRef]
9. Laoutid, F.; Bonnaud, L.; Alexandre, M.; Lopez-Cuesta, J.M.; Dubois, P. New Prospects in Flame Retardant Polymer Materials: From Fundamentals to Nanocomposites. *Mater. Sci. Eng. R Rep.* **2009**, *63*, 100–125. [CrossRef]
10. Altun, Y.; Doğan, M.; Bayramlı, E. The Effect of Red Phosphorus on the Fire Properties of Intumescent Pine Wood Flour—LDPE Composites. *Fire Mater.* **2016**, *40*, 697–703. [CrossRef]
11. Venier, M.; Salamova, A.; Hites, R.A. Halogenated Flame Retardants in the Great Lakes Environment. *Acc. Chem. Res.* **2015**, *48*, 1853–1861. [CrossRef]
12. Hobbs, C.E. Recent Advances in Bio-Based Flame Retardant Additives for Synthetic Polymeric Materials. *Polymers* **2019**, *11*, 224. [CrossRef]
13. Wang, M.; Yin, G.Z.; Yang, Y.; Fu, W.; Díaz Palencia, J.L.; Zhao, J.; Wang, N.; Jiang, Y.; Wang, D.Y. Bio-Based Flame Retardants to Polymers: A Review. *Adv. Ind. Eng. Polym. Res.* **2022**, *6*, 132–155. [CrossRef]
14. Alongi, J.; Poskovic, M.; Visakh, P.N.; Frache, A.; Malucelli, G. Cyclodextrin Nanosponges as Novel Green Flame Retardants for PP, LLDPE and PA6. *Carbohydr. Polym.* **2012**, *88*, 1387–1394. [CrossRef]
15. Prasad, V.S.; Pillai, C.K.S. Flame Retardation of Polyethylene: Effect of a Phosphorus Flame Retardant Having Both Hydrophobic and Hydrophilic Groups in the Same Molecule. *J. Appl. Polym. Sci.* **2000**, *77*, 2631–2640. [CrossRef]
16. Kormin, S.; Kormin, F.; Beg, M.D.H.; Piah, M.B.M. Physical and Mechanical Properties of LDPE Incorporated with Different Starch Sources. In Proceedings of the IOP Conference Series: Materials Science and Engineering, Melaka, Malaysia, 14 August 2017; Volume 226.
17. Nafchi, A.M.; Moradpour, M.; Saeidi, M.; Alias, A.K. Thermoplastic Starches: Properties, Challenges, and Prospects. *Starch/Staerke* **2013**, *65*, 61–72. [CrossRef]
18. Guarás, M.P.; Menossi, M.; Nicolini, A.T.; Alvarez, V.A.; Ludueña, L.N. Bio-Nanocomposites Films Based on Unmodified and Modified Thermoplastic Starch Reinforced with Chemically Modified Nanoclays. *J. Mater. Sci.* **2023**, *58*, 5456–5476. [CrossRef]
19. Radfar, R.; Hosseini, H.; Farhoodia, M.; Średnicka-Toberd, D.; Ghasemie, I.; Fathif, M.; Khaneghahg, A.M. Factors Affecting Mechanical Properties of Low-Density Polyethylene (LDPE)/Starch Blends: A Review. *Carpathian J. Food Sci. Technol.* **2020**, *12*, 6–26.
20. Kaboorani, A.; Gray, N.; Hamzeh, Y.; Abdulkhani, A.; Shirmohammadli, Y. Tailoring the Low-Density Polyethylene—Thermoplastic Starch Composites Using Cellulose Nanocrystals and Compatibilizer. *Polym. Test.* **2021**, *93*, 107007. [CrossRef]
21. *UL 94:2013*; Tests for Flammability of Plastic Materials for Parts in Devices and Appliances. Underwriters Laboratories: Northbrook, IL, USA, 2013.
22. *UNE EN ISO 4589-2:2017*; Plastics—Determination of Burning Behaviour by Oxygen Index—Part 2: Ambient-Temperature Test (ISO 4589-2:2017). Asociación Española de Normalización: Madrid, Spain, 2017.
23. *EN ISO 5660-1:2015*; Reaction-to-Fire Tests—Heat Release, Smoke Production and Mass Loss Rate—Part 1: Heat Release Rate (Cone Calorimeter Method) and Smoke Production Rate (Dynamic Measurement). International Organization for Standardization: Geneva, Switzerland, 2015.
24. Wang, D.; Wang, Y.; Li, T.; Zhang, S.; Ma, P.; Shi, D.; Chen, M.; Dong, W. A Bio-Based Flame-Retardant Starch Based on Phytic Acid. *ACS Sustain. Chem. Eng.* **2020**, *8*, 10265–10274. [CrossRef]
25. Pigłowska, M.; Kurc, B.; Rymaniak, Ł. Starch as the Flame Retardant for Electrolytes in Lithium-Ion Cells. *Materials* **2022**, *15*, 523. [CrossRef]

26. Xia, Y.; Chai, W.; Liu, Y.; Su, X.; Liao, C.; Gao, M.; Li, Y.; Zheng, Z. Facile Fabrication of Starch-Based, Synergistic Intumescent and Halogen-Free Flame Retardant Strategy with Expandable Graphite in Enhancing the Fire Safety of Polypropylene. *Ind. Crops Prod.* **2022**, *184*, 115002. [CrossRef]
27. Matzen, M.; Kandola, B.; Huth, C.; Schartel, B. Bernhard Schartel Influence of Flame Retardants on the Melt Dripping Behaviour of Thermoplastic Polymers. *Materials* **2015**, *8*, 5621–5646. [CrossRef] [PubMed]
28. Rajandas, H.; Parimannan, S.; Sathasivam, K.; Ravichandran, M.; Yin, L.S. A Novel FTIR-ATR Spectroscopy Based Technique for the Estimation of Low-Density Polyethylene Biodegradation. *Polym. Test.* **2012**, *31*, 1094–1099. [CrossRef]
29. Mendes, J.F.; Paschoalin, R.T.; Carmona, V.B.; Sena Neto, A.R.; Marques, A.C.P.; Marconcini, J.M.; Mattoso, L.H.C.; Medeiros, E.S.; Oliveira, J.E. Biodegradable Polymer Blends Based on Corn Starch and Thermoplastic Chitosan Processed by Extrusion. *Carbohydr. Polym.* **2016**, *137*, 452–458. [CrossRef] [PubMed]
30. Oromiehie, A.R.; Lari, T.T.; Rabiee, A. Physical and Thermal Mechanical Properties of Corn Starch/LDPE Composites. *J. Appl. Polym. Sci.* **2013**, *127*, 1128–1134. [CrossRef]
31. Oner, B.; Gokkurt, T.; Aytac, A. Studies On Compatibilization Of Recycled Polyethylene/Thermoplastic Starch Blends By Using Different Compatibilizer. *Open Chem.* **2019**, *17*, 557–563. [CrossRef]
32. Martins, A.B.; Cattelan, A.K.; Santana, R.M.C. How the Compatibility between Polyethylene and Thermoplastic Starch Can Be Improved by Adding Organic Acids? *Polym. Bull.* **2018**, *75*, 2197–2212. [CrossRef]
33. Schinazi, G.; Price, E.J.; Schiraldi, D.A. Chapter 3—Fire Testing Methods of Bio-Based Flame-Retardant Polymeric Materials. In *Bio-Based Flame-Retardant Technology for Polymeric Materials*; Hu, Y., Nabipour, H., Wang, X., Eds.; Elsevier: Amsterdam, The Netherlands, 2022; pp. 61–95, ISBN 978-0-323-90771-2.
34. Haurie, L.; Fernández, A.I.; Velasco, J.I.; Chimenos, J.M.; Lopez Cuesta, J.M.; Espiell, F. Thermal Stability and Flame Retardancy of LDPE/EVA Blends Filled with Synthetic Hydromagnesite/Aluminium Hydroxide/Montmorillonite and Magnesium Hydroxide/Aluminium Hydroxide/Montmorillonite Mixtures. *Polym. Degrad. Stab.* **2007**, *92*, 1082–1087. [CrossRef]
35. Chen, D.; Zhang, Y.; He, J.; Li, X. Making Polycarbonate Flame Retardant: Flame Retardant Selection and Calorimetric Analyses. *Polym. Test.* **2023**, *117*, 107876. [CrossRef]
36. Shang, S.; Yuan, B.; Sun, Y.; Chen, G.; Huang, C.; Yu, B.; He, S.; Dai, H.; Chen, X. Facile Preparation of Layered Melamine-Phytate Flame Retardant via Supramolecular Self-Assembly Technology. *J. Colloid. Interface Sci.* **2019**, *553*, 364–371. [CrossRef]
37. Vargün, E.; Sisli, Z.; Yilmaz, F.; Baş Sermenli, H. The Effects of Citric Acid and Stearic Acid Compatibilizers on the Degradation Behavior of TPS/LDPE Blends for Mulch Films Sitrik Asit ve Stearik Asit Uyumlaştırıcılarının TPN/AYPE Karışımı Malç Filmlerinin Bozunma Davranışı Üzerine Etkileri. *Hacet. J. Biol. Chem.* **2016**, *44*, 463–476. [CrossRef]

polymers

MDPI

Review

Guided Tissue and Bone Regeneration Membranes: A Review of Biomaterials and Techniques for Periodontal Treatments

Ali M. Alqahtani [1], Robert Moorehead [2] and Ilida Ortega Asencio [3,*]

1 Department of Restorative Dental Sciences, College of Dentistry, King Khalid University, Al Fara, Abha 62223, Saudi Arabia; aqahtani@kku.edu.sa
2 The Department of Materials Science and Engineering, The University of Sheffield, Sir Robert Hadfield Building, Sheffield S1 3JD, UK; r.moorehead@sheffield.ac.uk
3 The School of Clinical Dentistry, The University of Sheffield, 19 Claremont Crescent, Sheffield S10 2TA, UK
* Correspondence: i.ortega@sheffield.ac.uk; Tel.: +966-7-12345678

Abstract: This comprehensive review provides an in-depth analysis of the use of biomaterials in the processes of guided tissue and bone regeneration, and their indispensable role in dental therapeutic interventions. These interventions serve the critical function of restoring both structural integrity and functionality to the dentition that has been lost or damaged. The basis for this review is laid through the exploration of various relevant scientific databases such as Scopus, PubMed, Web of science and MEDLINE. From a meticulous selection, relevant literature was chosen. This review commences by examining the different types of membranes used in guided bone regeneration procedures and the spectrum of biomaterials employed in these operations. It then explores the manufacturing technologies for the scaffold, delving into their significant impact on tissue and bone regenerations. At the core of this review is the method of guided bone regeneration, which is a crucial technique for counteracting bone loss induced by tooth extraction or periodontal disease. The discussion advances by underscoring the latest innovations and strategies in the field of tissue regeneration. One key observation is the critical role that membranes play in guided reconstruction; they serve as a barrier, preventing the entry of non-ossifying cells, thereby promoting the successful growth and regeneration of bone and tissue. By reviewing the existing literature on biomaterials, membranes, and scaffold manufacturing technologies, this paper illustrates the vast potential for innovation and growth within the field of dental therapeutic interventions, particularly in guided tissue and bone regeneration.

Keywords: dental biomaterials; guided bone regeneration; guided tissue regeneration; tissue engineering; biocompatible polymers; membranes; scaffolds

Citation: Alqahtani, A.M.; Moorehead, R.; Asencio, I.O. Guided Tissue and Bone Regeneration Membranes: A Review of Biomaterials and Techniques for Periodontal Treatments. *Polymers* **2023**, *15*, 3355. https://doi.org/10.3390/polym15163355

Academic Editors: Vishal Gavande, Vasi Shaikh and Won-Ki Lee

Received: 25 July 2023
Revised: 5 August 2023
Accepted: 8 August 2023
Published: 10 August 2023
Corrected: 24 March 2025

1. Introduction

The field of periodontology has seen considerable advancements in recent years, with a keen focus on therapeutic strategies for restoring periodontal lesions and regenerating lost jawbone through cellular proliferation. Central to this endeavor is the availability of a substantial volume of hard bone tissue, the foundation for successful implant treatments [1].

The complex architecture of a healthy periodontium, with its multilayered structure and dynamic interplay of cells, tissues, and molecular factors, is fundamental to oral health. However, the periodontium can fall prey to a range of pathological conditions, leading to tooth loss and degenerative changes [2]. These conditions necessitate a host of diverse treatment modalities, from traditional periodontal therapies to more contemporary, regenerative procedures.

In the realm of regenerative therapies, guided bone regeneration (GBR) and guided tissue regeneration (GTR) have garnered significant attention. Primarily, GBR focuses on the regeneration of alveolar bone in edentulous regions, while GTR is tasked with repairing compromised periodontal tissues [3,4]. Both these techniques leverage the utility of a porous polymer membrane to physically impede the infiltration of undesirable tissues and cells into the lesion site [4]. This strategy aids in fostering an environment conducive for the proliferation of required cells [5].

GBR and GTR techniques play a wide array of roles, with GBR involved in maintaining and enhancing the alveolar ridge, correcting implant contractions or fenestrations, and promoting bone regeneration around implants [6]. Conversely, GTR is engaged in the regeneration of the periodontal ligament (PDL), bone, and cementum in proximity to the tooth [7].

However, successful bone and tissue regeneration are not merely reliant on the prevention of undesirable cell infiltration; they also demand the presence of osteogenic cells, alongside osteoconductive and osteoinductive materials [8]. Membranes, integral to GBR and GTR, should exhibit excellent biocompatibility and extended functional stability. They are also expected to ensure the spatial and biomechanical stability of the lesion site by filtering disruptive cells and tissues and protecting the emergent tissue [9].

Furthermore, these membranes are classified into several types based on their composition and bioactivity, such as bioabsorbable, non-resorbable, and metal and inorganic compound membranes [10]. In the latter part of this review, the pivotal role of scaffold manufacturing technologies, as a subset of these membrane technologies, in bone and tissue regenerations will be thoroughly discussed. This provides a unique vantage point to appreciate the wide spectrum of cutting-edge strategies that have been developed for the regeneration of bone and tissue in the periodontal context.

2. Materials and Methods

To conduct this comprehensive review, we sourced primary data from several established scientific databases, including Scopus, PubMed, and MEDLINE. Our objective was to unearth the most relevant and impactful literature relating to the use of biomaterials in guided tissue and bone regeneration procedures.

Our search strategy was developed with a focus on several key terms and phrases pertinent to our study. These include "biocompatible materials", "membrane", "bone regeneration", "tissue regeneration", and "dental biomaterials". We meticulously scanned all abstracts yielded by these search terms and selected full-text articles that aligned most significantly with our study's aims and objectives.

The review process involved a rigorous methodological approach. All selected articles underwent detailed evaluation, where data relating to the membrane types and range of biomaterials used for tissue and bone regeneration were extracted and scrutinized. In addition, we paid particular attention to articles that discussed the role of scaffold manufacturing technologies in these regenerative procedures.

Furthermore, we performed a narrative synthesis of the data obtained. The synthesis was aimed at providing a comprehensive overview of the current understanding and advancements in the use of biomaterials for guided tissue and bone regeneration, as well as the role of membranes in preventing the ingress of non-ossifying cells.

The exclusion criteria applied in the literature review process helped to maintain the focus and relevance of this study. Articles that did not focus on dental tissue and bone regeneration, used non-biocompatible materials, or did not discuss the use of scaffold manufacturing technologies were excluded from the review. Additionally, articles published in languages other than English were not considered.

This methodical approach ensured the selected literature was of high quality and relevant to our study, thereby supporting a more accurate and comprehensive review of the subject matter.

3. Results

Current periodontal treatment approaches are targeted at minimizing and/or removing inflamed tissues induced by bacterial plaque, repairing deficiencies or structural abnormalities, and regenerating new tissues in the region of lost tissues [11–13]. Various methods mentioned previously are only able to stop the progress of the problem, but are unable to reverse the damage or replace the lost tissue [14]. Bone grafting, enamel matrix derivative (EMD) and guided regeneration therapy are now used in the development of tissues that have been infected by periodontal diseases. To a certain degree, the overall structure and function of the damaged tissue can be restored [13].

3.1. Historical Viewpoint on Approaches to Periodontal Regeneration

The concept of placing a physical barrier along the tooth root surface after periodontal surgery to prevent epithelial downgrowth was first proposed in the 1970s [15] (Figure 1).

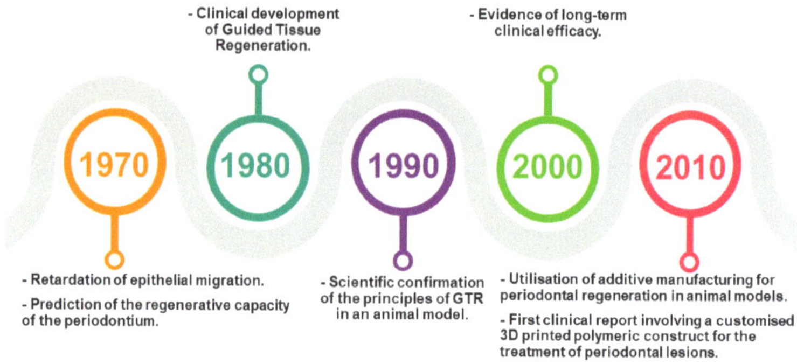

Figure 1. Timeline of periodontal regeneration approaches: from the original idea involving a free palatal graft for inhibiting epithelial migration to the most current developments involving additively engineered polymeric multiphasic scaffolds for periodontal tissue engineering.

Indeed, physicians had previously hypothesized that the collapse of gingival tissues into periodontal defects seriously impeded bone resorption [16,17]. Several early experiments suggested the placing of a harvested free palatal graft over the periodontal defect in order to delay or at least obstruct the downgrowth of epithelium around the tooth root surface [15]. Another popular procedure included the insertion of bone grafts (allogenic, autologous, or synthetic) obtained from the patient inside the periodontal defect to regenerate the missing bone [18]. However, neither of these methods is successful for periodontal recovery, and only periodontal healing is observed in the context of a fresh junctional epithelium. After researching the clinical and laboratory evidence, it can be hypothesized that the lack of compartmentalization between the periodontal defect and the underlying soft tissue was the cause of low regeneration rates. The problem of selective periodontal defect repopulation by tissues capable of fostering periodontal regeneration was presented in a series of pioneering papers by Nyman et al., which contributed to the development of the principle of guided tissue regeneration (GTR) [15,19].

3.1.1. Bone Graft Procedure

Bone grafts have been an option for a long time to successfully deal with the effects of periodontal disease, such as bone loss and damage. A bone graft is meant to fill the space that originated from the damaged tissue with a material that possesses certain qualities and characteristics (Figure 2). There are several types currently available, such as allografts, xenografts, alloplastic, and autograft materials. These kinds of graft materials are able to facilitate natural osseous repair through some mechanisms that have already been properly characterized [18,20,21]:

- Osteogenesis: The graft possesses cells that function as seeds for the continuous growth of the tissue by forming a bone matrix.
- Osteoinduction: The graft can release factors and biochemical signals that stimulate the formation of new bone by cells.
- Osteoconduction: The graft works as a scaffold on which the host bone develops.

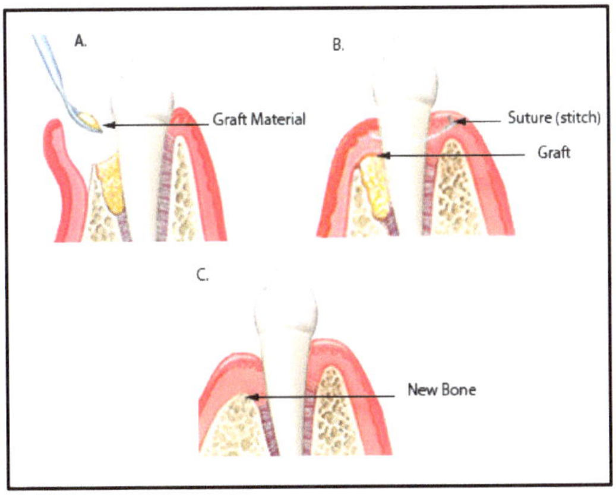

Figure 2. A schematic diagram of the management of periodontal defects by a bone graft technique. (**A**) Placing the graft. First, a gum flap is created. Growth factors may then be applied to the root. Graft material is packed into the area where bone was lost. (**B**) Closing up. The gum is closed and sewn together. (**C**) After the area heals. Stitches dissolve or are removed.

Diverse graft materials can be categorized into four general types that are shown in Table 1.

Table 1. Bone graft classification by material source. adapted from [20,21].

Type		Source	Benefit	Risk
Autograft		Patient	• Osteogenic, osteoinductive and osteoconductive • No immunological rejection living cells and matrices	• Morbidity at donor sites • Amount of bone volume is limited. • Rapid absorption
Allograft	Another human	Demineralized freeze-dried bone allograft (DFDBA)	• Osteoinductive and osteoconductive	• Potential of infection and immunological rejection
		Freeze-dried bone allograft (FDBA)	• Osteoinductive and osteoconductive	• Potential of infection and immunological rejection
Xenograft		Other species (Mostly bovine)	• Osteoconductive	• Potential of infection and immunological rejection • Slow resorption or non-resorbable
Alloplast	Synthetic	Sintered hydroxyapatite (HA)	• Osteoconductive	• Slow resorption or non-resorbable
		β-tricalcium phosphate (β-TCP)	• Osteoconductive	• Rapid resorption
		Natural products (coral, chitosan, etc.)	• Osteoconductive, low immunological rejection	• Slow resorption or non-resorbable

In the past, research publications have reported that at least 3.0 mm of bone height can be acquired, regardless of which material the graft is made of [21,22]. Probably the best option for osteogenesis, osteoconduction, and osteoinduction is autologous bone since it shares all of its properties with the surrounding bone. In this type, a part of the bone structure is extracted from a normal and un-damaged area of the patient who is getting the graft. The structure that is compatible and inherent to this type makes it very advantageous since it has the same vital bone structures in critical regions and includes nutrients, proteins, and cells as those found in the affected site. However, the autoimmune grafts come with some disadvantages, such as increasing the patient's pain at the same site of the excision In addition, only small amounts of bone can be extracted without incurring permanent damage to the patient.

Due to these drawbacks, several xenografts, bone grafts, as well as allogeneic materials have been designed and approved for commercial use [23]. It is important to note that clinicians must take into account the risks associated with these materials, such as infection, resorption, and immune responses. Currently, two types of allografts are available: freeze-dried bone allograft (FDBA) and demineralized freeze-dried bone allograft (DFDBA). These grafts are pre-treated chemically; they keep their osteoinductive capabilities due to the conservation of certain proteins, such as BMPs and TGF-s. These proteins work as powerful growth stimulants and induce the mobilization of cells from the mesenchyme into the implant. The overall process of demineralization enhances the excretion of these highly important factors into the extracellular medium [21,24].

The grafts that are extracted from animals, mainly farm cattle, are called xenografts. They go through chemical treatment in order to remove their inherent antigens, in order to avoid the human body's natural immune response. The greatest advantage in using these products lies in the fact that the general structure of the graft can be maintained, and this is because of the prolonged amounts of time required for these materials to be resorbed. These materials are capable of osteoconductivity and little else. However, the overall safeness and efficacy of them have been proved to be clinically relevant [21,25].

Due to the risks related to these types of grafts, scientists have also conducted investigations into the use of synthetic alternatives, which include composite grafts, polymer and inorganic materials, to repair osseous tissue [26]. In particular, materials such as hydroxyapatite and β-tricalcium phosphate (β-TCP) have become of increasing clinical importance when studying periodontal diseases. Furthermore, even though synthetic materials lack osteogenic and osteoconductive capabilities, their osteoconductive properties make them excellent options for bone regeneration. Recently, in a study by Schmidlin et al. (2013), it was found that polylactide-coated TCP was sufficient to repair problems in the rabbit's bone structure, all of this while retaining its biocompatibility [27].

Certain products of natural origin, such as coral, have been shown to be able to be used as bone grafts due to their similarity to human bone structure [28]. Similarly, to synthetic materials, numerous natural products have been proven safe for human use and are very cost-efficient. However, they are still only osteoconductive [21].

3.1.2. Guided Regeneration Therapy

The key aim of periodontal regeneration is to establish new cementum with periodontal ligament (PDL) fiber attached to the alveolar bone and promote new bone growth. Currently, there are two surgical approaches that have been used for regenerating periodontal tissues. These are guided tissue regeneration (GTR) and guided bone regeneration (GBR) [12,29,30]. The concept of guided tissue/bone regeneration began in the late 80s and was developed by Nyman et al. (1990) based on Melcher's theory. This theory hypotheses that when cells with regenerative capabilities are associated with damaged tissues, they can actually be used to aid in the regeneration of that tissue [19]. The hypothesis has been successfully proven in various animal experiments, and the principle of guided tissue regeneration (GTR) has been confirmed [12,15,29].

GTR refers to the procedure of regenerating periodontal tissue through the use of an occlusive barrier membrane between gingival (epithelial) and alveolar bone/PDL tissue. In this operation, an occlusive membrane will be inserted onto the surgical site in order to inhibit the migration of connective and epithelial tissue through the surgical site [12,15,29]. Progenitor cells that existed in the lining of the residual periodontal ligament, corresponding alveolar bone, or blood can then re-colonize the root region and divide into new periodontal supporting components [29]. The guided bone regeneration (GBR) approach is often used to repair defective alveolar ridges before or in conjunction with the placement of a dental implant at extraction sites. In GBR, a bone defect is protected by a membrane to stop fiber tissue intrusion into the site of the graft and to promote the development of a new bone. Intrabony abnormalities and furcations are also treated with GBR [12,15,29].

In order to promote the development of healthy bone structures surrounding bone defects, GBR permeable membranes can be utilized (Figure 3). GBR membranes can also be used in order to preserve the socket area that may be around the tooth due to the presence of periodontal disease, and can even be used to regenerate bone structure at the tooth site after it has been misplaced or extracted [31].

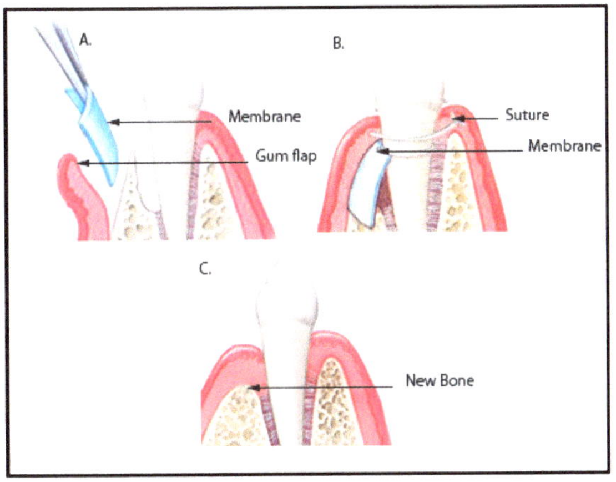

Figure 3. Schematic of GTR is a technique used to repair periodontal defects. (**A**) The gum is opened with a procedure known as a flap. Then, a membrane (with or without the bone graft material) is placed over the damaged bone. (**B**) Closing up. The gum is closed and sewn together. (**C**) After the area heals. Stitches dissolve or are removed.

The efficacy of this therapeutic approach was later verified by Gottlow et al., who effectively extended it to a large group of patients [32,33]. Subsequently, the theoretical and biological basis of GBR has been confirmed during the last three decades in several studies [34], and the effectiveness of the procedure has been shown in a multitude of clinical trials [35] and recorded in comprehensive reviews [15,36].

Guided regeneration has been shown to have many benefits when it comes to tissue regeneration over more conventional surgery approaches such as open-flap debridement, which is usually used to address intrabony defects and mild to moderate furcations [15,36].

The concept behind bone augmentation protocols mainly relates to the enhancement in function and aesthetics and can also be used to improve the functions of fixed dental prosthetics, including dental implants and fixed dentures. This procedure greatly helps in correcting contour deficiencies associated with the replacement of artificial teeth as well as offering a friendlier solution to altered speech patterns due to the uncovered areas frequently left between gingival tissues and the restoration [15,37].

3.1.3. Biologic Principles for Guided Regeneration Therapy

In order to successfully develop an engineered tissue, the following essential elements are required: properly defined levels and patterns of regulatory signals; an abundance of progenitor cells; a sufficient blood supply; and an appropriate biomaterial scaffold [38]. Whereas cells serve as the processing facility for the formation of newly formed tissue and the differentiation of cells. Cells require external stimuli in order to stimulate growth and matrix synthesis. These can be provided by growth factors or morphogens. New vascular networks are promoted as a result of angiogenic signals, which supply nutrients for tissue growth and maintenance. The three-dimensional architecture of scaffolds aids in directing cell regeneration [11,15].

The healing process for periodontal surgery wounds follows the same three stages as any other incisional wound. Initially, a fibrin clot is shaped along the flap's margin and the root surface. Then, a connective tissue matrix attached to the root surface takes the place of the fibrin clot [39]. By keeping the fibrin attached, a new connective tissue bond can form on the root's outer surface. However, a long-bonding epithelial connection forms if the limit of the fibrin clot's tensile strength is surpassed [15,40,41].

In general, the method of periodontal healing is more complicated than other wounds by considerations such as the involvement of various specific cell types and complexity of attachments; avascular root surfaces; different microbial flora; and stromal-cellular interfaces [42]. The first biological reaction that happens after the installation of the barrier membrane is the action of the tissue–membrane interface to absorb the plasma protein. Hence, the related growth factors and progenitor cells, which play an important role in tissue repair, are attracted to the surface of the membrane with the help of proteins [43]. In order to provide nourishment to the new tissue in the barrier membrane, which protects defects, much of the vascular supply comes from blood vessels that originate in the marrow [44]. This further explains why it is vital to plan multiple perforations in cortical bone (also called intra-marrow penetration), as this assists in the production of an excess of angiogenic and osteogenic cells as a means of creating new blood vessels and constructing new bone tissue. This serves two objectives: first, it induces bleeding or blood clotting around the grafts in order to induce bone formation around the grafts, and second, it increases the number of factors that raise the likelihood of bone growth [15,37,41].

There is also controversy about the effector cells in periodontal regeneration. Some reports indicate that PDL cells have the potential to behave as osteoblasts or cementoblasts when they are supplied with growth factors and allowed to proliferate. Other data indicate that PDL cells have the ability to regulate mineral formation, so this will help avoid ankylosis when undergoing regeneration [45]. In other studies, PDL cells in vivo and in vitro have been reported to exhibit minimal osteoblastic properties [42,45]. However, other studies argue that osteoblasts, and not PDL cells, are responsible for generating cementum-like material [17,46,47]. Such variations can be attributable to PDL cell heterogeneity, differing study designs, and/or loss of cell properties defined in vitro research. In summary, the majority of evidence points to PDL cells as the main source. Some reports also point to bone cells as the origin of regenerative cells [15,41,48].

3.2. *Requirements of GTR/GBR Membranes*

GTR and GBR membranes need to fulfil specific requirements to be most effective and successful [49]. These requirements can be summarized as follows:

- Cell exclusion: A growth guide membrane can be used to separate several types of unwanted tissue (e.g., epithelial cells) as well as get access to the site of interest [50].
- Framework: A more rigid framework is frequently necessary when clinical cases need more space maintenance in order to prevent membrane compression into the defect site. Bone grafts can provide this support [49,51].
- Porosity: In order to achieve appropriate cell growth and proliferate, the cells must have an underlying high-pore structure [9].
- Degradation: It is important to provide a degradation profile that suits the tissue regeneration, which takes approximately four to six weeks. Ideally, the membrane should fully degrade after it fulfils its purpose without leaving any residual materials [52].
- Stabilization: To prevent mechanical disturbances from the outside and overhanging of flap movement during the process of healing. Mini screws or sutures, can be used to keep the membrane in place [49].
- Clinical manageability: The membrane and the barrier need to possess physical characteristics that enable their handling by the clinician [53].
- Biocompatibility: Inflammation should be avoided at all costs in order to avoid increased morbidity and costs [49,54].

Owing to the large number of scaffolds that can be made from a wide range of materials, these materials have differing degrees of degradation and integrity. These scaffolds can often induce immune reactions in the hosts [55]. The following section discusses commercial membranes, whether non-absorbable or absorbable, in guided regeneration therapy.

3.3. *Types of Commercial Membranes Used in Guided Regeneration Therapy*

Barriers used in bone/tissue regeneration procedures have varying degrees of deterioration and properties and, therefore, have generally been divided into resorbable or non-absorbable membranes. Gottlow, (1993) was the first to divide these membranes into two generations depending on when they were created and developed; the first generation consists of non-resorbable membranes, while the second generation contains all resorbable membranes [56]. Before Elgali et al. (2017) reviewed this classification and added a new group, Third generation, its membranes rely on naturally derived sources combined with bone grafts and alternative materials to provide structural support to the defect site and to promote the intrinsic regenerative potential of the host tissue [5,15,41].

The following sections discuss the various commercially available periodontal membranes classified as non-resorbable or resorbable materials and are summarized in Table 2.

Table 3 includes a list of the primary biomaterials used in bone tissue engineering, along with their key characteristics.

3.3.1. First-Generation Membranes: Non-Resorbable Guided Membranes

In the 1960s and 1970s, the first generation of barrier membranes were developed with the goal of achieving a sufficient mix of physical qualities that would match those of the replaced tissue while also eliciting a low toxic response in the host [15,41].

In the initial GTR experiments, an occlusive membrane consisting of a bacterial filter made from cellulose acetate (Millipore, Burlington, MA, USA) was utilized. These experiments were conducted by Nyman et al. in 1982 [57]. Due to its toxicity, this form of membrane was not appropriate for clinical applications despite serving its goal. In later trials conducted in the 1990s, membranes of expanded polytetrafluoroethylene (e-PTFE) created specifically for periodontal regeneration were applied (Gore Tex Periodontal Material) [5,15,58].

E-PTFE has a dual-layered structure with pores measuring 5–20 microns in diameter. One side of this membrane is 1 mm thick and has an open microstructure that is 90% porous, preventing epithelial penetration; the other side is 0.15 mm thick and has a porous structure that is 30% thick, allowing space for new bone production [59]. Several investigations have shown that e-PTFE is effective, as described by Liu, J. and Kerns, D.G. [59] However, due to their very porous structural design, they have a high rate of exposure, which is seen as a major disadvantage, in addition to the need for additional surgery to remove them from the location of the newly created tissue.

A high-density d-PTFE membrane with hole sizes of less than 0.3 microns was created to counteract the drawbacks of e-PTFE [60]. In spite of the advantage of non-sticking of tissues to the membrane, which made its removal easy and simple, in addition to its ability to properly regenerate the bones even in exposed cases due to its modified transparency, However, the d-PTFE has limited flexibility, causing it to collapse into the site of the defect [59].

Titanium-reinforced e-PTFE and d-PTFE membranes were produced in order to address the lack of mechanical stiffness that appeared in the initial e-PTFE and d-PTFE membranes [61,62]. However, the requirement for a second surgery to remove the membrane is the most significant disadvantage, similar to other non-resorbable membranes, as well as the rigidity of titanium mesh can create some difficulties during removal due to the need for orthopedic fixation devices such as orthopedic screws. Ti-mesh also appears frequently, which restricts its applications, particularly in aesthetic applications. [63].

3.3.2. Second-Generation Membranes: Resorbable Guided Membranes

Regarding the several applications of GTR and GBR, an absorbable membrane has been proposed as a replacement to the membrane discussed in the previous section in order to minimize its limitations, most notably the requirement for extra surgery to remove the membrane. Based on the origin of the material used to produce the membrane, absorbable membranes are classified into two main groups: natural membranes and synthetic membranes [5,15].

Natural Resorbable Membranes

Numerous natural polymers have been shown to be useful in tissue engineering, which include polysaccharides (cellulose, alginate, starch, hyaluronic acid derivatives, chitin/chitosan), and proteins (soy, fibrin gels, collagen, silk) [26,64]. Natural polymers are also strongly coordinated and may include extracellular substances known as ligands that are essential for binding with cell receptors that can support cell adhesion and function. However, on either side, their medicinal use is constrained by their shortage and the complexity of their processing into scaffolds. In addition, they can induce an immune reaction since natural polymers can lead to cells growing at different developmental stages. Moreover, the rate of degradation varies between patients due to the enzymatic processes involved [5,15,65].

Collagen and chitosan appear to be the two main components of most natural membranes, which are naturally derived from many animal sources. Perhaps the most notable one is the use of bovine Achilles tendon (Cytoplast®), human skin (Alloderm®), or porcine skin (Bio-Gide®) to produce tissue-derived membranes based on collagen [66,67].

The presence of collagen in these membranes is a significant biological feature, as it contributes to many biological activities. Besides being biocompatible, biodegradable, and hemostatic, it also helps in attracting the gingival fibroblast and periodontal ligament (PDL) in addition to augmentation of the soft tissue. Using collagen type I, most of the commercially available collagen membranes are produced and developed, as well as a mixture of collagen types I and III [59]. In vivo experiments found that the collagen-dependent membrane showed some drawbacks, such as its modest efficiency, especially during degradation. Moreover, it may cause ethical and religious issues as well as be a cause of disease transmission [30]. Many biophysical characteristics and collagen framework stabilization can be improved by a number of methods that depend mainly on mechanical and chemical cross-linking, such as adding substances such as glutaraldehyde (GA), diphenyl-phosphoryl azide (DPPA), hexamethylene diisocyanate (HMDIC), and formaldehyde (FA), genipin (Gp), in addition to using ultraviolet light and irradiation [15,68,69].

Collagen structural integrity and mechanical properties are affected by the rehydration protocol, i.e., inserting a cross-linking agent that is natural, genipin into the AlloDerm® [68]. Studies have shown that extending the exposure time for genipin (Gp) to 6 h from 30 min significantly improves tensile strength in comparison with controls. Additionally, according to other studies, cross-linking is effective for controlling prolonged biological degradation, decreasing tissue amalgamation, and vascular depression, as well as for decreasing epithelial migration [70]. A biocompatible reaction of the membrane made of silk fibroin by osteoblast was also observed, which could be used for GBR as an alternate barrier membrane [71].

Synthetic Resorbable Membranes

Synthetic polymers have several advantages over natural polymers, including the ability to have their properties tuned, an infinite variety of forms, and well-established structures. The support that is provided by synthetic biomaterials can make it possible to restore the structural integrity and functional capacity of diseased or damaged tissues [72]. Synthetic polymers can be modified in terms of their molecular weight, molecular structure, and physical and chemical properties simply unlike polymers derived from natural sources, through the addition of certain functional groups and side chains, synthetic polymers may be self-cross-linked or cross-linked with enzymes or other bioactive molecules [73–75].

Synthetic biomaterials have the limitation of lacking cell attachment sites and requiring chemical alterations to improve cell adherence [76]. Physicochemical and mechanical properties of several commercially available synthetic polymers are close to those of biological tissues [77]. The mechanical and physical properties, such as stiffness, Elastic modulus, and degradation rate, are repeatable and predictable throughout a wide spectrum [5,15,76].

The most commonly investigated synthetic degradable materials are poly (-hydroxy esters), which include PCL, PGA, PLA, and their copolymer PLGA, and poly(ethers), which include PEO and PEG, PVA, and PU. These are perhaps the most common examples, however there are now many other synthetic materials being studied [72,76,77]. These polymers all have varying degrees of biodegradability, biocompatibility, and mechanical qualities; nevertheless, there is not a single polymer that possesses all three of these essential properties at the optimal amount [78].

PGA-Based Membranes

Polyglycolic acid is organically created through polycondensation of glycolic acid or ring-opening polymerization (ROP) of glycolide. PGA has a very high fusion point at around 226°. PGA can be processed through hydrolysis, and its by-products can be processed by the Krebs cycle and then eliminated. It is generally used as a suture, but it can also be used as a PLA co-polymer [79]. Resolut® is another commercially available product consisting of two layers: a PLGA compact layer that prevents epithelial cell penetration, and a porous network of polyglycolide fibers that promotes tissue integration. Histological studies showed similar effectiveness to non-resorbable membranes and complete resorption 5–6 months after placement [80,81].

Fibers of polyglactin 910, a copolymer of glycolide and L-lactide (9:1 *wt/wt*), were used to produce a woven mesh (Vicryl Periodontal Mesh®). The polyglactin 910 is inert (no reactions in the surrounding tissue during its adsorption were observed), not antigenic, and preserves its physicomechanical properties during the first 3–4 weeks [82]. Although animal studies indicated a lack of tissue integration and recession formation, clinical evaluation suggested a similar effectiveness as compared to that of other GBR membranes [56,80,81,83].

PCL-Based Membranes

Poly-ε-caprolactone is a polymer with some crystal-like properties that melts at approximately 60 °C. It possesses a relatively slow degradation rate, which makes it better suited for long-term applications such as drug delivery systems. A plethora of studies have evaluated this approach and determined that PCL is an effective delivery polymer. Additionally, its physical properties can be modified through the addition of materials such as PGA or PLA. It also possesses applications in osseous scaffolding [84–86]. Membranes based on copolymers of lactic acid and e-caprolactone have been produced, showing a lower degradation time as compared to pure PLA membranes. PCL is characterized by higher hydrophobicity and lower water solubility than PLA, PGA and their copolymers. A commercial product, called Vivosorb®, consisting of poly(DL-lactide-ecaprolactone), was found to be biocompatible, non-cytotoxic, occlusive and space maintaining [87].

PLA-Based Membranes

Polylactic acid is synthesized similarly to PGA, through ring-opening polymerization of its lactic acid (HOCHCH3COOH) [88]. Its structure can be seen in Figure 4.

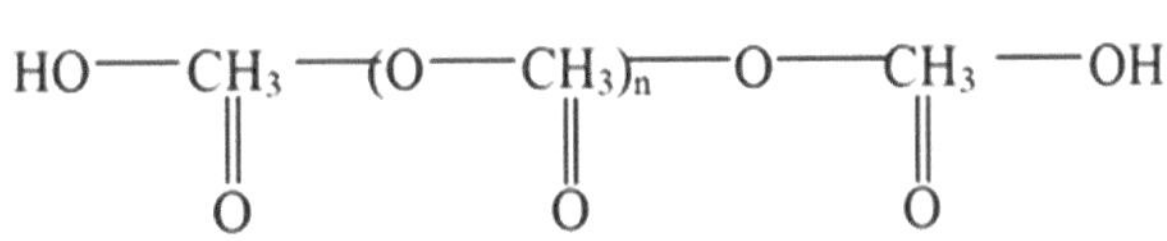

Figure 4. Chemical structure of PLA, where (n) denotes the central repeat unit [modified from [88,89].

PLA is one of the best biopolymers due to its biocompatibility and ease of biological degradation. Because of its properties, it has been used in various biomedical and clinical applications [90–92]. PLA exists in three optical isomers, specifically in its L-lactide form as (PLLA) and its D-lactide form as (PDLA). Additionally, it has a hybrid form (PDLLA) [93]. Because of its nature as an amorphous crystal, PDLLA degrades quicker than other forms of PLLA, in less than half a year [94,95].

The Guidor® Matrix Barrier is a bioresorbable membrane, first used for the regeneration of tissues in periodontology, consisting of polylactic acid treated with acetyltributylcitrate to achieve flexibility to guarantee close barrier adaptation to the bone defect. The Guidor® Matrix Barrier has a matrix with two differently perforated layers. The external layer, allowing integration of the overlying gingival flap, presents large pores (rectangular shape) to promote tissue integration and to enable gingival connective tissue to penetrate quickly into the matrix. The inner layer presents small pores (circular shape), able to retard tissue penetration while allowing nutrient permeation. The two layers are separated by many inner spacers, forming an interspace into which tissue can grow. According to the manufacturer, the barrier structure is not affected by the material degradation for at least the first 6 weeks, and a complete resorption takes place after one year due to hydrolysis [56,83].

Atrisorb® membrane is the first liquid product adapted directly at the surgical site: it consists of poly-DL-lactide acid dissolved in N-methyl-2- pyrrolidone. An irregular membrane is produced after polymer exposure to 0.9% saline solution for 4–6 min in a special cassette, in which it is possible to cut it into the desired shape. Membrane thickness is 600–750 μm, and it is positioned into the defect site by applying a moderate pressure. A histological complete resorption was observed 6–12 months after implantation [96]. Clinical studies reported its efficacy in the treatment of periodontal defects [97].

The Epi-Guide® Bioresorbable Barrier Matrix is a porous membrane consisting of D-L polylactic acid with a unique three-layer technology, used as an adjunct to periodontal restorative surgery. The Epi-Guide maintains its structure and functions for 5 months after implantation, with a complete bioresorption after one year [98]. The layer in contact with the gingiva is porous to promote fibroblast infiltration and attachment. On the contrary, the layer in contact with bone defects has a limited porosity that supports fluid uptake, helps adherence to the tooth surface, and inhibits fibroblast movement [98,99].

For successful periodontal tissue regeneration, the materials used must be compatible with living tissue and favorable in terms of mechanical properties. These specifications cannot be fulfilled by conventional single-component polymer materials. As a result, designing and preparing multicomponent polymer structures represents a promising approach for developing multifunctional biomaterials [100].

3.3.3. Third-Generation Membranes

By reviewing the previous absorbable and non-absorbable membranes, interests should arise in developing a new membrane which has a more advanced role as a barrier membrane and has an additional function such as releasing beneficial agents such as bioceramic, antibiotics, growth factors, and adhesion factors into the wound. The substance-releasing membrane should have a proper release time according to the environment of the graft site [9].

Resorbable Membranes Based on Polymer Composites

Polymer Blends

Polymer membranes must meet a few key criteria for successful guided bone and tissue regeneration (GBR and GTR, respectively), appropriated mechanical and physical properties, a suitable degradation profile, as well as the necessary strength to provide an effective barrier function and resist decomposition [101]. Due to a variety of requirements, a single polymer fails to meet all critical criteria. For instance, naturally occurring polymers cannot provide the required mechanical strength and suitable degradation profiles, while synthetic polymers are unable to interact with biological tissues. On the other hand, polyester membranes turn rigid and brittle after introduction to phosphate buffered saline or artificial saliva solution [102].

Therefore, the issue of developing membranes with the necessary mechanical properties, the expected rate of decomposition, as well as a structure similar to the natural extracellular matrix (ECM) remains topical [103]. A potential solution is to combine two or more polymers in order to offset their disadvantages and find a mutually reinforcing effect.

Natural Polymer and Synthetic Polymer Blends

Natural polymers are known for their increased biocompatibility and bioactive properties compared to synthetic counterparts. For instance, gelatin shows multiple integrin-binding sites to promote cellular adhesion and differentiation [104,105]. Mixing polymers of natural and synthetic origin should provide opportunities for taking advantage of both of them. For example, a material based on an amalgamation of gelatin with PCL has excellent biocompatibility as well as the essential mechanical, physical, and chemical qualities. Its unique properties allow it to be used in cartilage tissue engineering [106,107], neural tissue engineering [108], as well as GBR and GTR [104,105,109]. That being said, chemical segregation between PCL molecules and gelatin is a factor inhibiting the development of composites with the required characteristics.

It has been found that acetic acid can favorably affect the rate and strength of miscibility between PCL and gelatin. For this reason, it is effective for implementation when homogeneous nanofibers with improved performance are required [110,111]. The biodegradation period of such membranes is also appropriate for tissue regeneration [110].

The PLLA/chitosan multilayer membrane proposed by Ku et al. (2009) has shown excellent potential for utilization in GBR and GTR [112]. The membrane has external chitosan netting that promotes the adhesion of cells from nanoporous PLLA located in the middle layer. This layered structure allows for improved mechanical strength and integrity preservation for up to eight weeks.

Natural Polymers Blends:

Despite its natural origin, the bioactivity and mechanical properties of chitosan are inferior to those of protein polymers. In order to improve its properties, chitosan is often blended with other polymers. Due to the presence of free carboxyl groups in the structure of gelatin, it successfully blends with chitosan and forms a stable hydrogen bond with it. The ability of gelatin/chitosan membranes to maintain cellular adhesion and proliferation is better than that of gelatin and chitosan on their own [113]. Moreover, the enhancement by proanthocyanidin gives the gelatin/chitosan bond greater stability and improves its mechanical properties compared to membranes constructed from gelatin or chitosan and gelatin blend [113]. An example of the successful integration of natural polymers is a three-layer membrane with a chitosan interlayer sandwiched between two collagen membranes featuring 20 wt % HA [114]. Hunter and Ma, (2013) have shown that membranes based on hydroxyapatite/chitosan/gelatin can promote the growth of bone marrow mesenchymal stem cells (hBMSC) whilst improving the pace of osteogenic differentiation [115]. Research data assure that gelatin/chitosan or collagen/chitosan membranes possess adequate mechanical and structural properties to be implemented as a barrier membrane. Therefore, they demonstrate the potential to be used in bone and tissue regeneration.

Synthetic Polymer Blends

PLA, PLGA, PCL, and some other aliphatic polyesters are essential components for the production of fibrous scaffolds required for drug delivery systems and tissue regeneration [116]. At the same time, PLGA is characterized by reduced mechanical strength, which makes it impossible to maintain the scaffolding structure during in vitro and in vivo clinical trials. When PLGA was reinforced with other polymers such as PCL, applied in an equal ratio, the compressive strength of the PCL/PLGA scaffolding was far superior to the strength ensured by PLGA alone [117]. Cytological investigations have demonstrated that penetration of human embryonic kidney 293T cells can be prevented by using PDLLA/PLGA electrospinning devices with an appropriate degradation rate and effective cell occlusion for the purpose of GTR. In addition, implantation of a subcutaneous implant in rats demonstrated that PDLLA/PLGA membranes with a composite ratio of 70/30 and 50/50 are able to double as a physical barrier that stops cellular infiltration for a duration of 13 weeks [103]. These data suggest that PDLLA/PLGA membranes can become an effective barrier membrane for tissue regeneration purposes [103]. Along with this, composite membranes fabricated from PLA/PCL, PLGA/PCL, and other synthetic compounds may be deemed as a promising technology for GBR and GTR [117–119].

FloreonTM blend

Floreon is a new sustainable polymer blend created by Floreon-Transforming Packaging Limited in collaboration with the University of Sheffield and certified by the EN13432 standard [120,121]. Based on PLA, Floreon is composed of renewable components, which is likely to improve its mechanical and chemical properties [120,121].

In comparison to pure PLA, Floreon exhibits a remarkable four-fold increase in strength and is less susceptible to cracking and breakage during the manufacturing process and testing phases, as demonstrated by Floreon 3D (2014) and Floreon (2018). The compound has a maximum tensile strength of approximately 1.6 GPa while the elongation at break (fracture strain) is 14% [120–122]. Moreover, in comparison to PLA, Floreon exhibits enhanced thermal performance. It has a melting point of 210 °C [120,121,123], a crystallization temperature of 85 °C [123], and a glass transition temperature of 65 °C. Floreon is extruded at temperatures between 170 and 180 °C. However, since its destruction threshold is 250 °C [120,123], technological processes should not exceed 220 °C. In order to prevent moisture absorption, the material is dried at 65–90 °C after crystallization [120]. Floreon may undergo thermoforming, compounding, and injection molding processes in addition to extrusion (including film extrusion).

There are currently eight Floreon variants labelled in the range FL100–FL800 [120,121]. Due to its resilience to ultraviolet radiation, the Floreon blend is more effective than PLA for 3D printing and lithographic printing [120,121].

Although the Floreon was originally designed for the packaging industry, it has recently been investigated as a scaffold for musculoskeletal applications. The conclusion drawn is that the Floreon blend showed great promise for use in bone tissue regeneration [124].

Bio-Ceramic/Polymer Composites

The incorporation of polymer composites, bioceramic components, and the structural mimicry of bone extracellular matrix (ECM) can be advantageous for the development of biomaterials that are used for guided bone regeneration (GBR) and guided tissue regeneration (GTR) [9]. Hydroxyapatite (HA) [125], carbonated hydroxyapatite (CHA) [126], bioactive glass (BG) [127], β- tricalcium phosphate (β-TCP) and other bioceramics have been widely used in bone tissue engineering and shown to have excellent biocompatibility and osteoconduction properties [128,129].

The use of bioactive ceramics in GTR and GBR has a positive impact on mineralization and cell activity boost on polymer membranes, which suggests the required osteoconductivity and osteoinductivity [127,130–133]. On top of that, bioactive compounds are capable of affecting mechanical properties in a beneficial way [134]. While pure PLGA has a tensile strength of 0.49 MPa, the inclusion of 10–30 wt % nanoapatite into a membrane helps lift it to 0.61 MPa [135]. At the same time, the introduction of bioceramics is able to neutralize the acidic derivatives of PLA, chitosan, and other polymers formed due to their decomposition in an alkaline medium [130,136,137]. According to Khan et al. (2008), composite membranes have the ability to effectively and biomimicking preserve the structural and biological functions of damaged dense tissues [138].

Because hydroxyapatite is osteoinductive, it accelerates bone regeneration and allows the bio-ceramics component to connect directly to the regenerated bone, bypassing connective tissue. The composite has found wide application in orthopedic surgery and dentistry dealing with hard tissue restoration [26,139,140]. Inorganic–organic composites that emulate the structure of human bone offer increased toughness inherent in polymeric materials and the compressive strength characteristic of inorganic components. Their beneficial nature makes it possible to create bioactive materials with improved mechanical properties and degradation profiles. Such composites are stable enough since the alkalinity of the inorganic fraction (for example, hydroxyapatite) balances the acidic substances formed during the autocatalytic decomposition of polymers (such as PLA) [141]. Fabricated PCL/nHA nanocomposites possess properties characteristic of HA ceramics and simultaneously provide the qualities of synthetic polymer PCL, namely, osteoconductivity and biocompatibility [26,142]. Studies of poly (lactic acid) (PLA) nanofibers containing hydroxyapatite filler showed that HA contributes to the improvement of the mechanical and thermal features of the nanofibers [143]. In addition, testing of the β-chitin-HA composite membrane made it possible to detect inclusions of apatite on the surface of β-chitin membranes. This finding indicates increased biocompatibility and provides a suitable foundation for successful cell attachment, adhesion, and proliferation [144].

Bioactive glasses (BSs) are osteoconductive and osteoinductive silica biomaterials with a SiO_2-CaO-P_2O_5 structural grid. The introduction of BG stimulates osteogenesis and angiogenesis both in vitro and in vivo [145,146], and also generates high-performance collagen composites in imitation of bone mineralization. In particular, it is involved in the release of Ca, P, and Si and the subsequent deposition of Ca and P as well as amorphous Ca-P crystals on the implant surface. The following chemical dehydration reactions convert these crystals to hydroxycarbonate apatite (HCA) [147]. In a similar way, wollastonite ($CaSiO_3$) gives up Si and Ca ions, which induce the acceleration of osteogenic differentiation and cell multiplication. Simultaneously, this can lead to deposits of bone-like apatite on the surface of the implant after it has been introduced to simulated body fluids (SBF) [148,149]. Wollastonite exhibits the capability of increased structural mechanical strength, angiogenesis, and bone regenerative capacity. Despite this, it should be subject to further research to identify the bioactivity, osteogenic capacity, and immunogenicity of polymer composites when implanted in humans. These studies are driving the development of polymer/bioceramic based composites offering the advantages of both components [124].

Table 2. List of some commonly used barrier membranes for GTR/GBR therapy.

Resorbability	Barrier Membrane	Composition	Main Characterization	Comments	Refs.
(a) Non-resorbable	Gore-Tex®	Expanded polytetrafluoroethylene (e-PTFE).	- Good space maintainer. - Relatively stiff. - Handling.	- Longest clinical experience.	[150,151]
	Cytoplast® TXT-200	High-density polytetrafluoroethylene (d-PTFE).	- Pores with submicron (0.2 µm) size - Density precludes colonization of the host flora and prevents the infection.	- Avoids a second surgery.	[152,153]
	Gore-Tex-TI®	Titanium reinforced expanded polytetrafluoroethylene (Ti-e-PTFE).	- Most stable space maintainer, requires no filler material.	Titanium should not be exposed. For recession, ridge augmentation.	[154,155]
(b) Resorbable: natural	Bio-Gide®	Collagen derived from porcine skin (types I and III).	- Barrier function At least 6 weeks bioactive.	Usually employed in combination with filler substances	[156–159]
	BioMend Extend®	Collagen type I derived from bovine tendon.	- Resorption: 4–8 weeks - Collagen complexed with formaldehyde	Collagen network extends the resorption time	[160,161]
	AlloDerm®	Acellular dermal matrix human skin.	- Resorption: ~16 weeks. - promoting blood vessel growth, white cell movement, and cell growth.	Ethical concerns and health risks may be associated with the use of human skin.	[9,162]
	Cytoplast ® RTM collagen	Collagen type-I derived from bovine Achilles tendon.	- Resorption: 26–38 weeks. - Multilayered, long-lasting membrane.	Specifically oriented to enhance handling. Possesses good tensile strength.	[9,162]

Table 2. *Cont.*

Resorbability	Barrier Membrane	Composition	Main Characterization	Comments	Refs.
(c) Resorbable: synthetic	Guidor	Poly-DL-lactid/Poly-L-lactid + acetyltributylcitrate.	- Double-layered membrane. - Outer: large pores. - Inner: finer pores.	- No commercially available.	[56,83,163]
	Resolut	Poly-DL-lactid/Co-glycolid.	- Resorption: 10 weeks. - Functional integrity. - Good space maintainer.	- Good tissue integration-Separate suture material.	[80,81,157]
	Vicryl	Polyglactin 910: Polyglicolid/polylactid 9:1.	- Relatively soft. - Well adaptable. - Resorption: 4–12 weeks.	- Woven membrane. - Four prefabricated shapes.	[82,157, 164]
	Atrisorb	Poly-DL-lactide and solvent (N-methyl-2-pyrrolidone).	- Soft Well-adaptable. - Interesting resorptive characteristics.	- Customized membrane fabrication with "Barrier Kit".	[96,97]
	Epi-Guide	Poly-DL-lactic acid.	- 3-layer technology. - Bioresorption: after 6–12 months.	- Self-supporting, can be used without support from bone grafting materials.	[98,99]
	Vivosorb	Poly (DL-lactide-caprolactone) (PLCL).	- Anti-adhesive barrier. - Maintains its mechanical properties for up to eight weeks.	- Commercially available as a nerve guide.	[87]

Table 3. Types of biomaterials used in guided regeneration therapy.

Class	Example	Advantages	Disadvantages	Refs.
Polymers- Natural proteins	Collagen, fibrin, alginate, silk fibroin, hyaluronic acid	• Biocompatible • Biodegradable without inflammation bioactive	• Poor mechanical strength • Rapid resorption	[156,160,165,166]
Polymers- Natural Polysaccharides	Chitosan	• Biodegradable • Biocompatible • Has an antibacterial and bioadhesive properties • Promote wound healing	• Poor mechanical strength • Rapid resorption	[167–169]
Polymers- Synthetic	Polyglycolic acid (PGA)	• Versatile • Reproducible • Thermoplastic so it can be shaped easily	• Inflammatory or immune reaction due to acid release in enzymatic biodegradation • Mechanical stability is of limited duration • Less biocompatible than natural • Not bioactive • Rapid resorption • Low solubility in organic solvent	[73,170–172]
	poly-L-lactide acid (PLLA)	• Degrades slower and dissolves easier than PGA • Reproducible	• The potential to cause immune and foreign-body reactions because it does nor degrade completely • The mechanical stability is of limited duration	
	poly-ε-caprolactone (PCL)	• Slow degradation rate • Reproducible • Good workability	• Inflammatory or immune reaction • Mechanical stability is of limited duration	
Polymers- Synthetic	Hydrogel	• Modified easily • Biocompatible • Biodegradable	• Contracted • Lack stiffness	[73,170–172]
Metal	Titanium mesh	• High mechanical strength and fracture toughness • Biocompatible	• Corrosion may release toxic particles affecting the biocompatibility and induce an inflammatory reaction • Poor stimulation of new bone formation due to the elastic moduli which does not correspond with natural bone	[62,154,173]

Table 3. *Cont.*

Class	Example	Advantages	Disadvantages	Refs.
Ceramic	HA	• Biocompatible • Osteoconductive • Similar to the chemical structure of inorganic phase of bone	• Slow biodegradation • Difficult to shape due to hardness, fragility, and brittleness	[174,175]
	TCP	• Same to above	• Rigid and fragile • Faster resorption rate	[175,176]
	Bioglass	• Biocompatible • Osteoconductive • Bioactive • Promote angiogenesis • Enhance cell adhesion and proteins adsorption • Easy to control the chemical composition • Controlled degradation rate	• Brittleness • Low resistance to crack due to low strength and fracture toughness	[177,178]
Composite	PGA/β-TCP	• Better ability for osteogenesis, mineralization and biodegradation than HA	• Lack of osteoinductivity	[176]
	Bioglass 45S5 and poly (D, L-lactide) polymer	Improved mechanical properties and resorption rate	Reaction with polymer changes the bioglass surface properties and compromised its bioactivity	[177]
	Poly (b-hydroxybutyrate co-b-hydroxyvalerate) (PHBV) microsphere and poly (L-lactic-coglycolic acid) (PLGA)	• Supports drugs and growth factors delivery	• Changes in the surface topography and decrease porosity due to dehydration shrinkage	[179]
	Hyaluronic acid-gelatine	• Good mechanical propertyBiocompatible • High porosity Hydrophilic	• Suboptimal cell adhesion due to negative cell-scaffold interaction	[180]
	Nano HA/polymer	• Promote better cell adhesion and distribution No significant inflammatory response Biocompatible Improved mechanical properties	Unknown mechanism of cellular proliferation and differentiation	[50]

Multiphasic Scaffolds of Periodontal Tissues Regeneration

A multiphasic scaffold is defined by the differences in its architecture (porosity, pore organization, etc.) and its chemical composition, which usually mimics to some degree the structure or cellular and biochemical composition of the native tissue. Multiphasic scaffolds are designed to impart biomimetic functionality to tissue-engineered bone and soft tissue grafts have been recognized for some time as having the potential to facilitate clinical translation in the field of orthopedic tissue engineering, and more recently in the field of periodontal tissue regeneration [181].

In recent years, guided tissue regeneration and guided bone regeneration (GTR and GBR) approaches have been widely used to manage periodontitis. These membranes have separate functions on each side. The occlusive periodontal membrane acts as a barrier to inhibit the ingrowth of epithelial and undesirable tissues into the defective area during periodontal wound healing, whereas the opposite side promotes regeneration of periodontal tissues [181,182]. GTR/GBR membranes must have certain features, particularly those utilized in large-area repair, such as mechanical stability, osteoconductivity, and a balance between membrane degradation and tissue regeneration, all of which are required for the membranes to function [181]. In a number of studies, bilayer GTR/GBR membranes have been utilized as a treatment for periodontal diseases; here are a few examples from the last few years.

The Yoshimoto group has recently developed bilayer membranes based on PLGA or PCL [183,184]. These membranes consisted of a solid layer and a porous layer that, respectively, served as a barrier and provided cell support. By changing the freeze-drying temperature, they were able to control the thickness of each layer. These membranes were found to be more functional than monolayer membranes, with evidence suggesting that their porous structure aided in the osteogenic differentiation and proliferation of mesenchymal stem cells. In vivo studies also demonstrated that the PLGA bilayer membrane promoted bone regeneration with significantly increased bone formation compared to that with a monolayer membrane [184].

Requicha et al. (2016), in a related method, created a biphasic scaffold made of a porous fibrous PCL/starch scaffold for enabling bone ingrowth and an occlusive membrane developed using the same matter [185]. In this technique also, the occlusive membrane was devised to sustain periodontal ligament regeneration by inhibiting epithelial and gingival tissue invasion of the periodontal defect, hence carefully choosing osteoblast and periodontal fibroblast ingrowth as per the GTR law. After performing in vitro analyses, Requicha et al. (2014) and Requicha et al. (2016) discovered a high potential for osteogenesis, which is a key aspect of periodontal regeneration [185,186].

Park et al. (2010) suggested an approach that involves computer-assisted design and manufacturing (CAD/CAM), using two dissimilar sacrificial instruments to 3D print a mold with the negative imprint of the scaffold design. This method directly uses additive manufacturing technology to create a biphasic scaffold consisting of bone and ligament compartments [187]. Later, polymer solutions specific to each compartment (polyglycolic acid and polycaprolactone for bone and ligament compartments, respectively) were tossed into these molds. Consequently, the solvent evaporated before getting rid of the sacrificial material. The resultant porous scaffold had defined dimensions and shape and a definite internal pore architecture. In the process of developing the two compartments independently, they were consequently gathered by utilizing a thin PCL film, hence developing into a biphasic scaffold [187]. The researchers used fibrin to deliver BMP-7-transfected human gingival fibroblasts and human periodontal fibroblasts into the bone and periodontal ligament compartments, correspondingly. The usefulness of the cellularized biphasic scaffold was monitored by means of a murine ectopic model while a human dentin block was placed in the periodontal ligament compartment. This process showed that the presence of periodontal cells to a high degree enabled the attachment of a freshly developed ligament onto the dentine slice together with the sedimentation of cementum-like tissue 6 weeks before implantation [187].

Focusing on guided bone regeneration, which is the focal point of this project, Zhang et al. (2019) have recently studied the most commonly utilized GBR membrane, known as Bio-Gide, which is among the most commonly used commercial biodegradable membranes, and has a wide range of advantages [181]. Bio-Gide possesses a bilayer makeup in which one of the sides is structured to be compact and soft to inhibit epithelium and connective tissue interference on the other side of bone defects, and the opposite side is permeable and coarse to enable the bond of osteoblasts next to the bone defect.

The aforementioned experts described a unique form of multifunctional GBR membrane with similar design characteristics as those of the Bio-Gide membrane but including extra roles that the Bio-Gide membrane cannot accomplish. The unique GBR membrane is made up of a compact nacre-like coating and a permeable membrane. The function of the nacre-like layer is to give great mechanical properties and also to inhibit non-osteoblast interference. Conversely, the porous layer has been designed with the aim of necessitating osteoblast adhesion. For a number of reasons, they asserted that their multifunctional nanocomposite membrane was better than the other GBR membranes. These reasons include biocompatibility combination with the facial surface, high mechanical performance, sufficient rate of degradation, and efficacious bacteriostasis. For these reasons, this type of nanocomposite membrane qualifies to be considered as a perfect bioactive GBR membrane for medical use [181,188].

By combining the electrospinning technique with emulsion templating, a bilayer barrier membrane (BM) made of a biodegradable synthetic polymer, PCL, was effectively developed by [189]. Some of the qualities exhibited by the resultant BM included the absence of delamination, a qualitatively resistant structure to twisting and elongation, and simplicity in handling. The electrospun layer of the BM has been proven to possess the ability to act as a barrier, offering protection to the bone defect against soft tissue interference. On the other hand, the interconnected PCL polyHIPE layer has exhibited pivotal characteristics to be the bone-enhancing layer, supplying crucial needs including boosting collagen and mineral deposition and enhancing cellular infiltration and cell compatibility [189].

3.4. *Scaffold Manufacturing Technologies*

Scaffolding manufacturing is a highly nuanced and evolving field, encompassing a diverse array of techniques tailored to various applications and material requirements. The process of creating porous scaffolds is far from uniform, with methodologies varying significantly in complexity, cost, and final product quality. This section delves into a selection of prevalent manufacturing technologies, each catering to specific needs and producing results unique in structure and functionality. Distinct from existing studies, such as the one referenced in citation [190–200], this review aims to not only present an overview of established techniques but also emphasize recent advancements, novel applications, and critical evaluations. From the rapid yet modest-quality processes to intricate and time-intensive methods that yield superior structures, we provide an insightful and contemporary analysis, with a particular focus on how these methods contribute to various applications, including bone regeneration. This comprehensive examination serves as a valuable resource for researchers and practitioners seeking an up-to-date understanding of scaffolding manufacturing technologies.

3.4.1. Solid Free-Form Fabrication Technique

SFFT is a manufacturing technique also recognized as rapid prototyping (RP), which also refers to a type of fabrication process called additive manufacturing [191]. In which components are printed by depositing one cross-section layer over the other layer and assembled using a three-dimensional computer-aided design (CAD) model [190]. Three-dimensional scaffolds with complex geometries and dimensionally accurate structures can be manufactured using data obtained from medical scans and then adjusted to meet the needs of each individual patient [190,191].

This process is accomplished through several phases. The first phase is based on creating a computer-aided design (CAD) model, which is then sent to a file that can be manipulated with a stereolithography apparatus. Automatically, the STL file is divided into horizontal layers throughout the pre-production phase. Then, printing continues in this layered process. The final structure needs to be hardened and its surface treated before being used [191]. Through the use of sophisticated scanning techniques such as magnetic resonance imaging (MRI) or computer tomography (CT) [191]. These highly detailed 3D images can then be used to make the creation of precise [26], integrated scaffolds [192] and significantly reproducible [191]. This is particularly useful when making highly porous structures at approximately 90% or more of the total volume of the scaffold [192].

Scaffolds with sophisticated and controlled macro-and microporous structures can be provided by SFFT, potentially both within the same structure [26]. Table 4 compares the various SFFT types, which have been evaluated by different research groups [191–194]. This list includes their inherent advantages and disadvantages. SFFT is a modern development. It helps in creating solutions rather quickly, but not all types of SFFT can be used for scaffold manufacturing [191–194].

Table 4. Different types of SFFT with their advantages and disadvantages.

Techniques	Materials	Advantages	Disadvantages	Refs.
Stereolithography (SL)	PEG, PEGDA, PPF, PCL, PDLLA	High accuracy, complex 3D structure including agents and cells, easy removal of photopolymer by heating	Photo-polymerization of materials, photocurable materials, expensive materials and equipment	[191–194]
Fused deposition modelling (FDM)	Thermoplastic polymers and their composites (PVA, ABSP400)	High porosity, complete pore interconnectivity, possibility of controlling porosity and size of pores, macro shape control, good compressive strength, solvent-free	High processing temperature, limited material range, inconsistency in pores	
Selective laser sintering (SLS)	Polymer ceramics (PCL, HAp, TCP)	Complex structure, possibility of controlling porosity and size of pores independently, wide range of powder materials, solvent-free, any secondary binder system	High processing temperature, using only thermally stable polymers, limited to small pore size	
3D printing (3D-P)	Ceramics, polymers, metals	Easy process, high porosity, complete pore interconnectivity, possibility of controlling porosity and size of pores independently, macro shape control, wide range of materials	Use of toxic organic solvent, lack of mechanical strength, limited to small pore size	

3.4.2. Three-Dimensional Bioprinting Technique

Three-dimensional (3D) bioprinting is a sophisticated and intricate method of additive manufacturing that meticulously incorporates a range of biological materials to generate structures which resemble and function such as living tissues. This technique is known for its scalability, meaning it can be adjusted to create complex structures that meet individual patients' specific needs in terms of size and complexity. Furthermore, the technology enables the precise distribution of cellular components-including but not limited to growth factors, proteins, cells, and drug particles. These favorable conditions have spearheaded advancements in several medically and clinically relevant applications such as drug testing, high-throughput assays, tissue engineering, tissue regeneration, and cancer research [195].

One of the most challenging tasks in 3D bioprinting is the creation of blood vessels and organs. The complexity of this task arises from the need to integrate diverse cell types, the constraint of limited structural support, and the requirements of a concomitant capillary network, which is a typical characteristic of functional organs. Despite the considerable technological advances, these multifaceted requirements make the printing of such structures a significant challenge [196].

Nonetheless, relentless research efforts have led to some notable advancements. Researchers have successfully printed rudimentary structures, such as blood vessels, skin, and cartilage that does not require a blood supply. These achievements mark promising milestones in the field of 3D bioprinting [196]. Attempts have also been made to print bone tissue, specifically bone that comprises its natural constituents such as nerve and muscle tissue. Despite these endeavors, the structures produced are yet to match the functional superiority of their naturally occurring counterparts. Therefore, it is evident that continued research and development are needed to overcome the existing challenges in 3D bioprinting [197].

3.4.3. Gas-Foaming Technique

With this technique, there is no longer a need to use solvents that are normally present in the previously mentioned methods. This method creates a porous network through the dispersion of gas bubbles that, when the material is hardened, act as pores, as illustrated in Figure 5. A heated mold is used to heat the polymer material, which is usually made of polylactic-co-glycolic acid, which is then molded by compressing it to make rigid discs. After this, these molded structures are pumped with high pressure (5.5 MPa) CO_2 for 3 days at 25 °C. Afterward, gas pressure is reduced to atmospheric levels and, therefore, gas solubility is reduced. This process makes CO_2 gas create inner clumps, which then create the pores needed for the proper function of the implant. This method allows for the total number of porosities to reach up to 93% and sizes of approximately 100 mm. It is not trivial, however, to control pore size and interconnectivity with this technique [198,199].

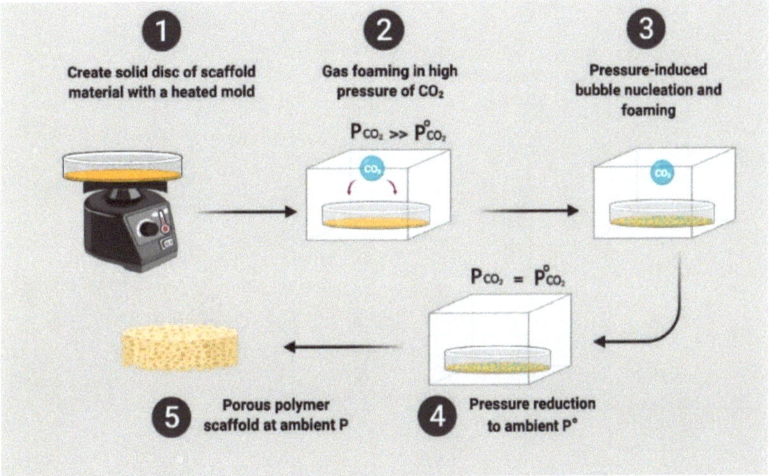

Figure 5. Schematic illustration of the gas-foaming technique, [source: [200], redesigned with copyright permission from Elsevier license number: 4724150945567, dated 8 December 2019].

3.4.4. Thermally Induced Phase Separation Technique

A procedure that allows for the fabrication of highly porous anisotropic scaffolds is called Thermally Induced Phase Separation (TIPS). These polymer scaffolds can be controlled with ease but have a low ability to be applied to affected tissues such as ligaments, muscles, nerves, intestines, and osseous structures [201]. Depending on the concentration of polymer used, certain characteristics will change, such as mechanical properties, pore shape, biological activity, and the rate of resorption. Furthermore, these properties will change depending on the volume of the phase separation [190]. A polymer phase fraction can be achieved by dissolving a polymer at a high degree of temperature in a certain solvent, then cooling the homogenous polymer/solvent solution to obtain a polymer porous scaffold. After this process is completed, a microporous scaffold can be obtained immediately after the solvent has evaporated, as schematically shown in Figure 6 [201].

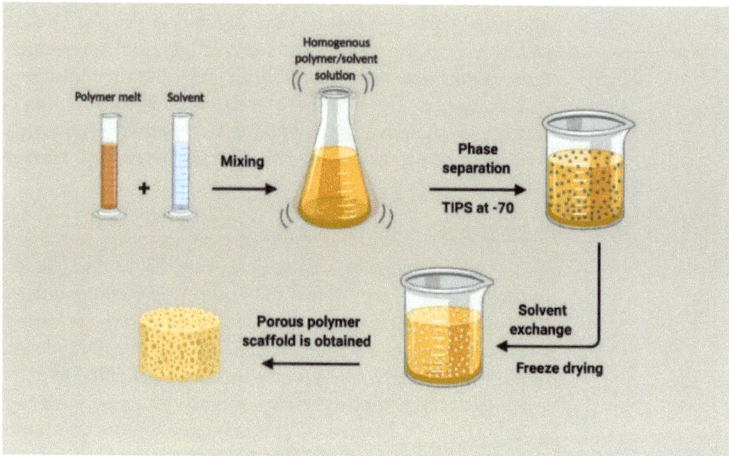

Figure 6. Schematic representation of the porous scaffold fabrication process with the Thermally Induced Phase Separation (TIPS), [source: [200], redesigned with copyright permission from Elsevier license number: 4724150945567, dated 8 December 2019].

3.4.5. Emulsion Freeze-Drying Technique

This technique is based on the phase fraction through the use of different physical properties of the fiber by emulsifying the solution and then drying it at a very low temperature [202], and producing a scaffold that has abundant pores, as illustrated in Figure 7.

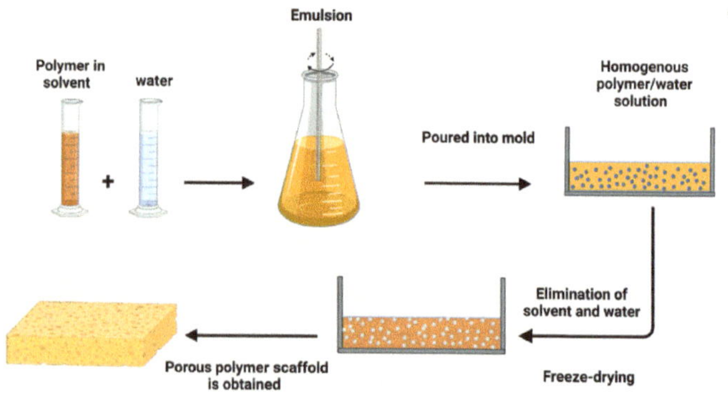

Figure 7. Schematic illustration of the freeze-drying process, [source: [200], redesigned with copyright permission from Elsevier license number: 4724150945567, dated 8 December 2019].

The first step in this process is the creation of the emulsion by homogenizing a polymer in a carbon-based solvent and water. This emulsion must be quickly frozen and the formed phases (solvent and water) are then eliminated by freeze-drying the sample. The resulting polymer scaffolds will have pores of between 20 and 200 μm [190]. This method could be combined with the third method that was mentioned, as well as adding crystal-forming polar compounds such as sucrose or NaCl, in order to further increase porosity. Once the sample has been dried, these particles can be cleared with the use of water [167].

3.4.6. Solvent Casting and Particulate Leaching Technique

Another common method for making scaffolds is solvent casing. This process starts with the deconstruction of a polymer in a carbon-based solvent, as schematically shown in Figure 8. The aforementioned method uses "porogens", a group of chemical compounds that can be distributed into a structure during the manufacturing process and then taken away through the use of water, leaving behind a porous structure. These porogens can create a coupled polymer-porogen structure when added to the overall solution. As soon as the polymer reaches its final form and starts hardening, and the original solvent evaporates away, water is then used to dissolve porogens, which is often a high polarity compound, such as NaCl. Although it is hard to control the final inner structure of the scaffold since it is difficult to predict and control where the porogen particles will be distributed and then dissolved, a three-dimensional porous polymer scaffold was obtained [199,203].

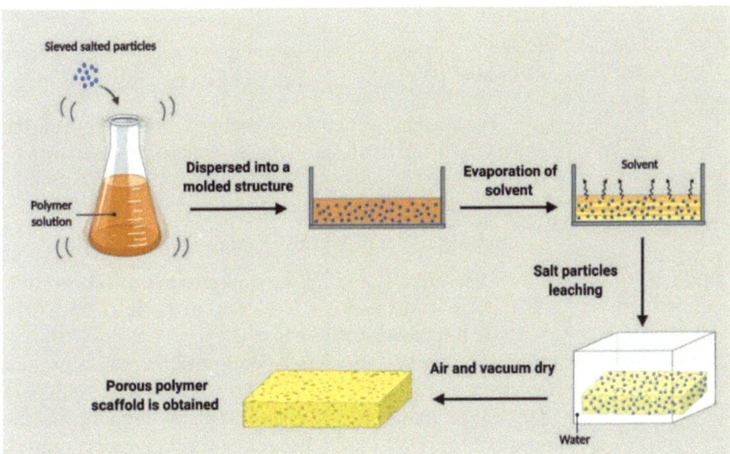

Figure 8. Schematic illustration of the solvent casting and particulate leaching technique, [source: [200], redesigned with copyright permission from Elsevier license number: 4724150945567, dated 8 December 2019].

3.4.7. Spin-Coating Technique

The spin coating technique is a widely used method for depositing thin films of materials onto a flat substrate. In this technique, a liquid solution containing the scaffold material is dispensed onto the substrate, which is then rotated at high speeds (typically in the range of 1000 to 4000 rpm). The centrifugal force generated by the spinning substrate causes the solution to spread out evenly over the substrate, forming a thin film. The speed and duration of the spinning process can be controlled to achieve a desired thickness and uniformity of the film.

In Figure 9, a schematic outline of the spin coating process is shown. The figure depicts the substrate, the spin coater, and the solution being applied to the substrate. As the substrate is spun, the centrifugal force pulls the solution towards the edges, creating a thin and uniform film on the substrate.

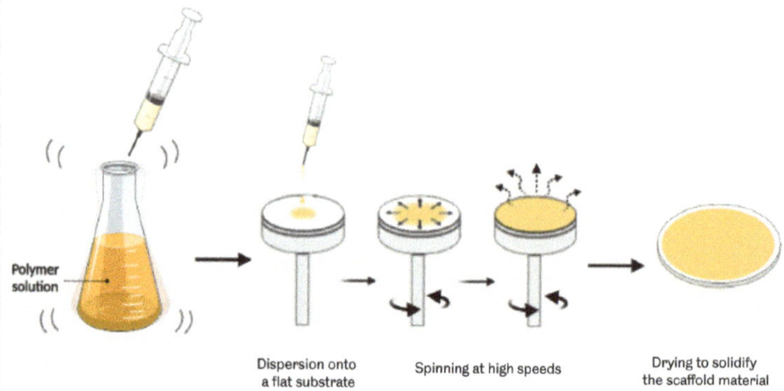

Figure 9. Schematic illustration of the spin coating technique, illustrating the steps involved in fabricating a scaffold.

The advantages of the spin coating technique include its simplicity and cost-effectiveness, as well as its ability to produce films with precise control over their thickness and uniformity. This technique can also be easily combined with other techniques to create complex multilayered structures, such as for periodontal tissue regeneration.

Limitations of the preceding manufacturing techniques

In practice, the techniques used to manufacture scaffolds are divided into solid free-form fabrication and conventional methods. Each of them produces various scaffolds with distinctive characteristics [204]. Even though SFFT provides a plethora of potential opportunities for tissue engineering and possesses undeniable advantages, there are some inherent drawbacks that must be considered. Firstly, each method uses a very specific fabrication material. SLS uses a fine powder, whereas the use of thermoplastics is more efficient for FDM. Even when the selected material is appropriate, if it is difficult to prepare, it can make the whole process much more challenging. Secondly, the fact that a material can be successfully printed does not guarantee its proper function, since successful scaffolds also require constructs to maintain their integrity throughout their layers. The material must be able to support itself after its fabrication, maintaining its integrity layer by layer. Thirdly, in the case of the printing of biological tissues, novel material solidification techniques that are used to preserve the fabricated scaffold integrity should be developed [205]. Lastly, when using materials that are cell-loaded, the flexibility of print parameters such as shear stress or temperature is restricted. This happens since cell environments are ever changing, and doing so would be deleterious for cell survival [206].

While conventional techniques of scaffolding fabrication include the construction of porous polymer structures such as substrates for cell adhesion, it is difficult to obtain complex structures with tunable microscale and macroscale using conventional methods [207]. In addition, some of these methods are manual based. Therefore, they are labor intensive and difficult to reproduce. Another limitation is the need for organic solvents and porogens, which are cytotoxic and their residues may cause inflammatory responses [208]. Benefits and limitations of conventional and Solid free-form manufacturing techniques are discussed and summarized in Table 5 [188,209].

Table 5. Comparison of different scaffold fabrication techniques: advantages and disadvantages, adapted from [188,209].

Manufacturing Method	Benefits	Potential Limitations
Gas foaming	- Eliminates use of chemical solvents.	- High pressures involved prohibits inclusion of cells and bioactive molecules directly into scaffolds - Temperature labile materials may be denatured during compression molding step - Difficult to control pore sizes and ensure interconnectivity
Emulsification freeze-drying	- Does not require use of solid porogen.	- Requires use of organic solvents - Small pore size and - Porosity often irregular - Long processing time
Phase separation	- Eliminates leaching step of porogen - Can be combined with other techniques easily.	- Small pore sizes limit use - Use of organic solvents inhibits use of bioactive molecules or cells during scaffold fabrication
3D Printing - SLA - Inkjet - SLS - Laser-assisted - FDM - Microvalve - Micro-extrusion	- Complex 3D shapes with high resolution, controlled pore size and morphology and controlled internal structures can be fabricated. Improved capacity to incorporate vascular structures into constructs. - Depending on technique used, cells may be included in high concentration directly in scaffold materials.	- Some techniques are limited by printable materials - Set up costs can be expensive for machinery
Solvent casting/ particulate leaching	- Relatively simple technique that allows creation of scaffolds with regular porosity, controlled composition and pore size.	- Use of organic solvents precludes cells and biomolecules being included directly in scaffolds - Can be difficult to control pore shape and interconnectivity - Limited thickness of structures and mechanical properties achievable
Spin-coating	- Simple and cost-effective method. - Precise control over the thickness and uniformity of the film.	- Requires optimization of parameters such as the spinning speed and duration for each specific material and substrate combination
Electrospinning	- Essential technique for developing nanofibrous scaffolds for the TE. - Homogeneous mixture made of fiber with high tensile strength. - Simple instrument. - Continuous process. - Cost effective compared to other existing methods. - Scalable. - Ability to fabricate fiber diameters few nm to several microns.	- Used solvents can be toxic - Jet instability - Packaging, shipping, handling

3.4.8. Electrospinning for Bone Regeneration

Over the last ten years, electrospinning technology has become one of the most interesting methods for creating scaffolds used for tissue engineering. The creation of these nanofiber scaffolds has become the focus of research for many investigators due to their many unique properties, especially those which are clinically relevant. In particular, this method is used in order to manufacture nanofibers used in different applications in dentistry such as tooth restoration, repair of oral mouth tissue, preventing tooth decay, and restoring other dental and periodontal tissues, such as the repair of dentin, endodontium, oral mucosa, periodontal tissue, as well as alveolar bone regeneration [210].

A type of material that has received a lot of attention recently has been biodegradable polymers, especially in biomedical areas such as bio-prosthetics, tissue engineering, and the application of drug delivery systems. Aliphatic polyesters are one of the most significant types of synthetic biodegradable polymers, owing in particular to their advantageous characteristics of biocompatibility and biodegradability. The main allure of these polymers (polyesters) is their biological compatibility and their ability to be degraded within the organ [211].

Electrostatic spinning, most commonly known as electrospinning, has been a focal point of research for the last 20 years due to the various potential uses of the created microfiber in both nanotechnologies and nanoscience [212]. Notable characteristics, such as high permeability, large surface-to-volume ratio, and excellent pore interconnection, have made electrospun microfiber ideal for normal cell functions, such as nutrient and cell transportation [213,214].

Additionally, the nanofibrous scaffolds manufactured through this method can provide excellent extracellular conditions, such as coupling, migration, and cell proliferation, especially for those cells in charge of hard tissue repair. Along with the simplicity of setup and cost efficiency, the opportunity to create microfiber with a large variety of physical and chemical properties is its own merit. This machine consists of a syringe needle, a grounded collector (metal plate), a high-voltage electrical source, and a syringe pump (Figure 10).

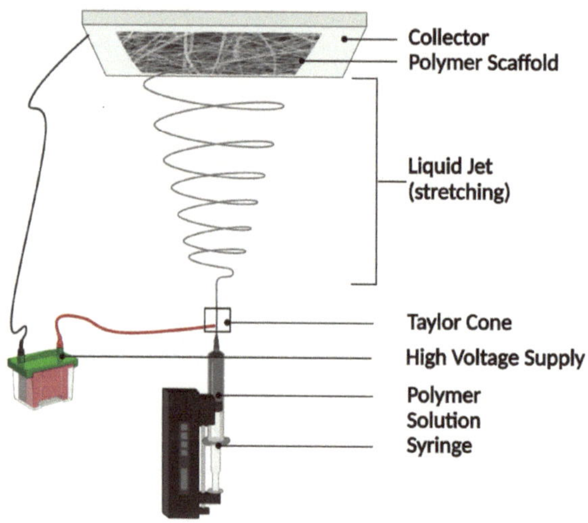

Figure 10. Example of electrospinning apparatus, source: [200], redesigned with copyright permission from Elsevier license number: 4724150945567, dated 8 December 2019.

The electrical source must carry around 10 to 30 kilovolts and is applied to solutions that are ejected via the syringe needle. When the electrical charge reaches the starting point, the surface tension of the charged solution begins to change, causing a deformation of the solution droplet into a conical droplet known as the Taylor cone. While the electrical force overcomes the surface tension of the charged solution, thin charged jets are ejected from the tip of the metallic needle in a nearly straight line towards the electrically inverse electrode. As the material is being extruded, the solvent is being evaporated away, resulting in the construction of continuous dry polymer fiber, which leads to the formation of a non-woven surface of the obtained fiber. The grounded collector surface is generally placed approximately 20 cm from the syringe's tip [215].

4. Future Directions and Conclusions

The dynamic landscape of guided tissue and bone regeneration is primarily driven by the interplay of biomaterials and scaffold manufacturing technologies, offering a diverse array of solutions for periodontal therapeutic interventions. As we move forward, several trends and directions can be envisioned that are expected to shape the future trajectory of this field.

One such direction is the exploration of more advanced, next-generation membranes, which could potentially surpass the benefits offered by the current third-generation membranes. The development of these membranes could focus on enhanced biocompatibility, optimized resorption rates, and improved mechanical strength. For instance, research on Floreon™ blend and other innovative polymer combinations that offer superior mechanical strength and enhanced bioresorbability may provide novel solutions for effective tissue regeneration [124]. A focus on personalized treatment approaches utilizing patient-specific 3D bioprinting technologies may also prove beneficial [195].

Furthermore, the integration of growth factors or other bioactive molecules within these membranes to promote bone and tissue regeneration is a promising area of exploration. With recent advances in drug delivery technologies, the controlled and localized delivery of these factors could significantly enhance therapeutic outcomes [49,170]. A notable recent development is the use of Zn-based biodegradable materials in biomedical applications. With their potential to serve as next-generation orthopedic implants, Zn-based materials may offer significant advantages over conventional alternatives, such as reducing the need for revision surgeries and minimizing biocompatibility issues [216]. These materials have demonstrated a significant role in bone metabolism and new cell growth, and they show medium degradation without the release of excessive hydrogen. The addition of alloying elements such as Mg, Zr, Mn, Ca, and Li into pure Zn has been found to enhance the mechanical properties of Zn alloys, making them a promising material for future guided tissue and bone regeneration applications [216]. Additionally, there is an increasing interest in nanotechnologies and their application in bone regeneration. The fabrication of nanocomposite scaffolds can potentially improve the cellular response and lead to better mimicking of the natural extracellular matrix. Such improvements could foster bone formation and promote better integration with host tissue [50,181].

In the realm of scaffold manufacturing technologies, there is considerable scope for enhancing precision and reducing production times, perhaps through the advancement of computer-aided design and manufacturing (CAD/CAM) techniques. Hybrid fabrication methods that combine the advantages of different techniques can also be explored, such as combining electrospinning with 3D printing, to create complex scaffold architectures with improved mechanical properties and porosity [187].

The future direction of guided tissue and bone regeneration is also increasingly being influenced by the emerging field of tissue engineering, specifically the use of stem cells. Combining biomaterials with stem cell therapy could lead to unprecedented advancements in periodontal tissue regeneration, offering high-potential treatment strategies that need to be explored extensively [217].

In conclusion, guided tissue and bone regeneration represents an exciting and rapidly advancing field with vast potential for further innovation and improvement. The progression in this field will rely heavily on a comprehensive understanding of biomaterials and their interactions with biological systems, as well as the continued refinement of manufacturing technologies. Embracing an interdisciplinary approach, encompassing materials science, biology, engineering, and clinical dentistry will be essential for the successful evolution of periodontal treatments.

Author Contributions: A.M.A.: Conceptualization, Methodology, Validation, Formal Analysis, Investigation, Resources, Data Curation, Writing—Original Draft, Writing—Review and Editing, Visualization, Supervision, Project Administration, Funding Acquisition. R.M.: Conceptualization, Methodology, Supervision, Validation, Visualization, Writing—Review and Editing. I.O.A.: Conceptualization, Methodology, Supervision, Validation, Visualization, Writing—Review and Editing. All authors have read and agreed to the published version of the manuscript.

Funding: This research received no external funding.

Institutional Review Board Statement: Not applicable.

Informed Consent Statement: Not applicable.

Data Availability Statement: All data analyzed during this study are included in this published article

Conflicts of Interest: The author declares no conflict of interest.

Abbreviations

%	Percentage
°C	Degree Celsius
BMP	Bone morphogenic protein
Ca	Calcium
CAD/CAM	Computer-aided design and manufacturing
$CaSiO_3$	Calcium silicate
CO	Carbon monoxide
conc	Concentration
CT	Computer tomography
DFDBA	Demineralized freeze-dried bone allograft
DPPA	Diphenyl-phosphoryl azide
EMD	Enamel matrix derivative
FA	Formaldehyde
FDBA	Freeze-dried bone allograft
FDM	Fused deposition modelling
GA	Glutaraldehyde
GBR	Guided bone regeneration
gm or g	Gram
Gp	Genipin
GTR	Guided tissue regeneration
HA	Hydroxyapatite
HMDIC	Hexamethylene diisocyanate
IGF	Insulin-like growth factor
mg	Milligram
mm	Millimeter
MPa	Megapascal
MRI	Magnetic resonance imaging
MSCs	Mesenchymal stem cells
Mw	Molecular weight
Na	Sodium
NOFs	Normal oral fibroblast cells
-OH	Hydroxyl
PCL	Poly-ε-caprolactone
PDGF	Platelet derived growth factor
PDL	Periodontal ligaments
PGA	Polyglycolic acid
PLA	Polylactic acid

PLGA	Poly (lactic-co-glycolic acid)
PO_4	Phosphate
PTFE	Polytetrafluoroethylene
ROP	Ring-opening polymerization
SBF	Simulated body fluid
SD	Standard deviation
SFFT	Solid free-form fabrication technique
SLS	Selective laser sintering
TCP	Tri calcium phosphate
TGF-beta	Transforming growth factor-beta
TIPS	Thermally induced phase separation
βTCP	Beta tricalcium phosphate

References

1. Huck, O.; Stutz, C.; Gegout, P.-Y.; Özçelik, H.; Benkirane-Jessel, N.; Petit, C.; Batool, F. Nanomedicine and Periodontal Regenerative Treatment. *Dent. Clin. N. Am.* **2022**, *66*, 131–155. [CrossRef]
2. Kinane, D.F.; Stathopoulou, P.G.; Papapanou, P.N. Periodontal diseases. *Nat. Rev. Dis. Prim.* **2017**, *3*, 17038. [CrossRef]
3. Liu, Y.; Guo, L.; Li, X.; Liu, S.; Du, J.; Xu, J.; Hu, J.; Liu, Y. Challenges and tissue engineering strategies of periodontal-guided tissue regeneration. *Tissue Eng. Part C Methods* **2022**, *28*, 405–419. [CrossRef] [PubMed]
4. Dimitriou, R.; Mataliotakis, G.I.; Calori, G.M.; Giannoudis, P.V. The role of barrier membranes for guided bone regeneration and restoration of large bone defects: Current experimental and clinical evidence. *BMC Med.* **2012**, *10*, 81. [CrossRef]
5. Elgali, I.; Omar, O.; Dahlin, C.; Thomsen, P. Guided bone regeneration: Materials and biological mechanisms revisited. *Eur. J. Oral Sci.* **2017**, *125*, 315–337. [CrossRef] [PubMed]
6. Esposito, M.; Grusovin, M.G.; Felice, P.; Karatzopoulos, G.; Worthington, H.V.; Coulthard, P. Interventions for replacing missing teeth: Horizontal and vertical bone augmentation techniques for dental implant treatment. *Cochrane Database Syst. Rev.* **2009**, *4*, CD003607. [CrossRef] [PubMed]
7. Chen, F.-M.; Zhang, J.; Zhang, M.; An, Y.; Chen, F.; Wu, Z.-F. A review on endogenous regenerative technology in periodontal regenerative medicine. *Biomaterials* **2010**, *31*, 7892–7927. [CrossRef]
8. Yuan, H.; Fernandes, H.; Habibovic, P.; De Boer, J.; Barradas, A.M.C.; De Ruiter, A.; Walsh, W.R.; Van Blitterswijk, C.A.; De Bruijn, J.D. Osteoinductive ceramics as a synthetic alternative to autologous bone grafting. *Proc. Natl. Acad. Sci. USA* **2010**, *107*, 13614–13619. [CrossRef]
9. Bottino, M.C.; Thomas, V.; Schmidt, G.; Vohra, Y.K.; Chu, T.-M.G.; Kowolik, M.J.; Janowski, G.M. Recent advances in the development of GTR/GBR membranes for periodontal regeneration—A materials perspective. *Dent. Mater.* **2012**, *28*, 703–721. [CrossRef]
10. Dahlin, C.; Linde, A.; Gottlow, J.; Nyman, S. Healing of bone defects by guided tissue regeneration. *Plast. Reconstr. Surg.* **1988**, *81*, 672–676. [CrossRef]
11. Taba, M., Jr.; Jin, Q.; Sugai, J.V.; Giannobile, W.V. Current concepts in periodontal bioengineering. *Orthod. Craniofac. Res.* **2005**, *8*, 292–302. [CrossRef]
12. Sculean, A.; Nikolidakis, D.; Schwarz, F. Regeneration of periodontal tissues: Combinations of barrier membranes and grafting materials–biological foundation and preclinical evidence: A systematic review. *J. Clin. Periodontol.* **2008**, *35*, 106–116. [CrossRef] [PubMed]
13. Liang, Y.; Luan, X.; Liu, X. Recent advances in periodontal regeneration: A biomaterial perspective. *Bioact. Mater.* **2020**, *5*, 297–308. [CrossRef] [PubMed]
14. Needleman, I.; Tucker, R.; Giedrys-Leeper, E.; Worthington, H. A systematic review of guided tissue regeneration for periodontal infrabony defects. *J. Periodont. Res.* **2002**, *37*, 380–388. [CrossRef] [PubMed]
15. Lee, H.-S.; Byun, S.-H.; Cho, S.-W.; Yang, B.-E. Past, present, and future of regeneration therapy in oral and periodontal tissue: A review. *Appl. Sci.* **2019**, *9*, 1046. [CrossRef]
16. Giannobile, W.V. Commentary: Treatment of periodontitis: Destroyed periodontal tissues can be regenerated under certain conditions. *J. Periodontol.* **2014**, *85*, 3–9. [CrossRef]
17. Melcher, A.H. On the repair potential of periodontal tissues. *J. Periodontol.* **1976**, *47*, 256–260. [CrossRef]
18. Dhruvakumar, D.; Arun Kumar, K.V.; Ayilavarapu, S. *Hand Book on Bone Regeneration: Materials, Techniques and Procedures: From Research to Clinical Practice*; Notion Press: Chennai, India, 2019; ISBN 1645878392.
19. Nyman, S.; Lang, N.P.; Buser, D.; Brägger, U. Bone regeneration adjacent to titanium dental implants using guided tissue regeneration: A report of two cases. *Int. J. Oral Maxillofac. Implants* **1990**, *5*, 94–105.
20. McAllister, B.S.; Haghighat, K. Bone augmentation techniques. *J. Periodontol.* **2007**, *78*, 377–396. [CrossRef]
21. Zhao, R.; Yang, R.; Cooper, P.R.; Khurshid, Z.; Shavandi, A.; Ratnayake, J. Bone grafts and substitutes in dentistry: A review of current trends and developments. *Molecules* **2021**, *26*, 3007. [CrossRef]
22. Mellonig, J.T. Autogenous and allogeneic bone grafts in periodontal therapy. *Crit. Rev. Oral Biol. Med.* **1992**, *3*, 333–352. [CrossRef]

23. Miron, R.J.; Zhang, Y.F. Osteoinduction: A review of old concepts with new standards. *J. Dent. Res.* **2012**, *91*, 736–744. [CrossRef] [PubMed]
24. Urist, M.R. Bone: Formation by autoinduction. *Science* **1965**, *150*, 893–899. [CrossRef] [PubMed]
25. Richardson, C.R.; Mellonig, J.T.; Brunsvold, M.A.; McDonnell, H.T.; Cochran, D.L. Clinical evaluation of Bio-Oss®: A bovine-derived xenograft for the treatment of periodontal osseous defects in humans. *J. Clin. Periodontol.* **1999**, *26*, 421–428. [CrossRef] [PubMed]
26. Rezwan, K.; Chen, Q.Z.; Blaker, J.J.; Boccaccini, A.R. Biodegradable and bioactive porous polymer/inorganic composite scaffolds for bone tissue engineering. *Biomaterials* **2006**, *27*, 3413–3431. [CrossRef]
27. Schmidlin, P.R.; Nicholls, F.; Kruse, A.; Zwahlen, R.A.; Weber, F.E. Evaluation of moldable, in situ hardening calcium phosphate bone graft substitutes. *Clin. Oral Implants Res.* **2013**, *24*, 149–157. [CrossRef]
28. Damien, E.; Revell, P.A. Coralline hydroxyapatite bone graft substitute: A review of experimental studies and biomedical applications. *J. Appl. Biomater. Biomech.* **2004**, *2*, 65–73.
29. Koo, K.-T.; Polimeni, G.; Pringle, G.A.; Agelan, A.; Safadi, F.F.; Wikesjö, U.M.E. Histopathological observations of a polylactic acid-based device intended for guided bone/tissue regeneration. *Clin. Implant. Dent. Relat. Res.* **2008**, *10*, 99–105.
30. Behring, J.; Junker, R.; Walboomers, X.F.; Chessnut, B.; Jansen, J.A. Toward guided tissue and bone regeneration: Morphology, attachment, proliferation, and migration of cells cultured on collagen barrier membranes. A systematic review. *Odontology* **2008**, *96*, 1–11. [CrossRef]
31. Agrawal, C.M.; Parr, J.E.; Lin, S.T. *Synthetic Bioabsorbable Polymers for Implants*; American Society for Testing and Materials: West Conshohocken, PA, USA, 2000.
32. Gottlow, J.; Nyman, S.; Lindhe, J.; Karring, T.; Wennström, J. New attachment formation in the human periodontium by guided tissue regeneration Case reports. *J. Clin. Periodontol.* **1986**, *13*, 604–616. [CrossRef]
33. Gottlow, J.; Nyman, S.; Karring, T.; Lindhe, J. New attachment formation as the result of controlled tissue regeneration. *J. Clin. Periodontol.* **1984**, *11*, 494–503. [CrossRef]
34. Needleman, I.; Worthington, H.V.; Giedrys-Leeper, E.; Tucker, R. Guided tissue regeneration for periodontal infra-bony defects. *Cochrane Database Syst. Rev.* **2006**, *19*, CD001724. [CrossRef] [PubMed]
35. Cortellini, P.; Prato, G.P.; Tonetti, M.S. Periodontal regeneration of human intrabony defects with bioresorbable membranes. A controlled clinical trial. *J. Periodontol.* **1996**, *67*, 217–223. [CrossRef] [PubMed]
36. Murphy, K.G.; Gunsolley, J.C. Guided tissue regeneration for the treatment of periodontal intrabony and furcation defects. A systematic review. *Ann. Periodontol.* **2003**, *8*, 266–302. [CrossRef] [PubMed]
37. Buser, D.; Dula, K.; Belser, U.; Hirt, H.-P.; Berthold, H. Localized ridge augmentation using guided bone regeneration. I. Surgical procedure in the maxilla. *Int. J. Periodontics Restor. Dent.* **1993**, *13*, 28–45.
38. Slavkin, H.C.; Bartold, P.M. Challenges and potential in tissue engineering. *Periodontology 2000* **2006**, *41*, 9–15. [CrossRef]
39. Wikesjö, U.M.E.; Crigger, M.; Nilvéus, R.; Selvig, K.A. Early healing events at the dentin-connective tissue interface. Light and transmission electron microscopy observations. *J. Periodontol.* **1991**, *62*, 5–14. [CrossRef]
40. Wikesjö, U.M.E.; Nilvéus, R.E.; Selvig, K.A. Significance of early healing events on periodontal repair: A review. *J. Periodontol.* **1992**, *63*, 158–165. [CrossRef]
41. Nuñez, J.; Vignoletti, F.; Caffesse, R.G.; Sanz, M. Cellular therapy in periodontal regeneration. *Periodontology 2000* **2019**, *79*, 107–116. [CrossRef]
42. Pitaru, S.; McCulloch, C.A.G.; Narayanan, S.A. Cellular origins and differentiation control mechanisms during periodontal development and wound healing. *J. Periodontal Res.* **1994**, *29*, 81–94. [CrossRef]
43. McClellan, S.J.; Franses, E.I. Adsorption of bovine serum albumin at solid/aqueous interfaces. *Colloids Surfaces A Physicochem. Eng. Asp.* **2005**, *260*, 265–275. [CrossRef]
44. Schenk, R.K.; Buser, D.; Hardwick, W.R.; Dahlin, C. Healing pattern of bone regeneration in membrane-protected defects: A histologic study in the canine mandible. *Int. J. Oral Maxillofac. Implants* **1994**, *9*, 13–29. [PubMed]
45. Lang, H.; Schiiler, N.; Arnhold, S.; Nolden, R.; Mertens, T. Formation of differentiated tissues in vivo by periodontal cell populations cultured in vitro. *J. Dent. Res.* **1995**, *74*, 1219–1225. [CrossRef] [PubMed]
46. Lin, W.; McCulloch, C.A.G.; Cho, M. Differentiation of periodontal ligament fibroblasts into osteoblasts during socket healing after tooth extraction in the rat. *Anat. Rec.* **1994**, *240*, 492–506. [CrossRef] [PubMed]
47. Boyko, G.A.; Melcher, A.H.; Brunette, D.M. Formation of new periodontalligament by periodontal ligament cells implanted in vivo after culture in vitro: A preliminary study of transplanted roots in the dog. *J. Periodontal Res.* **1981**, *16*, 73–88. [CrossRef]
48. Wang, H.-L.; MacNeil, R.L. Guided tissue regeneration. Absorbable barriers. *Dent. Clin. N. Am.* **1998**, *42*, 505–522. [CrossRef]
49. Gao, Y.; Wang, S.; Shi, B.; Wang, Y.; Chen, Y.; Wang, X.; Lee, E.-S.; Jiang, H.-B. Advances in modification methods based on biodegradable membranes in guided bone/tissue regeneration: A review. *Polymers* **2022**, *14*, 871. [CrossRef]
50. Basile, M.A.; d'Ayala, G.G.; Malinconico, M.; Laurienzo, P.; Coudane, J.; Nottelet, B.; Della Ragione, F.; Oliva, A. Functionalized PCL/HA nanocomposites as microporous membranes for bone regeneration. *Mater. Sci. Eng. C* **2015**, *48*, 457–468. [CrossRef]
51. Karring, T. Regenerative periodontal therapy. *J. Int. Acad. Periodontol.* **2000**, *2*, 101–109.
52. Sheikh, Z.; Khan, A.S.; Roohpour, N.; Glogauer, M.; u Rehman, I. Protein adsorption capability on polyurethane and modified-polyurethane membrane for periodontal guided tissue regeneration applications. *Mater. Sci. Eng. C* **2016**, *68*, 267–275. [CrossRef]

53. Kim, K.-H.; Jeong, L.; Park, H.-N.; Shin, S.-Y.; Park, W.-H.; Lee, S.-C.; Kim, T.-I.; Park, Y.-J.; Seol, Y.-J.; Lee, Y.-M. Biological efficacy of silk fibroin nanofiber membranes for guided bone regeneration. *J. Biotechnol.* **2005**, *120*, 327–339. [CrossRef] [PubMed]
54. Laycock, B.; Nikolić, M.; Colwell, J.M.; Gauthier, E.; Halley, P.; Bottle, S.; George, G. Lifetime prediction of biodegradable polymers. *Prog. Polym. Sci.* **2017**, *71*, 144–189. [CrossRef]
55. Brydone, A.S.; Meek, D.; Maclaine, S. Bone grafting, orthopaedic biomaterials, and the clinical need for bone engineering. *Proc. Inst. Mech. Eng. Part H J. Eng. Med.* **2010**, *224*, 1329–1343. [CrossRef]
56. Gottlow, J. Guided tissue regeneration using bioresorbable and non-resorbable devices: Initial healing and long-term results. *J. Periodontol.* **1993**, *64*, 1157–1165. [CrossRef]
57. Nyman, S.; Gottlow, J.; Karring, T.; Lindhe, J. The regenerative potential of the periodontal ligament: An experimental study in the monkey. *J. Clin. Periodontol.* **1982**, *9*, 257–265. [CrossRef]
58. Aaboe, M.; Pinholt, E.M.; Hjorting-Hansen, E. Healing of experimentally created defects: A review. *Br. J. Oral Maxillofac. Surg.* **1995**, *33*, 312–318. [CrossRef] [PubMed]
59. Liu, J.; Kerns, D.G. Suppl 1: Mechanisms of guided bone regeneration: A review. *Open Dent. J.* **2014**, *8*, 56. [CrossRef]
60. Madhuri, S.V. Membranes for Periodontal Regeneration. *Int. J. Pharm. Sci. Invent.* **2016**, *5*, 19–24.
61. Canullo, L.; Malagnino, V.A. Vertical Ridge Augmentation Around Implants by e-PTFE Titanium-Reinforced Membrane and Bovine Bone Matrix: A 24-to 54-Month Study of 10 Consecutive Cases. *Int. J. Oral Maxillofac. Implants* **2008**, *23*, 858–866. [PubMed]
62. Fontana, F.; Santoro, F.; Maiorana, C.; Iezzi, G.; Piattelli, A.; Simion, M. Clinical and Histologic Evaluation of Allogeneic Bone Matrix Versus Autogenous Bone Chips Associated with Titanium-Reinforced e-PTFE Membrane for Vertical Ridge Augmentation: A Prospective Pilot Study. *Int. J. Oral Maxillofac. Implants* **2008**, *23*, 1003–1012.
63. Sam, G.; Pillai, B.R.M. Evolution of barrier membranes in periodontal regeneration-"are the third generation membranes really here? *J. Clin. Diagn. Res.* **2014**, *8*, ZE14–ZE17. [CrossRef]
64. Peter, M.; Ganesh, N.; Selvamurugan, N.; Nair, S.V.; Furuike, T.; Tamura, H.; Jayakumar, R. Preparation and characterization of chitosan–gelatin/nanohydroxyapatite composite scaffolds for tissue engineering applications. *Carbohydr. Polym.* **2010**, *80*, 687–694. [CrossRef]
65. Cheung, H.-Y.; Lau, K.-T.; Lu, T.-P.; Hui, D. A critical review on polymer-based bio-engineered materials for scaffold development. *Compos. Part B Eng.* **2007**, *38*, 291–300. [CrossRef]
66. Felipe, M.E.M.C.; Andrade, P.F.; Grisi, M.F.M.; Souza, S.L.S.; Taba, M., Jr.; Palioto, D.B.; Novaes, A.B., Jr. Comparison of two surgical procedures for use of the acellular dermal matrix graft in the treatment of gingival recessions: A randomized controlled clinical study. *J. Periodontol.* **2007**, *78*, 1209–1217. [CrossRef]
67. Bottino, M.C.; Jose, M.V.; Thomas, V.; Dean, D.R.; Janowski, G.M. Freeze-dried acellular dermal matrix graft: Effects of rehydration on physical, chemical, and mechanical properties. *Dent. Mater.* **2009**, *25*, 1109–1115. [CrossRef]
68. Bottino, M.C.; Thomas, V.; Jose, M.V.; Dean, D.R.; Janowski, G.M. Acellular dermal matrix graft: Synergistic effect of rehydration and natural crosslinking on mechanical properties. *J. Biomed. Mater. Res. Part B Appl. Biomater.* **2010**, *95*, 276–282. [CrossRef]
69. Sundararaghavan, H.G.; Monteiro, G.A.; Lapin, N.A.; Chabal, Y.J.; Miksan, J.R.; Shreiber, D.I. Genipin-induced changes in collagen gels: Correlation of mechanical properties to fluorescence. *J. Biomed. Mater. Res. Part A Off. J. Soc. Biomater. Jpn. Soc. Biomater. Aust. Soc. Biomater. Korean Soc. Biomater.* **2008**, *87*, 308–320. [CrossRef]
70. Schwarz, F.; Rothamel, D.; Herten, M.; Sager, M.; Becker, J. Angiogenesis pattern of native and cross-linked collagen membranes: An immunohistochemical study in the rat. *Clin. Oral Implants Res.* **2006**, *17*, 403–409. [CrossRef] [PubMed]
71. Yoo, C.-K.; Jeon, J.-Y.; Kim, Y.-J.; Kim, S.-G.; Hwang, K.-G. Cell attachment and proliferation of osteoblast-like MG63 cells on silk fibroin membrane for guided bone regeneration. *Maxillofac. Plast. Reconstr. Surg.* **2016**, *38*, 17. [CrossRef]
72. Reddy, M.S.B.; Ponnamma, D.; Choudhary, R.; Sadasivuni, K.K. A comparative review of natural and synthetic biopolymer composite scaffolds. *Polymers* **2021**, *13*, 1105. [CrossRef]
73. Place, E.S.; George, J.H.; Williams, C.K.; Stevens, M.M. Synthetic polymer scaffolds for tissue engineering. *Chem. Soc. Rev.* **2009**, *38*, 1139–1151. [CrossRef] [PubMed]
74. Zhu, J. Bioactive modification of poly (ethylene glycol) hydrogels for tissue engineering. *Biomaterials* **2010**, *31*, 4639–4656. [CrossRef] [PubMed]
75. Geckil, H.; Xu, F.; Zhang, X.; Moon, S.; Demirci, U. Engineering hydrogels as extracellular matrix mimics. *Nanomedicine* **2010**, *5*, 469–484. [CrossRef] [PubMed]
76. Kluge, J.A.; Mauck, R.L. Synthetic/biopolymer nanofibrous composites as dynamic tissue engineering scaffolds. In *Biomedical Applications of Polymeric Nanofibers*; Springer: Berlin/Heidelberg, Germany, 2011; pp. 101–130.
77. Cascone, M.G.; Barbani, N.; Giusti, C.C.P.; Ciardelli, G.; Lazzeri, L. Bioartificial polymeric materials based on polysaccharides. *J. Biomater. Sci. Polym. Ed.* **2001**, *12*, 267–281. [CrossRef]
78. Ciardelli, G.; Chiono, V.; Vozzi, G.; Pracella, M.; Ahluwalia, A.; Barbani, N.; Cristallini, C.; Giusti, P. Blends of poly-(ε-caprolactone) and polysaccharides in tissue engineering applications. *Biomacromolecules* **2005**, *6*, 1961–1976. [CrossRef]
79. Liu, H.; Slamovich, E.B.; Webster, T.J. Increased osteoblast functions on nanophase titania dispersed in poly-lactic-co-glycolic acid composites. *Nanotechnology* **2005**, *16*, S601. [CrossRef]
80. Pontoriero, R.; Wennström, J.; Lindhe, J. The use of barrier membranes and enamel matrix proteins in the treatment of angular bone defects. A prospective controlled clinical study. *J. Clin. Periodontol.* **1999**, *26*, 833–840.

81. Donos, N.; Kostopoulos, L.; Karring, T. Alveolar ridge augmentation using a resorbable copolymer membrane and autogenous bone grafts: An experimental study in the rat. *Clin. Oral Implants Res.* **2002**, *13*, 203–213. [CrossRef]
82. De Sanctis, M.; Zucchelli, G. Guided tissue regeneration with a resorbable barrier membrane (Vicryl) for the management of buccal recession: A case report. *Int. J. Periodontics Restor. Dent.* **1996**, *16*, 434–441.
83. Araujo, M.G.; Berglundh, T.; Lindhe, J. GTR treatment of degree III furcation defects with 2 different resorbable barriers an experimental study in dogs. *J. Clin. Periodontol.* **1998**, *25*, 253–259. [CrossRef]
84. Marra, K.G.; Szem, J.W.; Kumta, P.N.; DiMilla, P.A.; Weiss, L.E. In vitro analysis of biodegradable polymer blend/hydroxyapatite composites for bone tissue engineering. *J. Biomed. Mater. Res. Off. J. Soc. Biomater. Jpn. Soc. Biomater. Aust. Soc. Biomater. Korean Soc. Biomater.* **1999**, *47*, 324–335. [CrossRef]
85. Li, W.; Danielson, K.G.; Alexander, P.G.; Tuan, R.S. Biological response of chondrocytes cultured in three-dimensional nanofibrous poly (ϵ-caprolactone) scaffolds. *J. Biomed. Mater. Res. Part A Off. J. Soc. Biomater. Jpn. Soc. Biomater. Aust. Soc. Biomater. Korean Soc. Biomater.* **2003**, *67*, 1105–1114.
86. Sinha, V.R.; Bansal, K.; Kaushik, R.; Kumria, R.; Trehan, A. Poly-ϵ-caprolactone microspheres and nanospheres: An overview. *Int. J. Pharm.* **2004**, *278*, 1–23. [CrossRef] [PubMed]
87. Meek, M.F.; Jansen, K.; Steendam, R.; Oeveren, W.V.; van Wachem, P.B.; van Luyn, M.J.A. In vitro degradation and biocompatibility of poly (dl-lactide-ϵ-caprolactone) nerve guides. *J. Biomed. Mater. Res. Part A An Off. J. Soc. Biomater. Japanese Soc. Biomater. Aust. Soc. Biomater. Korean Soc. Biomater.* **2004**, *68*, 43–51.
88. Conn, R.E.; Kolstad, J.J.; Borzelleca, J.F.; Dixler, D.S.; Filer, L.J., Jr.; LaDu, B.N., Jr.; Pariza, M.W. Safety assessment of polylactide (PLA) for use as a food-contact polymer. *Food Chem. Toxicol.* **1995**, *33*, 273–283. [CrossRef]
89. Garlotta, D. A literature review of poly (lactic acid). *J. Polym. Environ.* **2001**, *9*, 63–84. [CrossRef]
90. Rokkanen, P.U.; Böstman, O.; Hirvensalo, E.; Mäkelä, E.A.; Partio, E.K.; Pätiälä, H.; Vainionpää, S.; Vihtonen, K.; Törmälä, P. Bioabsorbable fixation in orthopaedic surgery and traumatology. *Biomaterials* **2000**, *21*, 2607–2613. [CrossRef]
91. Georgiou, G.; Mathieu, L.; Pioletti, D.P.; Bourban, P.; Månson, J.; Knowles, J.C.; Nazhat, S.N. Polylactic acid–phosphate glass composite foams as scaffolds for bone tissue engineering. *J. Biomed. Mater. Res. Part B Appl. Biomater. Off. J. Soc. Biomater. Jpn. Soc. Biomater. Aust. Soc. Biomater. Korean Soc. Biomater.* **2007**, *80*, 322–331. [CrossRef] [PubMed]
92. Gupta, B.; Revagade, N.; Hilborn, J. Poly (lactic acid) fiber: An overview. *Prog. Polym. Sci.* **2007**, *32*, 455–482. [CrossRef]
93. Kim, K.; Yu, M.; Zong, X.; Chiu, J.; Fang, D.; Seo, Y.-S.; Hsiao, B.S.; Chu, B.; Hadjiargyrou, M. Control of degradation rate and hydrophilicity in electrospun non-woven poly (D, L-lactide) nanofiber scaffolds for biomedical applications. *Biomaterials* **2003**, *24*, 4977–4985. [CrossRef]
94. Merolli, A.; Gabbi, C.; Cacchioli, A.; Ragionieri, L.; Caruso, L.; Giannotta, L.; Leali, P.T. Bone response to polymers based on poly-lactic acid and having different degradation times. *J. Mater. Sci. Mater. Med.* **2001**, *12*, 775–778. [CrossRef]
95. Chen, Y.-W.; Umeda, M.; Nagasawa, T.; Takeuchi, Y.; Huang, Y.; Inoue, Y.; Iwai, T.; Izumi, Y.; Ishikawa, I. Periodontitis may increase the risk of peripheral arterial disease. *Eur. J. Vasc. Endovasc. Surg.* **2008**, *35*, 153–158. [CrossRef]
96. Hou, L.; Yan, J.; Tsai, A.Y.; Lao, C.; Lin, S.; Liu, C. Polymer-assisted regeneration therapy with Atrisorb® barriers in human periodontal intrabony defects. *J. Clin. Periodontol.* **2004**, *31*, 68–74. [CrossRef] [PubMed]
97. Camargo, P.M.; Lekovic, V.; Weinlaender, M.; Vasilic, N.; Madzarevic, M.; Kenney, E.B. Platelet-rich plasma and bovine porous bone mineral combined with guided tissue regeneration in the treatment of intrabony defects in humans. *J. Periodontal Res.* **2002**, *37*, 300–306. [CrossRef]
98. Takata, T.; Miyauchi, M.; Wang, H. Migration of osteoblastic cells on various guided bone regeneration membranes. *Clin. Oral Implants Res.* **2001**, *12*, 332–338. [CrossRef] [PubMed]
99. Bilir, A.; Aybar, B.; Tanrikulu, S.H.; Issever, H.; Tuna, S. Biocompatibility of different barrier membranes in cultures of human CRL 11372 osteoblast-like cells: An immunohistochemical study. *Clin. Oral Implants Res.* **2007**, *18*, 46–52. [CrossRef]
100. Christenson, E.M.; Anseth, K.S.; van den Beucken, J.J.J.P.; Chan, C.K.; Ercan, B.; Jansen, J.A.; Laurencin, C.T.; Li, W.; Murugan, R.; Nair, L.S. Nanobiomaterial applications in orthopedics. *J. Orthop. Res.* **2007**, *25*, 11–22. [CrossRef] [PubMed]
101. Zupancic, S.; Kocbek, P.; Baumgartner, S.; Kristl, J. Contribution of nanotechnology to improved treatment of periodontal disease. *Curr. Pharm. Des.* **2015**, *21*, 3257–3271. [CrossRef] [PubMed]
102. Owen, G.R.; Jackson, J.K.; Chehroudi, B.; Brunette, D.M.; Burt, H.M. An in vitro study of plasticized poly (lactic-co-glycolic acid) films as possible guided tissue regeneration membranes: Material properties and drug release kinetics. *J. Biomed. Mater. Res. Part A* **2010**, *95*, 857–869. [CrossRef]
103. Zhang, E.; Zhu, C.; Yang, J.; Sun, H.; Zhang, X.; Li, S.; Wang, Y.; Sun, L.; Yao, F. Electrospun PDLLA/PLGA composite membranes for potential application in guided tissue regeneration. *Mater. Sci. Eng. C* **2016**, *58*, 278–285. [CrossRef]
104. Shi, R.; Xue, J.; He, M.; Chen, D.; Zhang, L.; Tian, W. Structure, physical properties, biocompatibility and in vitro/vivo degradation behavior of anti-infective polycaprolactone-based electrospun membranes for guided tissue/bone regeneration. *Polym. Degrad. Stab.* **2014**, *109*, 293–306. [CrossRef]
105. Xue, J.; He, M.; Liang, Y.; Crawford, A.; Coates, P.; Chen, D.; Shi, R.; Zhang, L. Fabrication and evaluation of electrospun PCL–gelatin micro-/nanofiber membranes for anti-infective GTR implants. *J. Mater. Chem. B* **2014**, *2*, 6867–6877. [CrossRef] [PubMed]
106. Xue, J.; Feng, B.; Zheng, R.; Lu, Y.; Zhou, G.; Liu, W.; Cao, Y.; Zhang, Y.; Zhang, W.J. Engineering ear-shaped cartilage using electrospun fibrous membranes of gelatin/polycaprolactone. *Biomaterials* **2013**, *34*, 2624–2631. [CrossRef] [PubMed]

107. Zheng, R.; Duan, H.; Xue, J.; Liu, Y.; Feng, B.; Zhao, S.; Zhu, Y.; Liu, Y.; He, A.; Zhang, W. The influence of Gelatin/PCL ratio and 3-D construct shape of electrospun membranes on cartilage regeneration. *Biomaterials* **2014**, *35*, 152–164. [CrossRef] [PubMed]
108. Gupta, D.; Venugopal, J.; Prabhakaran, M.P.; Dev, V.R.G.; Low, S.; Choon, A.T.; Ramakrishna, S. Aligned and random nanofibrous substrate for the in vitro culture of Schwann cells for neural tissue engineering. *Acta Biomater.* **2009**, *5*, 2560–2569. [CrossRef]
109. Ji, W.; Yang, F.; Ma, J.; Bouma, M.J.; Boerman, O.C.; Chen, Z.; van den Beucken, J.J.J.P.; Jansen, J.A. Incorporation of stromal cell-derived factor-1α in PCL/gelatin electrospun membranes for guided bone regeneration. *Biomaterials* **2013**, *34*, 735–745. [CrossRef] [PubMed]
110. Xue, J.; He, M.; Liu, H.; Niu, Y.; Crawford, A.; Coates, P.D.; Chen, D.; Shi, R.; Zhang, L. Drug loaded homogeneous electrospun PCL/gelatin hybrid nanofiber structures for anti-infective tissue regeneration membranes. *Biomaterials* **2014**, *35*, 9395–9405. [CrossRef]
111. Feng, B.; Tu, H.; Yuan, H.; Peng, H.; Zhang, Y. Acetic-acid-mediated miscibility toward electrospinning homogeneous composite nanofibers of GT/PCL. *Biomacromolecules* **2012**, *13*, 3917–3925. [CrossRef]
112. Ku, Y.; Shim, I.K.; Lee, J.Y.; Park, Y.J.; Rhee, S.; Nam, S.H.; Park, J.B.; Chung, C.P.; Lee, S.J. Chitosan/poly (l-lactic acid) multilayered membrane for guided tissue regeneration. *J. Biomed. Mater. Res. Part A Off. J. Soc. Biomater. Jpn. Soc. Biomater. Aust. Soc. Biomater. Korean Soc. Biomater.* **2009**, *90*, 766–772. [CrossRef]
113. Kim, S.; Nimni, M.E.; Yang, Z.; Han, B. Chitosan/gelatin–based films crosslinked by proanthocyanidin. *J. Biomed. Mater. Res. Part B Appl. Biomater. Off. J. Soc. Biomater. Jpn. Soc. Biomater. Aust. Soc. Biomater. Korean Soc. Biomater.* **2005**, *75*, 442–450. [CrossRef]
114. Teng, S.; Lee, E.; Wang, P.; Shin, D.; Kim, H. Three-layered membranes of collagen/hydroxyapatite and chitosan for guided bone regeneration. *J. Biomed. Mater. Res. Part B Appl. Biomater. Off. J. Soc. Biomater. Jpn. Soc. Biomater. Aust. Soc. Biomater. Korean Soc. Biomater.* **2008**, *87*, 132–138. [CrossRef]
115. Hunter, K.T.; Ma, T. In vitro evaluation of hydroxyapatite–chitosan–gelatin composite membrane in guided tissue regeneration. *J. Biomed. Mater. Res. Part A* **2013**, *101*, 1016–1025. [CrossRef] [PubMed]
116. Jiao, Y.-P.; Cui, F.-Z. Surface modification of polyester biomaterials for tissue engineering. *Biomed. Mater.* **2007**, *2*, R24. [CrossRef]
117. Kim, J.Y.; Yoon, J.J.; Park, E.K.; Kim, D.S.; Kim, S.-Y.; Cho, D.-W. Cell adhesion and proliferation evaluation of SFF-based biodegradable scaffolds fabricated using a multi-head deposition system. *Biofabrication* **2009**, *1*, 15002. [CrossRef] [PubMed]
118. Vaquette, C.; Cooper-White, J. A simple method for fabricating 3-D multilayered composite scaffolds. *Acta Biomater.* **2013**, *9*, 4599–4608. [CrossRef]
119. Xu, H.; Cui, W.; Chang, J. Fabrication of patterned PDLLA/PCL composite scaffold by electrospinning. *J. Appl. Polym. Sci.* **2013**, *127*, 1550–1554. [CrossRef]
120. Floreon 3D. "Floreon 3D." 2014. Available online: https://www.floreon.com/floreon3d (accessed on 17 January 2020).
121. Floreon. Ltd. Floreon-Transforming Packag. 2018. Available online: https://www.floreon.com (accessed on 17 January 2020).
122. *ISO 527-3:2018*; Plastics—Determination of Tensile Properties—Part 3: Test Conditions for Films and Sheets. International Organization for Standardization: Geneva, Switzerland, 2018.
123. Lee, C.H.; Sapuan, S.M.; Hassan, M.R. Thermal analysis of kenaf fiber reinforced floreon biocomposites with magnesium hydroxide flame retardant filler. *Polym. Compos.* **2018**, *39*, 869–875. [CrossRef]
124. Ramos-Rodriguez, D.H.; Pashneh-Tala, S.; Bains, A.K.; Moorehead, R.D.; Kassos, N.; Kelly, A.L.; Paterson, T.E.; Orozco-Diaz, C.A.; Gill, A.A.; Ortega Asencio, I. Demonstrating the Potential of Using Bio-Based Sustainable Polyester Blends for Bone Tissue Engineering Applications. *Bioengineering* **2022**, *9*, 163. [CrossRef] [PubMed]
125. Bottino, M.C.; Thomas, V.; Janowski, G.M. A novel spatially designed and functionally graded electrospun membrane for periodontal regeneration. *Acta Biomater.* **2011**, *7*, 216–224. [CrossRef]
126. Liao, S.; Wang, W.; Uo, M.; Ohkawa, S.; Akasaka, T.; Tamura, K.; Cui, F.; Watari, F. A three-layered nano-carbonated hydroxyapatite/collagen/PLGA composite membrane for guided tissue regeneration. *Biomaterials* **2005**, *26*, 7564–7571. [CrossRef]
127. Rowe, M.J.; Kamocki, K.; Pankajakshan, D.; Li, D.; Bruzzaniti, A.; Thomas, V.; Blanchard, S.B.; Bottino, M.C. Dimensionally stable and bioactive membrane for guided bone regeneration: An in vitro study. *J. Biomed. Mater. Res. Part B Appl. Biomater.* **2016**, *104*, 594–605. [CrossRef]
128. Shim, J.-H.; Yoon, M.-C.; Jeong, C.-M.; Jang, J.; Jeong, S.-I.; Cho, D.-W.; Huh, J.-B. Efficacy of rhBMP-2 loaded PCL/PLGA/β-TCP guided bone regeneration membrane fabricated by 3D printing technology for reconstruction of calvaria defects in rabbit. *Biomed. Mater.* **2014**, *9*, 65006. [CrossRef]
129. Shim, J.-H.; Huh, J.-B.; Park, J.Y.; Jeon, Y.-C.; Kang, S.S.; Kim, J.Y.; Rhie, J.-W.; Cho, D.-W. Fabrication of blended polycaprolactone/poly (lactic-co-glycolic acid)/β-tricalcium phosphate thin membrane using solid freeform fabrication technology for guided bone regeneration. *Tissue Eng. Part A* **2013**, *19*, 317–328. [CrossRef] [PubMed]
130. Mota, J.; Yu, N.; Caridade, S.G.; Luz, G.M.; Gomes, M.E.; Reis, R.L.; Jansen, J.A.; Walboomers, X.F.; Mano, J.F. Chitosan/bioactive glass nanoparticle composite membranes for periodontal regeneration. *Acta Biomater.* **2012**, *8*, 4173–4180. [CrossRef] [PubMed]
131. Yang, F.; Both, S.K.; Yang, X.; Walboomers, X.F.; Jansen, J.A. Development of an electrospun nano-apatite/PCL composite membrane for GTR/GBR application. *Acta Biomater.* **2009**, *5*, 3295–3304. [CrossRef] [PubMed]
132. Leal, A.I.; Caridade, S.G.; Ma, J.; Yu, N.; Gomes, M.E.; Reis, R.L.; Jansen, J.A.; Walboomers, X.F.; Mano, J.F. Asymmetric PDLLA membranes containing Bioglass® for guided tissue regeneration: Characterization and in vitro biological behavior. *Dent. Mater.* **2013**, *29*, 427–436. [CrossRef]

133. Zhao, X.; Wu, Y.; Du, Y.; Chen, X.; Lei, B.; Xue, Y.; Ma, P.X. A highly bioactive and biodegradable poly (glycerol sebacate)–silica glass hybrid elastomer with tailored mechanical properties for bone tissue regeneration. *J. Mater. Chem. B* **2015**, *3*, 3222–3233. [CrossRef]
134. Li, W.; Ding, Y.; Yu, S.; Yao, Q.; Boccaccini, A.R. Multifunctional chitosan-45S5 bioactive glass-poly (3-hydroxybutyrate-co-3-hydroxyvalerate) microsphere composite membranes for guided tissue/bone regeneration. *ACS Appl. Mater. Interfaces* **2015**, *7*, 20845–20854. [CrossRef]
135. Jamuna-Thevi, K.; Saarani, N.N.; Kadir, M.R.A.; Hermawan, H. Triple-layered PLGA/nanoapatite/lauric acid graded composite membrane for periodontal guided bone regeneration. *Mater. Sci. Eng. C* **2014**, *43*, 253–263. [CrossRef]
136. Peter, M.; Binulal, N.S.; Nair, S.V.; Selvamurugan, N.; Tamura, H.; Jayakumar, R. Novel biodegradable chitosan–gelatin/nano-bioactive glass ceramic composite scaffolds for alveolar bone tissue engineering. *Chem. Eng. J.* **2010**, *158*, 353–361. [CrossRef]
137. Qasim, S.B.; Delaine-Smith, R.M.; Fey, T.; Rawlinson, A.; Rehman, I.U. Freeze gelated porous membranes for periodontal tissue regeneration. *Acta Biomater.* **2015**, *23*, 317–328. [CrossRef]
138. Khan, A.S.; Ahmed, Z.; Edirisinghe, M.J.; Wong, F.S.L.; Rehman, I.U. Preparation and characterization of a novel bioactive restorative composite based on covalently coupled polyurethane–nanohydroxyapatite fibres. *Acta Biomater.* **2008**, *4*, 1275–1287. [CrossRef] [PubMed]
139. Bianco, A.; Cacciotti, I.; Lombardi, M.; Montanaro, L.; Gusmano, G. Thermal stability and sintering behaviour of hydroxyapatite nanopowders. *J. Therm. Anal. Calorim.* **2007**, *88*, 237–243. [CrossRef]
140. Porter, A.E.; Patel, N.; Skepper, J.N.; Best, S.M.; Bonfield, W. Effect of sintered silicate-substituted hydroxyapatite on remodelling processes at the bone–implant interface. *Biomaterials* **2004**, *25*, 3303–3314. [CrossRef] [PubMed]
141. Ferraz, M.P.; Monteiro, F.J.; Manuel, C.M. Hydroxyapatite nanoparticles: A review of preparation methodologies. *J. Appl. Biomater. Biomech.* **2004**, *2*, 74–80.
142. Bianco, A.; Di Federico, E.; Moscatelli, I.; Camaioni, A.; Armentano, I.; Campagnolo, L.; Dottori, M.; Kenny, J.M.; Siracusa, G.; Gusmano, G. Electrospun poly (ε-caprolactone)/Ca-deficient hydroxyapatite nanohybrids: Microstructure, mechanical properties and cell response by murine embryonic stem cells. *Mater Sci. Eng. C* **2009**, *29*, 2063–2071. [CrossRef]
143. Sonseca, A.; Peponi, L.; Sahuquillo, O.; Kenny, J.M.; Giménez, E. Electrospinning of biodegradable polylactide/hydroxyapatite nanofibers: Study on the morphology, crystallinity structure and thermal stability. *Polym. Degrad. Stab.* **2012**, *97*, 2052–2059. [CrossRef]
144. Madhumathi, K.; Binulal, N.S.; Nagahama, H.; Tamura, H.; Shalumon, K.T.; Selvamurugan, N.; Nair, S.V.; Jayakumar, R. Preparation and characterization of novel β-chitin–hydroxyapatite composite membranes for tissue engineering applications. *Int. J. Biol. Macromol.* **2009**, *44*, 1–5. [CrossRef]
145. Gorustovich, A.A.; Roether, J.A.; Boccaccini, A.R. Effect of bioactive glasses on angiogenesis: A review of in vitro and in vivo evidences. *Tissue Eng. Part B Rev.* **2010**, *16*, 199–207. [CrossRef]
146. Xynos, I.D.; Edgar, A.J.; Buttery, L.D.K.; Hench, L.L.; Polak, J.M. Gene-expression profiling of human osteoblasts following treatment with the ionic products of Bioglass® 45S5 dissolution. *J. Biomed. Mater. Res. Off. J. Soc. Biomater. Jpn. Soc. Biomater. Aust. Soc. Biomater. Korean Soc. Biomater.* **2001**, *55*, 151–157. [CrossRef]
147. Izquierdo-Barba, I.; Arcos, D.; Sakamoto, Y.; Terasaki, O.; López-Noriega, A.; Vallet-Regí, M. High-performance mesoporous bioceramics mimicking bone mineralization. *Chem. Mater.* **2008**, *20*, 3191–3198. [CrossRef]
148. Li, X.; Chang, J. Preparation and characterization of bioactive collagen/wollastonite composite scaffolds. *J. Mater. Sci. Mater. Med.* **2005**, *16*, 361–365. [CrossRef]
149. Wang, X.; Zhou, Y.; Xia, L.; Zhao, C.; Chen, L.; Yi, D.; Chang, J.; Huang, L.; Zheng, X.; Zhu, H. Fabrication of nano-structured calcium silicate coatings with enhanced stability, bioactivity and osteogenic and angiogenic activity. *Colloids Surf. B Biointerfaces* **2015**, *126*, 358–366. [CrossRef] [PubMed]
150. Villar, C.C.; Cochran, D.L. Regeneration of periodontal tissues: Guided tissue regeneration. *Dent. Clin.* **2010**, *54*, 73–92. [CrossRef] [PubMed]
151. Becker, W.; Becker, B.E.; Handelsman, M.; Ochsenbein, C.; Albrektsson, T. Guided tissue regeneration for implants placed into extraction sockets: A study in dogs. *J. Periodontol.* **1991**, *62*, 703–709. [CrossRef] [PubMed]
152. Bartee, B.K. The use of high-density polytetrafluoroethylene membrane to treat osseous defects: Clinical reports. *Implant Dent.* **1995**, *4*, 21–31. [CrossRef] [PubMed]
153. Barber, H.D.; Lignelli, J.; Smith, B.M.; Bartee, B.K. Using a dense PTFE membrane without primary closure to achieve bone and tissue regeneration. *J. Oral Maxillofac. Surg.* **2007**, *65*, 748–752. [CrossRef]
154. Jovanovic, S.A.; Nevins, M. Bone formation utilizing titanium-reinforced barrier membranes. *Int. J. Periodontics Restor. Dent.* **1995**, *15*, 56–69. [CrossRef]
155. Lindfors, L.T.; Tervonen, E.A.T.; Sándor, G.K.B.; Ylikontiola, L.P. Guided bone regeneration using a titanium-reinforced ePTFE membrane and particulate autogenous bone: The effect of smoking and membrane exposure. *Oral Surg. Oral Med. Oral Pathol. Oral Radiol. Endodontology* **2010**, *109*, 825–830. [CrossRef]
156. Taguchi, Y.; Amizuka, N.; Nakadate, M.; Ohnishi, H.; Fujii, N.; Oda, K.; Nomura, S.; Maeda, T. A histological evaluation for guided bone regeneration induced by a collagenous membrane. *Biomaterials* **2005**, *26*, 6158–6166. [CrossRef]
157. Zhao, S.; Pinholt, E.M.; Madsen, J.E.; Donath, K. Histological evaluation of different biodegradable and non-biodegradable membranes implanted subcutaneously in rats. *J. Cranio-Maxillofac. Surg.* **2000**, *28*, 116–122. [CrossRef] [PubMed]

158. Zwahlen, R.A.; Cheung, L.K.; Zheng, L.; Chow, R.L.K.; Li, T.; Schuknecht, B.; Grätz, K.W.; Weber, F.E. Comparison of two resorbable membrane systems in bone regeneration after removal of wisdom teeth: A randomized-controlled clinical pilot study. *Clin. Oral Implants Res.* **2009**, *20*, 1084–1091. [CrossRef] [PubMed]
159. Gielkens, P.F.M.; Schortinghuis, J.; De Jong, J.R.; Raghoebar, G.M.; Stegenga, B.; Bos, R.R.M. Vivosorb®, Bio-Gide®, and Gore-Tex® as barrier membranes in rat mandibular defects: An evaluation by microradiography and micro-CT. *Clin. Oral Implants Res.* **2008**, *19*, 516–521. [CrossRef] [PubMed]
160. Sela, M.N.; Kohavi, D.; Krausz, E.; Steinberg, D.; Rosen, G. Enzymatic degradation of collagen-guided tissue regeneration membranes by periodontal bacteria. *Clin. Oral Implants Res.* **2003**, *14*, 263–268. [CrossRef]
161. Maksoud, M.A. Immediate implants in fresh posterior extraction sockets: Report of two cases. *J. Oral Implantol.* **2001**, *27*, 123–126. [CrossRef] [PubMed]
162. Sheikh, Z.; Abdallah, M.N.; Hamdan, N.; Javaid, M.A.; Khurshid, Z.; Matilinna, K. Barrier membranes for tissue regeneration and bone augmentation techniques in dentistry. In *Handbook of Oral Biomaterials*; Pan Stanford Publishing: Singapore, 2014; pp. 605–636.
163. Miller, N.; Penaud, J.; Foliguet, B.; Membre, H.; Ambrosini, P.; Plombas, M. Resorption rates of 2 commercially available bioresorbable membranes: A histomorphometric study in a rabbit model. *J. Clin. Periodontol.* **1996**, *23*, 1051–1059. [CrossRef]
164. Thoma, D.S.; Halg, G.; Dard, M.M.; Seibl, R.; Hammerle, C.H.F.; Jung, R.E. Evaluation of a new biodegradable membrane to prevent gingival ingrowth into mandibular bone defects in minipigs. *Clin. Oral Implants Res.* **2009**, *20*, 7–16. [CrossRef]
165. Thomas, V.; Dean, D.R.; Jose, M.V.; Mathew, B.; Chowdhury, S.; Vohra, Y.K. Nanostructured biocomposite scaffolds based on collagen coelectrospun with nanohydroxyapatite. *Biomacromolecules* **2007**, *8*, 631–637. [CrossRef]
166. Wikesjö, U.M.E.; Qahash, M.; Huang, Y.; Xiropaidis, A.; Polimeni, G.; Susin, C. Bone morphogenetic proteins for periodontal and alveolar indications; biological observations–clinical implications. *Orthod. Craniofac. Res.* **2009**, *12*, 263–270. [CrossRef]
167. Alizadeh, M.; Abbasi, F.; Khoshfetrat, A.B.; Ghaleh, H. Microstructure and characteristic properties of gelatin/chitosan scaffold prepared by a combined freeze-drying/leaching method. *Mater. Sci. Eng. C* **2013**, *33*, 3958–3967. [CrossRef]
168. Pighinelli, L.; Kucharska, M. Chitosan–hydroxyapatite composites. *Carbohydr. Polym.* **2013**, *93*, 256–262. [CrossRef] [PubMed]
169. Thein-Han, W.W.; Kitiyanant, Y. Chitosan scaffolds for in vitro buffalo embryonic stem-like cell culture: An approach to tissue engineering. *J. Biomed. Mater. Res. Part B Appl. Biomater. Off. J. Soc. Biomater. Jpn. Soc. Biomater. Aust. Soc. Biomater. Korean Soc. Biomater.* **2007**, *80*, 92–101. [CrossRef] [PubMed]
170. Nair, L.S.; Laurencin, C.T. Biodegradable polymers as biomaterials. *Prog. Polym. Sci.* **2007**, *32*, 762–798. [CrossRef]
171. Agrawal, C.M.; McKinney, J.S.; Huang, D.; Athanasiou, K.A. The use of the vibrating particle technique to fabricate highly porous and permeable biodegradable scaffolds. In *Synthetic Bioabsorbable Polymers for Implants*; ASTM International: West Conshohocken, PA, USA, 2000.
172. Bhattarai, D.; Aguilar, L.; Park, C.; Kim, C. A review on properties of natural and synthetic based electrospun fibrous materials for bone tissue engineering. *Membranes* **2018**, *8*, 62. [CrossRef] [PubMed]
173. Kedem, S.; Schmidt, J.; Paz, Y.; Cohen, Y. Composite polymer nanofibers with carbon nanotubes and titanium dioxide particles. *Langmuir* **2005**, *21*, 5600–5604. [CrossRef]
174. Sun, F.; Zhou, H.; Lee, J. Various preparation methods of highly porous hydroxyapatite/polymer nanoscale biocomposites for bone regeneration. *Acta Biomater.* **2011**, *7*, 3813–3828. [CrossRef]
175. Sulaiman, S.B.; Keong, T.K.; Cheng, C.H.; Saim, A.B.; Idrus, R.B.H. Tricalcium phosphate/hydroxyapatite (TCP-HA) bone scaffold as potential candidate for the formation of tissue engineered bone. *Indian J. Med. Res.* **2013**, *137*, 1093–1101.
176. Cao, H.; Kuboyama, N. A biodegradable porous composite scaffold of PGA/β-TCP for bone tissue engineering. *Bone* **2010**, *46*, 386–395. [CrossRef]
177. Fabbri, P.; Cannillo, V.; Sola, A.; Dorigato, A.; Chiellini, F. Highly porous polycaprolactone-45S5 Bioglass® scaffolds for bone tissue engineering. *Compos. Sci. Technol.* **2010**, *70*, 1869–1878. [CrossRef]
178. Abdollahi, S.; Ma, A.C.C.; Cerruti, M. Surface transformations of Bioglass 45S5 during scaffold synthesis for bone tissue engineering. *Langmuir* **2013**, *29*, 1466–1474. [CrossRef]
179. Huang, W.; Shi, X.; Ren, L.; Du, C.; Wang, Y. PHBV microspheres–PLGA matrix composite scaffold for bone tissue engineering. *Biomaterials* **2010**, *31*, 4278–4285. [CrossRef] [PubMed]
180. Linh, N.T.B.; Min, Y.K.; Lee, B.-T. Fabrication and in vitro evaluations with osteoblast-like MG-63 cells of porous hyaluronic acid-gelatin blend scaffold for bone tissue engineering applications. *J. Mater. Sci.* **2013**, *48*, 4233–4242. [CrossRef]
181. Zhang, K.-R.; Gao, H.-L.; Pan, X.-F.; Zhou, P.; Xing, X.; Xu, R.; Pan, Z.; Wang, S.; Zhu, Y.; Hu, B. Multifunctional bilayer nanocomposite guided bone regeneration membrane. *Matter* **2019**, *1*, 770–781. [CrossRef]
182. Retzepi, M.; Donos, N. Guided bone regeneration: Biological principle and therapeutic applications. *Clin. Oral Implants Res.* **2010**, *21*, 567–576. [CrossRef]
183. Abe, G.L.; Sasaki, J.-I.; Katata, C.; Kohno, T.; Tsuboi, R.; Kitagawa, H.; Imazato, S. Fabrication of novel poly (lactic acid/caprolactone) bilayer membrane for GBR application. *Dent. Mater.* **2020**, *36*, 626–634. [CrossRef]
184. Yoshimoto, I.; Sasaki, J.-I.; Tsuboi, R.; Yamaguchi, S.; Kitagawa, H.; Imazato, S. Development of layered PLGA membranes for periodontal tissue regeneration. *Dent. Mater.* **2018**, *34*, 538–550. [CrossRef] [PubMed]
185. Requicha, J.F.; Viegas, C.A.; Hede, S.; Leonor, I.B.; Reis, R.L.; Gomes, M.E. Design and characterization of a biodegradable double-layer scaffold aimed at periodontal tissue-engineering applications. *J. Tissue Eng. Regen. Med.* **2016**, *10*, 392–403. [CrossRef]

186. Requicha, J.F.; Viegas, C.A.; Muñoz, F.; Azevedo, J.M.; Leonor, I.B.; Reis, R.L.; Gomes, M.E. A tissue engineering approach for periodontal regeneration based on a biodegradable double-layer scaffold and adipose-derived stem cells. *Tissue Eng. Part A* **2014**, *20*, 2483–2492. [CrossRef]
187. Park, C.H.; Rios, H.F.; Jin, Q.; Bland, M.E.; Flanagan, C.L.; Hollister, S.J.; Giannobile, W.V. Biomimetic hybrid scaffolds for engineering human tooth-ligament interfaces. *Biomaterials* **2010**, *31*, 5945–5952. [CrossRef]
188. Eltom, A.; Zhong, G.; Muhammad, A. Scaffold Techniques and Designs in Tissue Engineering Functions and Purposes: A Review. *Adv. Mater. Sci. Eng.* **2019**, *2019*, 3429527. [CrossRef]
189. Aldemir Dikici, B.; Dikici, S.; Reilly, G.C.; MacNeil, S.; Claeyssens, F. A novel bilayer polycaprolactone membrane for guided bone regeneration: Combining electrospinning and emulsion templating. *Materials* **2019**, *12*, 2643. [CrossRef] [PubMed]
190. Liu, X.; Ma, P.X. Polymeric scaffolds for bone tissue engineering. *Ann. Biomed. Eng.* **2004**, *32*, 477–486. [CrossRef] [PubMed]
191. Yuan, B.; Zhou, S.; Chen, X. Rapid prototyping technology and its application in bone tissue engineering. *J. Zhejiang Univ. B* **2017**, *18*, 303–315. [CrossRef]
192. Leong, K.F.; Cheah, C.M.; Chua, C.K. Solid freeform fabrication of three-dimensional scaffolds for engineering replacement tissues and organs. *Biomaterials* **2003**, *24*, 2363–2378. [CrossRef] [PubMed]
193. Lee, J.W.; Kim, J.Y.; Cho, D.-W. Solid free-form fabrication technology and its application to bone tissue engineering. *Int. J. Stem Cells* **2010**, *3*, 85–95. [CrossRef]
194. Stevens, B.; Yang, Y.; Mohandas, A.; Stucker, B.; Nguyen, K.T. A review of materials, fabrication methods, and strategies used to enhance bone regeneration in engineered bone tissues. *J. Biomed. Mater. Res. Part B Appl. Biomater. Off. J. Soc. Biomater. Jpn. Soc. Biomater. Aust. Soc. Biomater. Korean Soc. Biomater.* **2008**, *85*, 573–582. [CrossRef]
195. Ozbolat, I.T.; Peng, W.; Ozbolat, V. Application areas of 3D bioprinting. *Drug Discov. Today* **2016**, *21*, 1257–1271. [CrossRef]
196. Ozbolat, I.T.; Yu, Y. Bioprinting toward organ fabrication: Challenges and future trends. *IEEE Trans. Biomed. Eng.* **2013**, *60*, 691–699. [CrossRef]
197. Datta, P.; Ozbolat, V.; Ayan, B.; Dhawan, A.; Ozbolat, I.T. Bone tissue bioprinting for craniofacial reconstruction. *Biotechnol. Bioeng.* **2017**, *114*, 2424–2431. [CrossRef]
198. Dehghani, F.; Annabi, N. Engineering porous scaffolds using gas-based techniques. *Curr. Opin. Biotechnol.* **2011**, *22*, 661–666. [CrossRef] [PubMed]
199. Liao, C.; Chen, C.; Chen, J.; Chiang, S.; Lin, Y.; Chang, K. Fabrication of porous biodegradable polymer scaffolds using a solvent merging/particulate leaching method. *J. Biomed. Mater. Res. Off. J. Soc. Biomater. Jpn. Soc. Biomater. Aust. Soc. Biomater. Korean Soc. Biomater.* **2002**, *59*, 676–681. [CrossRef] [PubMed]
200. Puppi, D.; Chiellini, F.; Piras, A.M.; Chiellini, E. Polymeric materials for bone and cartilage repair. *Prog. Polym. Sci.* **2010**, *35*, 403–440. [CrossRef]
201. Liang, H.-Q.; Wu, Q.-Y.; Wan, L.-S.; Huang, X.-J.; Xu, Z.-K. Polar polymer membranes via thermally induced phase separation using a universal crystallizable diluent. *J. Memb. Sci.* **2013**, *446*, 482–491. [CrossRef]
202. Mikos, A.G.; Temenoff, J.S. Formation of highly porous biodegradable scaffolds for tissue engineering. *Electron. J. Biotechnol.* **2000**, *3*, 23–24. [CrossRef]
203. Johnson, T.; Bahrampourian, R.; Patel, A.; Mequanint, K. Fabrication of highly porous tissue-engineering scaffolds using selective spherical porogens. *Biomed. Mater. Eng.* **2010**, *20*, 107–118. [CrossRef]
204. Roseti, L.; Parisi, V.; Petretta, M.; Cavallo, C.; Desando, G.; Bartolotti, I.; Grigolo, B. Scaffolds for bone tissue engineering: State of the art and new perspectives. *Mater. Sci. Eng. C* **2017**, *78*, 1246–1262. [CrossRef]
205. An, J.; Teoh, J.E.M.; Suntornnond, R.; Chua, C.K. Design and 3D printing of scaffolds and tissues. *Engineering* **2015**, *1*, 261–268. [CrossRef]
206. Nair, K.; Gandhi, M.; Khalil, S.; Yan, K.C.; Marcolongo, M.; Barbee, K.; Sun, W. Characterization of cell viability during bioprinting processes. *Biotechnol. J. Healthc. Nutr. Technol.* **2009**, *4*, 1168–1177. [CrossRef]
207. Abdelaal, O.A.M.; Darwish, S.M.H. Review of rapid prototyping techniques for tissue engineering scaffolds fabrication. In *Characterization and Development of Biosystems and Biomaterials*; Springer: Berlin/Heidelberg, Germany, 2013; pp. 33–54.
208. Hutmacher, D.W. Scaffolds in tissue engineering bone and cartilage. In *The Biomaterials: Silver Jubilee Compendium*; Elsevier: Amsterdam, The Netherlands, 2000; pp. 175–189.
209. Kumbar, S.G.; James, R.; Nukavarapu, S.P.; Laurencin, C.T. Electrospun nanofiber scaffolds: Engineering soft tissues. *Biomed. Mater.* **2008**, *3*, 34002. [CrossRef]
210. Zafar, M.; Najeeb, S.; Khurshid, Z.; Vazirzadeh, M.; Zohaib, S.; Najeeb, B.; Sefat, F. Potential of electrospun nanofibers for biomedical and dental applications. *Materials* **2016**, *9*, 73. [CrossRef]
211. Bikiaris, D.N.; Papageorgiou, G.Z.; Achilias, D.S.; Pavlidou, E.; Stergiou, A. Miscibility and enzymatic degradation studies of poly (ε-caprolactone)/poly (propylene succinate) blends. *Eur. Polym. J.* **2007**, *43*, 2491–2503. [CrossRef]
212. Ramakrishna, S.; Fujihara, K.; Teo, W.-E.; Yong, T.; Ma, Z.; Ramaseshan, R. Electrospun nanofibers: Solving global issues. *Mater. Today* **2006**, *9*, 40–50. [CrossRef]
213. Jang, J.-H.; Castano, O.; Kim, H.-W. Electrospun materials as potential platforms for bone tissue engineering. *Adv. Drug Deliv. Rev.* **2009**, *61*, 1065–1083. [CrossRef] [PubMed]
214. Agarwal, S.; Wendorff, J.H.; Greiner, A. Progress in the field of electrospinning for tissue engineering applications. *Adv. Mater.* **2009**, *21*, 3343–3351. [CrossRef] [PubMed]

215. Jiang, T.; Carbone, E.J.; Lo, K.W.-H.; Laurencin, C.T. Electrospinning of polymer nanofibers for tissue regeneration. *Prog. Polym. Sci.* **2015**, *46*, 1–24. [CrossRef]
216. Hussain, M.; Ullah, S.; Raza, M.R.; Abbas, N.; Ali, A. Recent Developments in Zn-Based Biodegradable Materials for Biomedical Applications. *J. Funct. Biomater.* **2022**, *14*, 1. [CrossRef]
217. Howard, D.; Buttery, L.D.; Shakesheff, K.M.; Roberts, S.J. Tissue engineering: Strategies, stem cells and scaffolds. *J. Anat.* **2008**, *213*, 66–72. [CrossRef]

MDPI AG
Grosspeteranlage 5
4052 Basel
Switzerland
Tel.: +41 61 683 77 34

Polymers Editorial Office
E-mail: polymers@mdpi.com
www.mdpi.com/journal/polymers

www.ingramcontent.com/pod-product-compliance
Lightning Source LLC
Chambersburg PA
CBHW041407210926
51850CB00013B/3198
* 9 7 8 3 7 2 5 8 3 8 6 0 8 *